Evolutionary Economics and Social Complexity Science

Volume 7

The Japanese Association for Evolutionary Economics (JAFEE) always has adhered to its original aim of taking an explicit "integrated" approach. This path has been followed steadfastly since the Association's establishment in 1997 and, as well, since the inauguration of our international journal in 2004. We have deployed an agenda encompassing a contemporary array of subjects including but not limited to: foundations of institutional and evolutionary economics, criticism of mainstream views in the social sciences, knowledge and learning in socio-economic life, development and innovation of technologies, transformation of industrial organizations and economic systems, experimental studies in economics, agent-based modeling of socio-economic systems, evolution of the governance structure of firms and other organizations, comparison of dynamically changing institutions of the world, and policy proposals in the transformational process of economic life. In short, our starting point is an "integrative science" of evolutionary and institutional views. Furthermore, we always endeavor to stay abreast of newly established methods such as agent-based modeling, socio/econo-physics, and network analysis as part of our integrative links.

More fundamentally, "evolution" in social science is interpreted as an essential key word, i.e., an integrative and /or communicative link to understand and re-domain various preceding dichotomies in the sciences: ontological or epistemological, subjective or objective, homogeneous or heterogeneous, natural or artificial, selfish or altruistic, individualistic or collective, rational or irrational, axiomatic or psychological-based, causal nexus or cyclic networked, optimal or adaptive, micro- or macroscopic, deterministic or stochastic, historical or theoretical, mathematical or computational, experimental or empirical, agent-based or socio/econo-physical, institutional or evolutionary, regional or global, and so on. The conventional meanings adhering to various traditional dichotomies may be more or less obsolete, to be replaced with more current ones vis-à-vis contemporary academic trends. Thus we are strongly encouraged to integrate some of the conventional dichotomies.

These attempts are not limited to the field of economic sciences, including management sciences, but also include social science in general. In that way, understanding the social profiles of complex science may then be within our reach. In the meantime, contemporary society appears to be evolving into a newly emerging phase, chiefly characterized by an information and communication technology (ICT) mode of production and a service network system replacing the earlier established factory system with a new one that is suited to actual observations. In the face of these changes we are urgently compelled to explore a set of new properties for a new socio/economic system by implementing new ideas. We thus are keen to look for "integrated principles" common to the above-mentioned dichotomies throughout our serial compilation of publications. We are also encouraged to create a new, broader spectrum for establishing a specific method positively integrated in our own original way.

More information about this series at http://www.springer.com/series/11930

Yoshinori Shiozawa • Tosihiro Oka • Taichi Tabuchi
Editors

A New Construction of Ricardian Theory of International Values

Analytical and Historical Approach

Springer

Editors
Yoshinori Shiozawa
Osaka City University
Osaka, Japan

Tosihiro Oka
Faculty of Economics
Fukui Prefectural University
Fukui, Japan

Taichi Tabuchi
Faculty of Commerce
Doshisha University
Kyoto, Japan

ISSN 2198-4204 ISSN 2198-4212 (electronic)
Evolutionary Economics and Social Complexity Science
ISBN 978-981-10-0190-1 ISBN 978-981-10-0191-8 (eBook)
DOI 10.1007/978-981-10-0191-8

Library of Congress Control Number: 2017936973

Printed on acid-free paper

This Springer imprint is published by Springer Nature
The registered company is Springer Nature Singapore Pte Ltd.
The registered company address is: 152 Beach Road, #21-01/04 Gateway East, Singapore 189721, Singapore

Preface

The classical theory of value is characterized as value determined by production costs. Ricardo based many of his propositions (propositions on distribution, taxation, and dynamics) on the labor theory of value, a special form of the production cost theory of value. He was aware of difficulties with the labor theory, but did not find a consistent formulation of the production cost theory that could uphold the whole of his propositions. It was in the latter half of the twentieth century that the classical vision of value was reestablished by using simultaneous equations without recourse to the labor theory. It was established that relative prices were regulated by production costs determined by technological, as well as distributional, variables. This development of the value theory was, however, confined to economies without international trade.

Ricardo himself left the question of how relative prices are determined in economies with international trade unresolved, whereas neoclassical explanations developed after J. S. Mill's declaration that international values should be determined by the law of supply and demand. Attempts to formulate international values in accordance with the classical vision of value faced difficulties in dealing with multiple countries and commodities, including intermediate commodities traded internationally.

A breakthrough was made by Shiozawa. He established that there is a combination of international prices and wages that enables a set of techniques to be carried out competitively as to produce any points on a facet that is part of the maximal boundary of the production possibility set. The prices and wages are determined by production costs in the sense that they are uniquely determined once a facet of the production possibility set is chosen, and the law of supply and demand has no role. Technologies are assumed to be different among countries, and the difference in wages enables multiple countries to produce a commodity by using different methods of production.

This is the first book written in English on the new theory of international values. The volume is divided into three parts. Part I consists of Chap. 1, in which Shiozawa gives a basic framework of the new theory of international values and some of its consequences. After presenting a short history of international trade theory, he provides definitions for describing the Ricardo-Sraffa trade economy and presents the fundamental theorem of the new theory with its mathematical proof. Shiozawa discusses gains from trade, possibility of trade conflicts, and some subjects concerning extensions and generalizations of the new theory. It is shown that the capital-labor ratio has no relation to the development pattern of less developed countries or to the fragmentations of production processes; rather, the wage disparity is important. These facts are in contrast to the neoclassical theory of trade. Implications to some other issues are presented: international input-output table, classical theory of value, international political economy, and development economics.

Part II consists of five chapters. In Chap. 2, Oka gives an introduction to the new theory of international values by putting it in the context of the Ricardo-Sraffian theory of value and distribution. After giving a sketch of the development of the classical theory from Ricardo to Sraffa, Oka describes how J. S. Mill addressed the question of international values that had been left unsolved by Ricardo. By using one of Mill's numerical examples, it is shown that Mill's solution is not valid, also showing how his example would be properly dealt with by the new theory of international values.

Chapter 3 addresses the question of the role of demands in determining international value. As explained in Chap. 1, the fundamental theorem of the new theory is founded on the "equivalent economy," in which each input coefficient is defined as the physical input coefficient multiplied by $(1 + m)$, where m is the markup rate. Net outputs in the equivalent economy should be interpreted as the quantities of products that can be consumed after deducting the amounts necessary as investment for growth at rate m. The maximal boundary of the production possibility set should also be understood for the net outputs in this sense. As such, it does not represent the real maximal boundary, because the growth rate of each industry is not necessarily equal to its markup rate. In the new theory, international prices are connected with the net outputs and with the final demand through the relation that the price vector is perpendicular to a facet of the maximal boundary. Therefore, the fact that this boundary is not real complicates the relation between demand and value. In this chapter, Oka presents a precise relation of the value to real demand by distinguishing three types of production possibility frontiers: R-efficient locus, physical maximal frontier, and capitalistically feasible frontier.

In Chap. 4, Ogawa makes an attempt at translations from "evolutionary economics" to "modern economics" with regard to the new theory of international trade. Ogawa attributes the unpopularity of the results of Shiozawa in the mainstream economics to the style of writing too inclined to evolutionary economics. He gives expressions of the results in terms of the mainstream economics and discusses their meanings.

In Chap. 5, Takamasu surveys the development of the trade theory focusing on the Neo-Ricardian tradition. He shows how Ricardo's comparative advantage theory becomes invalid when the number of commodities or countries is larger than two or when there are intermediate goods. He also presents the arguments on the possibility of losses from trade when the rate of profit is positive, and also the concept of intertemporal efficiency, on the basis of which transition from autarky to trade turns out to give rise to gains for every country, taking time preference into account. He considers new findings of the new theory of international values from the perspective of the Neo-Ricardian tradition and provides a proof that the theorem on intertemporal efficiency of international trade is also valid in the framework of the new theory.

In Chap. 6, Hirano addresses the issue of international trade and unemployment indicated by the new theory. He explores how this issue can be analyzed by extrapolating the Ricardo-Sraffa trade economy to the national self-sufficiency vision put forward by Keynes. After comparing the views of Keynes, Parrinello, and Shiozawa on the market mechanisms not eliminating unemployment, he comes to the notion of a nation's economic competitiveness, with regard to which he focuses on the industrial sector's long-term development decisions that are not supported by current cost calculations.

Part III consists of four chapters. Chapter 7, written by Shiozawa, gives an "internalist explanation" for the turn in the history of the value theory from classical to neoclassical. By "internalist explanation" he means "explanation from logical necessity to solve theoretical deficiencies." Shiozawa argues that, when John Stuart Mill tried to solve the international value problem, he was forced to revert from classical principles of cost determination of value to a special form of the law of demand and supply and that this marked a crucial turning point. To prove this contention, he presents how the fathers of British neoclassical value theory were influenced by Mill's misleading solutions.

In Chap. 8, Yoshii draws attention to the fact that J. S. Mill's price theory is not the one of supply-demand equilibrium, that is, a system consisting of supply and demand equations that determines a quantity and a price simultaneously, but rather a sequential process model with time. Yoshii attributes the misunderstanding of Mill's theory as a supply-demand equilibrium one to Jenkin's interpretation. He also remarks Marshall's contribution to the development of the misunderstanding through purification of economic theory and introduction of the theme of the stability of equilibrium. Yoshii identifies the turning point of the history of economic thought to these events.

In Chap. 9, Tabuchi examines how the doctrine of comparative costs has developed from Ricardo to modern economics in the light of value theories and asks who was responsible for the reconstruction of Ricardo's theory. He focuses especially on the controversy between Viner and Haberler in the 1930s and concludes that the modern theory formulated by Samuelson and others was based on some unrealistic assumptions derived from Haberler's opportunity cost approach.

In Chap. 10, Sato gives an overview of research into international values in Japan. He divided the postwar period of research in Japan into two parts; one is until the 1980s and the other is since the 1990s. He regards the former period as being led by Marxian economists; the purpose of research is identified as how Marx's labor theory of value should be modified to determine international values, and Toichi Nawa is taken as a pioneering scholar, who argued that the difference in the productivity of a key commodity between countries determined the exchange rate of labor in the two countries. A consensus of researchers involved in the argument, however, is described as that the average productivity of individual sectors, rather than the productivity of the key commodity, matters. After pointing out the difficulties in the notion of the average productivity, Sato discusses the limits, as well as advantages, of the research in this period. He identifies the latter period (i.e., since the 1990s) as based on Graham and Sraffa. He first describes the rediscovery of Graham's theory, in which "link commodities" link the opportunity costs in countries that produce them in common, so that relative prices of all the commodities produced in those countries are determined uniquely, irrespective of reciprocal demands. He also mentions the arguments on international values, based on the Sraffian framework. He identifies the new theory of international values launched by Shiozawa as a development of Graham's theory toward the inclusion of intermediate goods and profits, as well as a development of Sraffa's theory as to include international trade.

This volume is a collection of some of the achievements of the Workshop on the Theory of International Values started in 2014. There are many domains that would benefit from research based on the new theory of international values, under-employment, development and growth, international division of labor, international input-output analysis, and so on, as remarked in several chapters. We hope this volume gives an impetus for such research endeavors.

In celebration of the bicentennial year of the first edition of Ricardo's *Principles.*

Osaka, Japan — Yoshinori Shiozawa
Fukui, Japan — Tosihiro Oka
Kyoto, Japan — Taichi Tabuchi

Contents

Part I General Introduction

The New Theory of International Values: An Overview 3
Yoshinori Shiozawa

Part II Theoretical Topics in and about the New Theory

The New Theory of International Values in the Context of the Ricardo-Sraffian Theory of Value and Distribution 77
Tosihiro Oka

The Relation Between Value and Demand in the New Theory of International Values 99
Tosihiro Oka

Analysis of Production-Efficient Patterns of Specialization Allowing Intermediate Inputs: The Meaning of Shiozawa's Model from the Viewpoint of Modern Economics 123
Takeshi Ogawa

The Neo-Ricardian Trade Theory and the New Theory of International Values 149
Akira Takamasu

Application of Normal Prices to Trade Analysis: National Self-Sufficiency and Factors of Competition 175
Yoshitaka Hirano

Part III Re-Examining the History of International Trade Theory

An Origin of the Neoclassical Revolution: Mill's "Reversion" and Its Consequences 191
Yoshinori Shiozawa

An Extinction of Adjustment Time and an Introduction of Stability Condition in Economics through Misunderstandings to J.S. Mill's Law of Supply and Demand and International Value Theory 245
Satoshi Yoshii

Comparative Advantage in the Light of the *Old* Value Theories 265
Taichi Tabuchi

An Overview of Research into International Values in Japan 281
Hideo Sato

Part I
General Introduction

The New Theory of International Values: An Overview

Yoshinori Shiozawa

Abstract This chapter is a general introduction of the new theory of international values, which is an extension of the cost-of-production theory of value to the international trade situations. Within a general framework comprising input trade and choice of production techniques, the new theory analyzes the international values (i.e., the system that consists of wages of each country and prices of goods), gains and losses from trade, and the patterns of specializations. It is the first theory to treat traded input goods in this general form. It facilitates the analysis of recent conspicuous trade aspects, such as rapidly increasing trade volume of intermediate goods, fragmentations of production processes, and the complex network of global value chains. Besides the Introduction (Division I), this work is divided into four parts: (1) the presentation of the theory (Division II), (2) extensions of the theory to more complex situations (Division III), (3) some examples of applications of the theory (Division IV), and (4) possible implications to three neighboring fields (Division V). Terms and concepts are explained in detail and theorems are fully stated. A mathematical proof of the fundamental theorem is given in the Appendix.

Keywords Classical theory of value • International trade • Development economics • International input–output tables • Globalization

Division I: Introduction

Following a short introduction, I provide a brief history of international trade (Sect. 2). I consider this particular section necessary, because the new theory emerged from a very thin research strand that had remained underutilized for a considerable length of time. Its entire history requires many more pages than can be used here, and so I focus here on two points: (1) I look to situate the new theory in a long history of trade theory. The crucial bifurcation point goes back to John Stuart

Y. Shiozawa (✉)
Osaka City University, Osaka, Japan
e-mail: y@shiozawa.net

Y. Shiozawa et al. (eds.), *A New Construction of Ricardian Theory of International Values*, Evolutionary Economics and Social Complexity Science 7,
DOI 10.1007/978-981-10-0191-8_1

Mill, who thought he had solved the unsettled problem left by Ricardo. (2) I pay special attention to why intermediate (or input) goods have not been implemented in the theory, despite awareness of their importance.

1 General Introduction

The theory of international values presented in this chapter lies in a research strand that remained underutilized for quite some time, and with only a few exceptions, it differs markedly from most trade theories. The theory both resurrects and belongs to the classical tradition, wherein production plays a major role in determining wages and prices.

The new theory provides a value theory for a wide class of Ricardo–Sraffa (RS) trade economies. The model includes a many-country, many-commodity trade economy in which input goods are traded and the choice of techniques is explicitly incorporated. There is no comparable general theory in trade theory, save for general equilibrium theory (GET) *à la* Arrow and Debreu (1954). Heckscher–Ohlin–Samuelson (HOS) theory and its generalization Heckscher–Ohlin–Vanek (HOV) theory are both completely contingent on GET. Krugman's (1979, 1980) trade theory (or new trade theory) explains why intra-industry trade occurs only in an extremely symmetrical situation and thus depends implicitly on a generalized GET that permits increasing returns. Melitz's (2003) new new trade theory is originally formulated in GET, but can be easily incorporated into the new theory, because each firm can have its own (and different) production techniques. However, trade theories based on GET all bear a common weakness: GET generally excludes corner solutions and is not well suited to analyzing those cases, while specialization is but a typical case of corner solutions. The new theory, developed specifically to analyze international specialization, does not bear this weakness.

The merits of the new theory do not lie solely in its generality. One of its fundamental innovations is general treatments of input goods (or intermediate goods). The world economy is becoming more and more global in nature, with concurrent reductions in information and transportation costs. Trade in input goods is rapidly increasing in terms of both volume and proportion. Firms are obliged to adopt global optimal procurement policy, and engineers are designing the optimal fragmentation of production processes. A great unbundling is now taking place, and trade in tasks is common everywhere (Baldwin 2006, 2014). The trade theory for RS economies provides a sound analytical tool, because it assumes many different and distinct production techniques, and production processes can be fragmented into a series of different production techniques.

One result of trade and production globalization is that trade is no longer solely in (final) goods, but a complex network of value-adding processes (i.e., trade in value-added). The new theory of international values provides a basic framework for analyzing this complexity. The new theory typically assumes that each firm's production and procurements are based on global optimal procurement policy.

One of the biggest merits of the new theory is that an RS economy has a good representation in the form of international input–output tables (IIOTs). Although there are some differences in viewpoints, an RS economy and IIOTs have a similar vision vis-à-vis how economies work. As workable statistical tables, IIOTs have various restrictions in terms of information collection and are obliged to condense information at a more or less aggregate level. An RS economy is an ideal virtual entity that is constructed abductively from preceding economic theories and observations; however, it is rather easy to transcribe ideas from an RS economy into IIOTs and argue complex processes that are taking place in the current globalized economy.

We can easily compare in this regard the new theory and some other trade theories, such as the HOV model. The latter sees trade as an exchange of factor contents and considers that factor endowments among countries serve as the main driving force of international trade. This vision may have been true when transportation costs were high, and the location of primary resources may have played a major role in determining the pattern of production and trade. However, reduced transportation costs have undermined the *raison d'être* of this kind of labor division, and trade within the globalized economy works under a variety of principles other than the locations and proportions of primary resources. RS economies and IIOTs have a common vision—namely, that the production of commodities is a production by means of commodities (Sraffa 1960). The driving forces of globalization and trade in input goods and tasks are now differences in production techniques; these techniques differ by country and even by firm. This is the common starting point of both RS economies and IIOTs. We can expect that the new theory will provide some theoretical basis for analysis based on IIOTs and will profit in turn from developments in those IIOTs. Such a trial has already started.[1]

Although its formal formulation is highly mathematical and the theorems require knowledge of nonelementary mathematics, the essence of the new theory is quite simple. The world production possibility set forms a polytope. In the N-commodity case, the production possibility set is a polytope of dimension N, and it is covered by facets of dimension $N-1$. The simplest example (i.e., minimal model) is seen in Fig. 1. The production possibility set is a three-dimensional body whose positive boundary comprises two triangles and one parallelogram, as facets. These three polygons are the facets (three two-dimensional faces) in which we have interest, and the interior of these facets is called regular domains. At any point in a regular domain, a unique value exists (up to scalar multiplication) whose price component is perpendicular to the domain.[2] As far as the demand remains within the same domain, the international value remains constant.

[1] One example is Escaith and Miroudot (2016).

[2] An international value $\mathbf{v} = (\mathbf{w}, \mathbf{p})$ comprises two parts: wage vector $\mathbf{w}$ and price vector $\mathbf{p}$. A wage vector expresses the set of wages w_i in country i, which may differ from country to country. The core of the fundamental theorem (Theorem 4.4) is to prove the existence of a wage vector $\mathbf{w}$ that forms an admissible international value together with price vector $\mathbf{p}$.

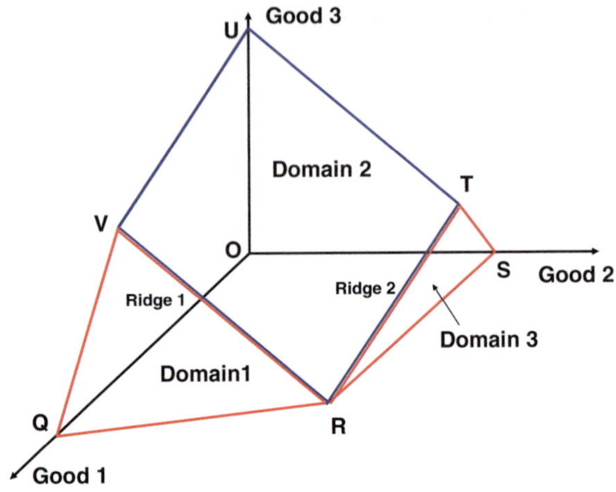

Fig. 1 A minimal model of the Ricardian trade theory (two-country three-good case)

At an intersection of two facets (i.e., at a ridge in the three-dimensional case), the values are not uniquely determined, but we can ignore those points as cases of low probability. It is important to note that in the interior of a facet, price adjustment fails to work. Inside such a domain, the main mechanism that works to adjust demand and production is the change in production levels for each product. This is why the new theory of international values basically aligns with the classical theory of value.[3]

A major limitation of the present study is that it assumes that the set of production techniques is given and fixed. In this sense, the new theory is still strictly static in nature, but this does not mean that we cannot develop a more dynamic theory at some point in the future. In fact, the new theory already makes it possible to deal with changes in production techniques. Suppose that we are given a set of production techniques S_0, and some new techniques are added. We now have two sets of production techniques, S_0 and S_1. Normally, we can suppose $S_0 \subset S_1$.[4] Then, a transition in the state of technology, from S_0 to S_1, raises the question of choice of production techniques.[5] This logic is incorporated into the new theory of international values. Two questions now arise. (1) How do production techniques evolve? (2) What are the effects of technological changes? These are underdeveloped questions, even in neoclassical micro- and macro-level economics. Various notions of neutral technical change (e.g., Solow, Harrod, and Hicks neutralities, among others) have been introduced only to facilitate

[3] As for the characterization and understanding of classical theory of value, see Shiozawa (2016).

[4] This relation holds inasmuch as old production techniques remain possible and socially permissible. On the other hand, some old techniques have become inadmissible, for ecological and other reasons.

[5] Other important choices of techniques include the adoption of new articles and the abandonment of old ones. However, I do not enter into this important but difficult issue.

macroeconomic analysis. More frequently used labor and capital-saving technical changes are rough notions that can loosely indicate a long-term trend, but with no firm reasons, as we see in Sect. 13. Analyzing the effects of technological change is more difficult; in principle, we can analyze changes in specialization patterns. If in-use production techniques change, the values change. We still know very little about these changes and mutual relations. Obtaining this knowledge is central to the next step in developing the new theory.

If a set of production techniques is fixed, one will encounter no great challenge in developing growth theory where quantities increase proportionally. Analysis of a closed economy can be easily generalized to the international trade case. However, every sincere economist knows that real economic development differs markedly from such proportional growth: prices, wage rates, consumption, people's lives, and technology all change with time, and we are not yet ready to analyze all these changes.

Despite all the difficulties and challenges we face, I believe that the new theory of international values provides a firm basis for further study of the interconnected world economy. The new theory will also contribute to studies of any country's economic development, which should be considered in the global context.

2 A Short History of International Trade Theory

The new theory of international values resurrects an old tradition that goes back to Ricardo. Many people may think it too trivial to mention. Although the first form of trade theory really started with Ricardo, the history of international trade theory is disoriented and quite sinuous. I have no intention of covering all currents. This section mainly addresses why the new theory of international values was so late to appear. It is true that a general theory could not appear before the 1950s, because the new theory required a certain level of mathematics. The general theory of linear inequality—the essential tool for the new theory—was mainly developed around the 1950s in connection with the arrival of linear programming. However, this was not the major obstruction. A much graver obstacle was the mode of thinking in economics.

The first turning point came when Mill tried to solve the problem that Ricardo had left unsettled. His "solution" is now referred to as the theory of reciprocal demand. It is important to note that Mill's theory is based fully on supply and demand relations. Mill wanted to be Ricardo's loyal disciple, but when he poised himself to solve the unsettled problem in international trade, he was obliged to abandon the cost-of-production theory of value, and he returned to the "antecedent law"—namely, the law of demand and supply (Mill 1848, III.18.4), which is "more fundamental" and "anterior to [the] cost of production" (Mill 1848, III.16.5). Following a discussion of the logical status of the cost of production and the law of demand and supply, Mill concludes that

> This law of International Values is but an extension of the more general law of Value, which we called the Equation of Supply and Demand. (Mill 1848, III.18.24)

Because I argue in chap. 7 of this book (Shiozawa 2017), just how and why Mill was guided to this conclusion, I will not argue these points here. However, it is inevitable to emphasize that a change to the *problématique* occurred in economics.

Classical economics was an economics of production (*plutology*, after Hicks (1976)). In examining a situation in which both countries enjoy gains from trade, Mill was guided to consider complete specialization—namely, a scenario in which each country produces just one commodity. In such a scenario, the production of each country is completely determined, including the commodity that a country produces and the quantity it produces, because the labor force and the labor input coefficient are given. Nominally, there are "productions," but the commodity and the quantity that a country can supply are uniquely determined. This is equivalent to a pure exchange economy. When countries set out to trade, each country has its own commodity in hand and tries to obtain a bundle of commodities that maximizes its satisfaction. The "solution" Mill obtains is but an economics of exchange (*catallactics*, after Hicks (1976)). Thus, Mill inaugurated the long tradition of catallactics in international trade. This was the real change to the *problématique* in economics, and it triggered the explosion of neoclassical economics.

The problem set by Mill was refined and formulated in a more mathematical way by Alfred Marshall and Francis Ysidro Edgeworth, near the end of the nineteenth century. They paved the way to a basis for international trade theory. The core logic of their analysis was that of a pure exchange economy.[6] After Marshall and Edgeworth, there were many contributions in this field, but only a few economists deserve special mention for the emergence of a Ricardian theory of international values.

In the 1930s, works such as those of Haberler (1933), Ohlin (1933), and Viner (1937) appeared. There was a tripartite dispute among them, but they all belonged to Mill's tradition. Among them, Ohlin deserves special note; he introduced in his book (Ohlin 1933) a formulation that later became the Heckscher–Ohlin theory of international trade. It was Paul Samuelson (1948, 1949, 1953) who transformed Ohlin's observation into more precise formulations and produced a series of theorems, among which the factor price equalization theorem was included. Chipman (1965–1966) distinguished "neoclassical" and "modern" approaches by virtue of the HOS theory. While reference to the factor endowments was a new orientation and more modern tools were employed, the "modern approach" intensified its neoclassical character, and with the appearance of Arrow and Debreu (1954), it was absorbed into GET. International trade theory became a part of standard microeconomics, and it was interpreted that a special feature of the theory lies only in the special situation's setting and assumptions, such as trade among nations and the immobility of factors. It is not accidental that Krugman's (1979, 1980) new trade theory supposes an

---

[6] I also explain this history in some detail in my chapter "An Origin of the Neoclassical Economics: Mill's Reversion and His Followers" (chap. 7 in this volume).

extremely high symmetry for its situation setting. His basic presumptions must be that the general case is supported by GET.

At the margins of mainstream international trade theory, there was a thin undercurrent that paid attention to the special features of Ricardian theory. The most remarkable proponent was Frank Dunstone Graham (1948) who endeavored to correct the misdirected orientation started by Mill and set Ricardian theory on the right path. However, following his publication of a book full of numerical examples (Graham 1948), he died in an unexpected accident. Lionel W. McKenzie was one of Graham's students at Princeton, and he partially took up Graham's research program; he developed it into a more modern style. McKenzie founded a new graduate course in Rochester and recruited Ronald Jones.[7] They produced a series of papers that can be called "Ricardian trade theory," in a broad sense. Its culmination was Jones (1961).

Praising this work, Ethier (1999, p. 764) comments in this manner:

> The contribution was so definitive that the Ricardian model has since been used almost entirely as a tool of other purposes and not as a subject of research in its own right. The main exception is the extension, by Samuelson (1964) and by Dorbusch et al. (1977) to the model of a continuum of commodities.

Ethier is correct in his conveyance of the general atmosphere in Rochester and elsewhere, but was incorrect on two critical points.

First, Jones indicated that his theory was extended to include the trade of intermediate products, but what he did was a study of a symmetrical case: in other words, he succeeded only in providing a general theory, wherein all countries have an identical matrix of material input coefficients (RII economy in the end of this section). The extension to a wider situation (i.e., an asymmetrical case) was not pursued, save in some sporadic studies in Japan and elsewhere. However, building a trade theory by which we can analyze the trade of intermediate (or input) products is a crucial problem, because all issues—from the importation of primary materials, processing trade (*kakō bōeki*, in Japanese), and outsourcing to fragmentation—are concerned with trade in intermediate products.[8] As McKenzie (1954, p. 179) points out, "Lancashire would be unlikely to produce cotton cloth if the cotton has to be grown in England." McKenzie (1954, p. 180) concludes his paper with this warning: "we have found that this simplicity [of the theory] is bought at the expense of prohibiting all trade in intermediate products (with a slight exception), which is indeed a heavy price." Ethier should have known this fact. Just after Ethier (1999), Samuelson (2001) gave an example that shows that gains from trade are

[7]The two mentored many Japanese economists, with the majority of them being specialists in international trade theory. This helped pave the way to a strong tradition in Japan of Ricardian trade theory.

[8]All primary materials are intermediate products, because they are extracted and processed. The difference between primary material and intermediate product does not matter. What makes analysis difficult is that the cost of a product is dependent on other countries' product prices, if imported products are used as inputs.

multiplied when two countries have strongly asymmetrical production techniques. The assumption of symmetry is not naïve. It is introduced to make analysis tractable, but excludes an important mechanism of gains from trade. It is a condition that must be removed, if we truly want to understand gains from input trade.

The real difficulty inherent in value theory in introducing input trade lies in the fact that a country's cost of production depends on the price and wages of other countries through the importation of input goods (We are now ignoring many other cost factors that comprise tariffs and transportation costs). The cost of imported inputs depends on the prices and wages of other countries, because material inputs have a kind of fractal structure: if a product comprises a part, that part comprises other parts.[9] Simply stated, the cost of a product depends on the wages of all countries. Herein lies the essential difference between value theory without input trade and that with input trade. The fact that two trade theories have different mathematical structures will be explained at the end of this section.

Second, Jones (1961) was more interested in the case where prices can move freely, even within a certain range. In a world where neoclassical thinking dominated, it was natural for Jones to be interested in this situation; the Ricardian framework had opened another possibility, but he could not find it. Sraffa's (1960) seminal book had appeared just before Jones's paper, and so Jones had not been able to read the book beforehand and consider its consequences. After all, Jones remained in the tradition or the *problématique* initiated by Mill. As I argue in Shiozawa (2017, chap. 7 of this book), Mill concentrated his analysis on the case where two countries enjoy gains from trade; this analysis was conducted to examine an economy that corresponds to the internal vertex (or internal extreme point) of the production frontier (point C in Fig. 1, Chap. 7 of this book).[10] When Jones studied many-commodity cases, he had no necessity to confine himself to the examination of an internal vertex. But he expended much effort in characterizing internal vertices of the world production frontier, and he was ultimately rewarded with his beautiful theorem. Jones's formula provides a complete characterization of the internal vertices. He proved that internal vertices on the frontier are in fact unique, if they exist, and he provided us with a way of knowing the possible specialization pattern.[11] However, knowing the complete specialization is not the end of international trade theory. A price vector can be determined in a manner similar to that of Mill, Marshall, and Edgeworth, but the production specified by a complete specialization pattern is uniquely determined, as was the case with Mill's

[9] As an illustration of the fractal structure of the international division of labor, see Figure 5.1 of Escaith and Inomata (2013).

[10] The extreme point of a convex set is defined as the point that cannot be the middle point of a segment contained in the set. Normal vectors at a boundary point of a convex polytope can have full-dimensional freedom (i.e., $N - 1$ dimension) only when the point is a vertex. The adjective "internal" here means that the point is in the interior of the positive orthant.

[11] Jones provides a necessary and sufficient condition for the existence of an internal vertex. He provides a proof of the necessary part, but does not prove the sufficient part. See Shiozawa (2015), Sect. 10.

case. What happens if world final demand is not proportional to its net production? What prices emerge outside the internal vertex, and how do they change when demand changes? A deficiency of relevant theory is manifest, but Jones and his followers did not pursue these questions.

More conspicuous is how Jones and McKenzie neglected to examine the case where the number of commodities exceeds the number of countries. This neglect is exorbitant, because if we observe the real world, the number of commodities far exceeds the number of countries and economic areas. The first is in the order of 10–100 million, while the latter is, at most, in the vicinity of 200. If they had examined the situation where the number of commodities is greater than the number of countries, they must have found it quite inconvenient, because in such a case, there is no internal vertex on the frontier. In fact a Ricardian economy has no internal vertex on the production frontier when the number of commodities is greater than the number of countries. This is easily understood. An internal vertex on the frontier represents a complete specialization, where "complete specialization" means that each country produces only one product. If the number of products exceeds the number of countries, how can this economy produce all kinds of commodities?[12] If Jones had ever examined the case where the number of products exceeds the number of countries, he would have noticed that their research program based on the existence of an internal vertex cannot be justified, because no such point exists.

Curiously, McKenzie (1954) examines cases where the number of countries is greater than that of commodities. It is possible that he too was preoccupied by the assertion that prices must move freely. When I started to study McKenzie and Jones in the 1980s, I was already a Sraffian and had the idea to separate price determination and quantity determination. I was critical of the general equilibrium (GE) framework. In spite of this, I sought for quite some time to characterize the Mill–Jones points (i.e., internal vertex of the world production possibility set). A major part of my struggle was the need to escape from such preoccupation, i.e. to escape from cases where prices move freely.

My first paper in international trade was published in Shiozawa (1985). Its main thrust was a generalization of the minimal price theory for a two-country case. I assumed the relative wage rate to be given. In the two-country case, it was sufficient to move the wage rate in one country relative to that of another, from a very low to a very high wage rate. By the intermediate value theorem, I can easily prove the existence of a wage ratio in which each country can have at least one competitive good. This result was entirely obvious, and I did not create an English version of that paper. It is, I believe, in the post-Keynesian tradition. I discussed the proportional growth path, among other things.

When I published that paper in 1985, I was planning to generalize it to many-country (i.e., three- or more-country) cases, and I was thinking that this could be done in relatively short order. However, the general argument with the many-country case was much more difficult than I had imagined. I tried many methods (e.g., the

[12] I omit from here a formal proof. A simple example of nonexistence is found in Fig. 1.

theory of linear inequalities, various forms of the fixed-point theorem, combinatorial geometry, matroid theory, and convex cones, inter alia); none worked, and time passed. My research was interrupted many times. After some time, I almost gave up. The main difficulty was in characterizing the Mill–Jones point. I wanted, at least, to prove its existence. I was also trapped in the thinking that the first task was to find that characterization. I pursued in vain, for more than 20 years, a sufficient condition for the existence of a Mill–Jones point or an open cone where the relative prices can freely move.[13] I eventually succeeded, but it was not a very satisfactory result, because the sufficient condition required too much, and was in a sense tautological. Half satisfied and half dissatisfied, I wrote two papers, one in Japanese and another in English. The latter became Shiozawa (2007).

In writing these papers, I came to know that there is a beautiful one-to-one correspondence between the modal decomposition of the wage or price simplex and the modal decomposition of the production possibility set frontier. This result was more important than the existence of a Mill–Jones point. It was evident that the Mill–Jones point is merely a point on a frontier; if the net world demand was not at that point, however, what happens? I came to know that the situation was very different from that of the Mill–Jones point. If we consider an economy of N-commodities, the production frontier is a set of faces of less than or equal to the N – 1 dimension. The faces of the greatest dimension are in particular called facets; the faces of lower dimension are expressed as the common set of several facets. Then, what happens in (the interior of) a facet must represent a more general situation. With this acknowledgment, all became clear and simple; it was even obvious. There was no need to emphasize that the prices remain constant while the demand moves in the interior of a facet.

With Shiozawa (2007) in hand, I looked for opportunities to discuss these new interpretations and succeeded in talking at 16 seminars and workshops, including the Ricardo Society's 14th Seminar, on January 12, 2008, at Meiji University, Tokyo. This experience gave me the confidence that my idea was headed in the right direction. Summarizing my idea, I formulated a fundamental theorem. A rough result was reported in the International Conference on Structural Economic Dynamics, on September 3, 2012, at Meiji University. A young economist, Yasuaki Tsukamoto, taught me that the history of doctrines can be a good weapon in persuading people to embrace a new idea. I compiled what I had come to know from the long pursuit of the international theory of values; that work became the essential parts of chap. 4 of my book (Shiozawa 2014). Hiroshi Takahashi, the editor of the book, recommended that I newly compose the major part of the book and I reformulated the plan. In Chap. 3, I provide only the core concept; the mathematical

[13] See Theorem 4.3 in Shiozawa (2007), which proves the existence in the wage simplex of an open face in which all countries have at least one competitive product (i.e., a strongly shared pattern of specialization). By the duality theorem (Shiozawa 2007, Theorem 5.7), this corresponds to internal vertex of the production frontier.

parts were concentrated in Chap. 5. This reformulation provided good results, as Chap. 5 became more rigorously reconstructed.[14]

The theoretical situation after Jones (1961) can be summarized by Ethier's two misunderstandings. While the general preoccupation in price adjustment was more epistemological (in the sense that it is more concerned with the focus of interest), the construction of a general theory of international trade with input trade was more substantial, because the omission of input trade changes world trade, both effectively and tremendously. As McKenzie put it, input trade was the vital condition that made the cotton industry revolution possible in Lancashire.[15] In the age of globalization, trade in tasks (or the fragmentation of the production process) and global optimal procurement or global supply chain management are more and more ubiquitous. Any analysis of value-added trade is theoretically impossible if we do not have a theory of input trade. However, generalization to input trade raises a mathematical problem that is structurally different from theory that excludes input trade.

Although I have no intention of entering into the details of mathematical questions, let me distinguish four different types of Ricardian trade models and add some remarks on their mutual relationships. R0 is a pure labor input economy; in this model, all products are produced purely by the labor, and no goods are input for the production. As there are no input goods, there is no input trade. When production is undertaken by labor with the aid of material input, we can distinguish RI and RII. In RI, only final demand goods (or consumer goods) are permitted for export. In RII, input goods are internationally traded, but it is supposed that all countries have the same material input coefficient matrices. The fourth type is the RS economy; in this model, input matrices can differ from country to country, and input products are traded freely (under the same title as other final demand goods).

Simple evidence that the RS economy is mathematically different from other Ricardian economies is given by Jones's aforementioned formula. The internal vertex of an N-country, N-commodity pure labor input economy with labor input coefficients a_{ij} is characterized by Jones's formula as a point that has a specialization pattern that attains the strict minimum among all permutation products:

$$\Pi_{\sigma \in P(N)}\ a_{1\sigma(1)} \cdot a_{2\sigma(2)} \cdot \quad \ldots \quad \cdot a_{N\sigma(N)}.$$

This means that an R0 economy has at most one internal vertex. However, as Higashida (2005) shows, a three-country, three-commodity RS economy can have three internal vertices. This cannot occur if an RS economy has the same mathematical structure as an R0-type Ricardian economy. To define it more concretely, there is no isomorphism that converts an RS economy into an R0 economy via a suitable transformation.

---

[14]Section 3 is a digest of Chaps. 3 and 5 of Shiozawa (2014). Contents of Division IV are not found in Shiozawa (2014).

[15]From a different angle, Samuelson (2001) emphasizes that input trade comprises a new logic of gains from trade.

In comparison to this fact, two extended forms of Ricardian trade economy are structurally identical to a pure labor input economy R0. In fact, in the RI economy case, the prices in the closed economy can act as substitute of labor input coefficients. I do not argue here the equivalence of mathematical structures between R0 and RII, but the hint is the common material net input coefficient matrix. We can choose a square submatrix A that is invertible. Using the inverse matrix A^{-1}, we can transform RII into R0.[16]

If input goods are traded, the cost price of a product of a country is dependent now on the prices of input goods and thus on the wage level of another country. This interdependence makes the analysis of RS economy much more difficult than other types of Ricardian economies.

The equivalences between three types of Ricardian economies were known at the time of McKenzie and Jones. In fact, McKenzie (1954, p. 166) noted that RI is structurally identical (or can be reduced) to R0. Jones (1961, Section 4) reduces RII to R0. However, an RS economy cannot be reduced to R0 or to either RI or RII.

The structural differences among R0, RI, and RII in one part and RS in another are the fact that we should bear in mind when we examine RS economies, because theorems found for R0 or others cannot necessarily be generalized to RS economies. With these structural relations in mind, we can group R0, RI, and RII into a single group of Ricardian economies, whereas the RS economy must constitute an independent class. The formulations and results in Division II are always concerned with the RS economy. As RS economy is a general class that comprises three Ricardian economies as special cases, all results obtained for an RS economy are valid for any Ricardian economy.

As a mathematical entity, Ricardian trade economy (either R0, RI, or RII) has an interesting mathematical structure. The theory of Ricardian trade economy can be interpreted as subtropical convex geometry, based on min-times algebra (Shiozawa 2015). Whether or not this interpretation can be generalized to an RS economy is an open problem. The following exposition is completely independent of this interpretation.

Division II: The Theory

This division (Sects. 3, 4, 5 and 6) presents the new theory of international values. In Sect. 3, the basic assumptions are defined and a fundamental result (i.e., the fundamental theorem) is given. Section 4 shows the light and shadow of international trade. Section 5 explains how to incorporate markup pricing into the theory. Section 6 deals with a delicate question that we cannot ignore, if we wish to implement the theory in an actual economy.

[16]Shiozawa (2014) provides a concrete procedure for converting the RII model (which is there referred to as the Ricardo–Jones trade economy) to R0; see Chapter 4, Subsection 8.2 (pp. 286–287).

3 A Short Summary of the New Theory of International Values[17]

The new theory of international values is constructed on a model that is highly general and permits the trade of intermediate goods. The word "general" here means that the theory does not depend on a special hypothesis vis-à-vis numbers of countries or commodities, and there must be no assumptions regarding symmetry. Such a model is called an RS economy. The new theory redresses the theory of international values, which predates Mill. It is a theory of value that Ricardo would have imagined to construct, but for which he could not provide even a rough design. It contains an account of how wage disparities occur between countries. Discussions of types of specializations are omitted below, but are implicitly contained in the fundamental theorem.

As the theory is explained elsewhere in detail (Shiozawa 2007, 2014), only essential and minimal information is provided here.

We assume the following conditions.

(a) There are M countries.
(b) There are N goods that are traded freely among countries.
(c) The labor of each country is assumed to be homogeneous.
(d) Production is a transformation of inputs into outputs. Input comprises labor and produced goods, and outputs comprise a set of produced goods.
(e) A possible production is a positive combination of productions, each of which belongs to a production technique.
(f) Productions that belong to a production technique are simple and proportional with each one another. ("Simple" here means that the output of a production has only one positive net output.)
(g) The production of a good requires a positive amount of labor.
(h) Any production technique belongs to a country. Labor used in a single technique must be that of the single country to which the production technique belongs.
(i) Goods are transported without cost within a country and from one country to another.
(j) Each country has at least one productive system of production techniques.

Although we assume a finite number of production techniques, there are in general many production techniques that will each produce the same good. Some techniques are in operation and some others are not. Thus, we naturally consider a choice of techniques, and input substitutions are built in as an internal logic of the theory. We see later that a production technique is expressed by a set of input–output

[17]This section is highly mathematical and can be skipped if you understand that new theory international values determine a wage price system $(\mathbf{w}, \mathbf{p})$, where $\mathbf{w} = (w_i)$ gives the wages of countries and $\mathbf{p} = (p_i)$ gives the prices of goods.

(IO) coefficients. These are expressed by physical units that differ from those in the IO tables (IOTs). In Sect. 15, we examine how these two different sets of coefficients are related.

The above conditions (a)–(j) are chosen to give the core framework of the theory. Various generalizations are possible. Condition (f) implies that joint productions are excluded; as a consequence, durable capital goods are excluded. Extensions to include durable capital goods (or fixed capital) are explained in Sect. 8. Condition (g) excludes any production process that requires many production periods (e.g., wine production). This condition can be eliminated by assuming that labor is directly or indirectly necessary; we assume that a production technique can be divided into a series of production techniques and that at least one production technique requires a labor input. As this is rather a classical treatment, we do not explain this process explicitly.

Condition (h) is the crucial property that permits us to construct the whole of the theory. By condition (i), there is no need to distinguish the place where the commodities exist. Of course, this is a strong assumption. Generalization to the case of positive transportation costs is given in Sect. 9. Condition (i) is equivalent to assuming that products are freely traded, and thus intermediate or input trade is incorporated into the new theory. "Productive system" is precisely described in Definition 3.1.

Condition (j) excludes the case where some country cannot produce some products, such as petrol or rare earths; this may seem too restrictive as a real-world model, as many countries cannot produce petrol or rare earths. However, condition (j) is not as strong as it seems. We may suppose that each country has a productive system, but some of its production techniques are extremely inefficient. For example, it can be possible to synthesize petrol, but at a cost that is not economically feasible. Another method is to weaken condition (j) to (j′): the world as a whole contains at least one productive system of production techniques. In this case, it is sufficient that *some* countries can produce petrol or rare earths. In the following, however, this generalization is avoided, because with it, expositions become overly long and complicated.

Labor mobility among countries is not normally considered. Condition (c) implies, in effect, that labor moves freely within a country. In some cases, we can consider a migration of labor force from one country to another. As we see in Sect. 11, this is possible if the migrating labor force can be assimilated into the labor force of the host country. Condition (c) also excludes the case where a country has different categories of labor (e.g., skilled and unskilled labor). Even in this case, most of the results in this paper can be generalized if we can assume that the relative wage rates of the different classes are fixed. However, a new theory of international values cannot treat variation in wage discrepancies among labor classes.

We use the following notations. A set of different goods is denoted by an N-row vector $\mathbf{x}$ and is called the commodity vector. As goods can be transported freely without cost (condition (g)), we can treat goods of the same kind as the same one, independent of the country in which it is found. In the same way, the price of a good is treated as the same everywhere in the world. A price vector will be denoted by an N-column vector $\mathbf{p} = (p(i))$, where $p(i)$ is the price of a good i. The wage rate of a country i is denoted by $w(i)$. A set of wage rates for all countries is denoted by an M-column vector $\mathbf{w} = (w(i))$. A value vector $\mathbf{v}$ is a couple $(\mathbf{w}, \mathbf{p})$ that is also deemed as a column vector of $M + N$ entries. Each entry indicates either the wage of a country or the price of a good.

An industry is a set of activities that produces product i; it is called by the same index i.[18] As we assume that production techniques are simple (condition (d)), each technique belongs by definition to an industry that produces a single positive product. A country is the place where the production takes place (condition (h)). Each country has at least one production process or one production technique for any good. Production processes that produce the same product but belong to (or are operating in) different countries are treated as different techniques. We suppose there are in total H different techniques in the world, where H must be greater than $M \times N$. H is finite if it can be as large as we can imagine.[19] Techniques are numbered in a certain order, but there is no need to enter into this detail. It is sufficient to suppose that this order is preserved for all expressions.

A production technique h is expressed by a net production vector $\mathbf{a}(h)$ that requires one unit of labor input. The set of all production techniques is expressed by an $H \times N$ matrix A comprising vectors of net output vectors $\mathbf{a}(h)$. The set of all labor input is expressed by an $H \times M$ matrix J whose entries are either 0 or 1. The (h, i) entry of matrix J takes the value 1 if and only if h is the production technique of country i. Note that labor is assumed to be different when it belongs to a different country; this is only because labor in country A cannot be used as an input in production that takes place in any other country B. Each row vector of J contains only one entry with value 1, which indicates the country in which the production takes place.

Each country has a certain quantity of labor force $q(i)$. The set of labor forces of the world is denoted by the M-row vector $\mathbf{q}$. A demand vector $\mathbf{d}$ is a set of demand for each product; it is an N-dimensional row vector. Activity vector $\mathbf{s} = (s(h))$ is given by a set of the activities $s(h)$ for each production technique h. It is an H-dimensional row vector. Then, the net material production of the world is $\mathbf{s}A$, and the total labor input of the world is $\mathbf{s}J$. When $s(h)$ is positive, we say that production technique h is in operation.

---

[18]This implies that industry and product correspond on a one-to-one basis. Another possible treatment is to suppose that an industry includes a group of products. Here, we work as if an industry can be divided into product levels. The same explanations apply to relations between firms and products (i.e., firm can produce a variety of products). In the following, we suppose that each firm produces only one product.

[19]There is sometimes the criticism that the Ricardian framework ignores input substitutions. This is a misunderstanding, because substitution occurs between different production techniques that produce the same products. What is excluded is the differentiability of the "production function." Note that this last property is too strong to assume—except in cases such as agriculture, in which one can choose input ratios quite arbitrarily.

A production technique has two affiliations—namely, its country of production and the good it produces. In the following, these are denoted as $c(h)$ and $g(h)$, respectively.

The production possibility set $P(\Gamma, \mathbf{q})$ for a set of techniques Γ is defined as the set of vectors $\{\mathbf{s}A \mid \mathbf{s}J \leqq \mathbf{q}, \mathbf{s} \geqq \mathbf{0}\}$. To examine the production possibility set, we need some basic concepts of the theory of (convex) polytopes. A polytope P is a set of a vector space $\mathbb{R}^N$ that is spanned by a finite set of points of vector space $\mathbb{R}^N$. A face of a polytope P is a subset of P that is the common set of P and the hyperplane of a half-space that contains P. A facet of polytope P is a face of codimension 1; in our case, a facet has the dimension $N-1$. A point $\mathbf{x}$ of a set P is maximal when there are no points z in P that satisfy $\mathbf{z} \geqq \mathbf{x}$ and $\mathbf{z} \neq \mathbf{x}$. By this terminology, $P(\Gamma, \mathbf{q})$ is a polytope in R^N. We are normally concerned with a nonnegative subset of $P(\Gamma, \mathbf{q})$, because such a point can represent only an economy that reproduces itself materially. The frontier of the production possibility set $P(\Gamma, \mathbf{q})$, or the maximal boundary of $P(\Gamma, \mathbf{q})$, is a set of maximal points of $P(\Gamma, \mathbf{q})$. The boundary points of $P(\Gamma, \mathbf{q})$ are covered by a finite number of facets.

Definition 3.1 (Productive System)
A system of production techniques is by definition productive when there exists a nonnegative vector $\mathbf{s}$ such that $\mathbf{s}A > 0$. □

Definition 3.2 (Ricardo–Sraffa Trade Economy)
An economy that satisfies conditions (a)–(h) is called an RS trade economy (or more simply RS economy). □

As noted above, condition (h) can be weakened to (h′), but we assume condition (h) so as to avoid making the propositions overly complicated.

Definition 3.3 (Regular Domain)
The frontier or nonnegative boundary of production possibility set $\mathrm{P}(\Gamma, \mathbf{q})$ comprises a finite number of facets. The interior of any such facet is called the regular domain. □

Theorem 3.4 (Fundamental Theorem for the Ricardo–Sraffa Trade Economy)
Let $\mathcal{E}$ be an RS economy with A, J, and $\mathbf{q}$ as described above. For any final demand vector $\mathbf{d}$ that belongs to the production possibility set, there are a production activity vector $\mathbf{s}$ and an international value vector $\mathbf{v} = (\mathbf{w}, \mathbf{p})$ that satisfies the following conditions:

(i) $\mathbf{s}A = \mathbf{d}$.
(ii) $\mathbf{s}J \leqq \mathbf{q}$.
(iii) $J\mathbf{w} \geqq A\,\mathbf{p}$.
(iv) $\langle \mathbf{q}, \mathbf{w} \rangle = \langle \mathbf{d}, \mathbf{p} \rangle$.

The value vector $\mathbf{v} = (\mathbf{w}, \mathbf{p})$ is unique up to scale if the final demand $\mathbf{d}$ is in a regular domain of the production possibility set, and it remains constant as long as $\mathbf{d}$ stays in the regular domain. □

The proof of the fundamental theorem requires brief preparation. As it is purely mathematical in nature, the proof is given in the Appendix to this chapter.

Remark 3.5 (Nonuniqueness Out of Full Employment)
When demand is not sufficient and the state of the economy is not in full employment, the uniqueness of value does not generally hold. However, if the net production is sufficiently close to a point in the interior of a facet of the frontier, the international value that satisfies conditions (iii) and (iv) is unique up to scalar multiplication. □

Remark 3.6 (Trade Flows)
Even if demand vector of each country $\mathbf{d}(i)$ satisfies the condition

$$\mathbf{d}(1) + \mathbf{d}(2) + \cdots + \mathbf{d}(M) = \mathbf{d},$$

the trade flows from one country to another are not determined. Suppose that a product is produced in two countries $i = 1, 2$, and other countries $i = 3$ consume the product along with the producing countries. Suppose also that country j consumes the product in the amount of z_i. Then, in order for the allocation problem to have a solution, it is necessary that production y_i $(i = 1, 2)$ satisfies the equation:

$$y_1 + y_2 = z_1 + z_2 + z_3.$$

Let x_{ij} $(i = 1, 2, j = 1,2, 3,)$ be the quantity of product that is transported from country i to country j; then, x_{ij} must satisfy the conditions:

$$\begin{aligned}
&x_{11} + x_{21} = y_1\\
&x_{21} + x_{22} = y_2\\
&x_{11} + x_{12} + x_{13} = z_1\\
&x_{12} + \mathrm{x}_{22} + x_{13} = z_2\\
&x_{13} + \mathrm{x}_{23} + x_{33} = z_3\\
&x_{11}, x_{12}, x_{13}, x_{21}, x_{22}, x_{23} \geqq 0.
\end{aligned}$$

This is a simple example of the classical *transportation problem*, and it has solutions that form the *transportation polytope*. In the above case, the polytope has the dimension 2 (i.e., is of degree of freedom 2). In the case of m producer countries and n consumer countries (the same country can simultaneously be a producer and a consumer), the transportation problem has a polytope whose dimension is $(m - 1)(n - 1)$. □

Let us apply suitable names to those international values for which Theorem 3.4 shows the existence.

Definition 3.7 (Admissible and Regular Values)
The international value $\mathbf{v} = (\mathbf{w}, \mathbf{p})$ that satisfies conditions (i)–(iv) for some couple of positive $\mathbf{d}$ and $\mathbf{q}$ is called an *admissible value*; one that corresponds to a final demand on a regular domain when $\mathbf{q}$ is fixed is called a *regular value*.[20]

Definition 3.8 (Competitive Technique)
A production technique h is considered competitive when

$$w\,(\mathrm{c}(h)) = \langle \mathbf{a}(h), \mathbf{p} \rangle$$

for an international value $\mathbf{v} = (\mathbf{w}, \mathbf{p})$ that satisfies condition (iii).

If international value $\mathbf{v} = (\mathbf{w}, \mathbf{p})$ is admissible, any country has at least one competitive production technique, and any commodity has at least one competitive production technique that produces it. This is equivalent to saying that the competitive type associated with the international value $\mathbf{v} = (\mathbf{w}, \mathbf{p})$ is spanning. Regular value has a maximal spanning competitive type.[21]

The number of regular values is always finite, as they correspond one to one with facets of the frontier. The price part $\mathbf{p}$ of the regular value is perpendicular to the facet that includes the final demand. The proof of the theorem is given in Shiozawa (2014; Chap. 5, Theorem 44).[22] See also Theorems 5.2 and 5.7 of Shiozawa (2007). Shiozawa (2007) uses more geometric intuition, whereas Shiozawa (2014) gives a straightforward and algebraic proof.

The conditions of Theorem 3.4 have concrete economic meanings. Suppose that an economy is in a self-replacing state; then, each of four conditions stands for the following propositions:

(i) The supply is equal to the demand.
(ii) The labor force of each country is fully employed.
(iii) No production technique is running with excess profit.
(iv) The value of the net product is equal to the total sum of wages.

The term "excess profit" in (iii) may require an explanation. As I will discuss in Sect. 5, we normally assume a normal markup rate for each industry, and input coefficients are modified into equivalent ones. "Excess profit" here refers to a profit margin that exceeds this markup rate. We are here following Ricardo, who contended that the cost of production should be understood to include (normal)

[20] Japanese readers are asked to note that I changed the definitions of these two notions. In Shiozawa (2014), I provide different notions in Definitions 18 and 19 in Chap. 3 and Definitions 38 and 39 in Chap. 5.

[21] I do not enter into the details of competitive types that have a possible specialization pattern for the set of techniques. The competitive pattern is spanning when it has a link that is connected to any given vertex. We can define an international value as one that has a spanning competitive pattern. It is expected that this definition makes it possible to define admissible and regular values completely independent from demand.

[22] Theorem 17 in Chap. 3 of Shiozawa (2014) provides an equivalent theorem in different expressions.

profits (Ricardo 1951, p. 47, footnote to the 3rd edition). Note that the competitive production technique operates with normal profit margins, even if it is not producing excess profit.

Condition (iv) combined with condition (iii) implies that those production techniques with positive activities are all competitive—that is, they satisfy condition (iii) with equality. The proof of this property is not difficult. Indeed, suppose that there is a production technique with positive $y(h)$ with $w(h) > \langle \mathbf{a}(h)\ , \mathbf{p}\rangle$; then,

$$\begin{aligned}\langle \mathbf{q}, \mathbf{w}\rangle - \langle \mathbf{d}, \mathbf{p}\rangle &\geqq \langle \mathbf{s}J, \mathbf{w}\rangle - \langle \mathbf{s}A, \mathbf{p}\rangle = \langle \mathbf{s}, J\mathbf{w}\text{–}A\mathbf{p}\rangle \\ &\geqq s(c(h)) \cdot \{w\,(\mathrm{c}(h)) - \langle \mathbf{a}(h), \mathbf{p}\rangle\} > 0.\end{aligned}$$

This is a contradiction, because from condition (iv) the leftest member of the equations is 0. By consequence, it follows that no technique with negative excess profit is operating.

We have omitted from the above formulation any reference to markups. In other words, we assume that all markup rates are 0. In such a case, no profit is produced by production. Cases of positive markup rates are addressed in Sect. 5. The definitions of "net output matrix" and "final demand" require some modifications; with these modifications, we can reinterpret Theorem 3.4 as proving the existence of a proportionally growing economy with positive profits.

Let us remember that the new theory of international values is based on a wide set of circumstances, where each country has its own set of production techniques. Although we do not explicitly mention firms, such circumstances can also comprise cases where a firm holds several production techniques that produce the same products (see also footnote 18). Accordingly, Melitz (2003) and others' "new new trade theory" can easily be incorporated into the framework of the new theory of international values. Questions of the choices of techniques and input substitutions are incorporated into and solved in this framework.

Another important characteristic is that the new theory is a natural generalization of the classical value theory. We stated above that the new theory of international values "is a theory of value that Ricardo would have imagined to construct, but for which he could not provide even a rough design." To support this, the new theory must satisfy the essential characteristics of the classical theory of value or the cost-of-production theory of value. The most important point is that regular value remains constant whenever the final demand changes within the same regular domain. Although it is necessary to add the supplementary condition "within the same regular domain," international value defined in this way conserves the basic property of the classical theory—namely, that value is primarily independent of demand. This contrasts sharply with the neoclassical theory of value, in which small variations in demand crucially change prices. In the new theory, there is no need to appeal to concepts such as marginal product or marginal cost.[23]

[23]This is in accordance with Sraffa's opinion, as expressed in the preface to Sraffa (1960).

In the classical theory of value—including the new theory of international values—the values and quantities are primarily separated. Thus, quantity variables such as production scale and number of employment moves (within a certain range) are independent of value variables. The most conspicuous effect of this independence is that we can examine the circumstances in which unemployment exists. We discuss this point in the next section.

It is also important to note that the classical theory is not based on the GE framework. For example, Theorem 3.4 describes the existence of a self-replacing state, but it does not assume or affirm that the economy converges to such a state. On the contrary, the theorem can be interpreted as showing how difficult it is to bring about a state of full employment.

In the next section, we will first see the gains from trade and then the possibility of trade conflicts, including unemployment problems.

4 Gains from Trade and Possibility of Trade Conflicts

Suppose an RS economy $\mathcal{E}$ with A, J, and $\mathbf{q}$ as defined in Sect. 3. We can imagine each country's closed economy $\mathcal{E}(i)$, with production techniques and a labor force belonging to country i. Suppose that each country has at least a productive system of techniques. The economy $\mathcal{E}(i)$ with matrices $A(i)$, $I(i)$, and $\mathbf{q}(i)$ composes country i's closed economy. In a closed economy, or an economy in one country, we have the minimal price theory, which can be expressed in various forms.[24] The lemma that follows is one of them.

Lemma 4.1 (Minimal Price Theorem)
Let $\mathcal{E}$ be a one-country economy that satisfies conditions from (c) to (g), and (j) in the definition of an RS economy. Then, there exists a system of production techniques that gives the minimal price for all goods when the wage rate is fixed at w.

Lemma 4.1 can be paraphrased as follows. A system of production techniques is a set of production techniques that includes exactly one production technique that produces each of all goods. If a system of production techniques γ is productive, its associated input coefficient matrix $\mathrm{A}(\gamma)$ is a square matrix and nonnegatively invertible—that is, $\mathrm{A}(\gamma)^{-1}$ exists and is nonnegative. Thus, a price vector $\mathbf{p}$ associated to system γ with wage w can be expressed by the formula

$$\mathbf{p} = w \cdot \mathrm{A}(\gamma)^{-1}\mathbb{1},$$

[24]This lemma was discovered by Samuelson and named the nonsubstitution theorem. Samuelson proved the two-good case and Koopmans (1951b) the three-good case. The general N-good case was proved by Arrow (1951). It is known that the lemma was discovered simultaneously and independently by Samuelson and Georgescu–Roegen in 1949. See Samuelson (1951) and Georgescu–Roegen (1951).

where $\mathbb{1}$ is an N-dimensional column vector whose entries are all 1. Lemma 4.1 tells us that there exists a system of production techniques γ^*, such that

$$\mathbf{p}_* = w \cdot \mathrm{A}(\gamma *)^{-1}\mathbb{1} \leqq \mathbf{p} = w \cdot \mathrm{A}(\gamma)^{-1}\mathbb{1}$$

for all system γ of production techniques.

As a corollary to Lemma 4.1, we obtain Theorem 4.2.

Theorem 4.2 (Gains from Trade)
Let $\mathcal{E}$ be an RS economy with A, J, and $\mathbf{q}$. Let $\mathbf{v} = (\mathbf{w}, \mathbf{p})$ be an admissible value and $\mathbf{p}^*(i)$ be the minimal price vector with wage w_i. Then, we have

$$\mathbf{p} \leqq \mathbf{p}_* (i) \,. \tag{1}$$

If $\mathbf{p}$ is not proportional to $\mathbf{p}^*(i)$ for a country index i, an inequality of (1) is strict for some product j. □

The proof of Theorem 4.2 is easy. If $\mathbf{v} = (\mathbf{w}, \mathbf{p})$ is admissible, we have condition (iii) of Theorem 3.5. Then,

$$J\mathbf{w} \geqq A\mathbf{p}. \tag{2}$$

Taking production techniques that belong to country i, (2) can be expressed as

$$w_i \mathbb{1}\,(i) \geqq A\,(i)\,\mathbf{p} \tag{3}$$

for all h belonging to i, where $\mathbb{1}(i)$ is the $H(i)$-column vector comprising only 1 and $A(i)$ and $H(i)$ are respectively the matrix and the number of production techniques belonging to country i. Now let γ be the system of techniques that gives the minimal price of country i. Restricting inequalities (3) to the production techniques that belong to γ, we obtain

$$w_i \mathbb{1} \geqq A\,(\gamma)\,\mathbf{p}.$$

In applying to this inequality the nonnegative matrix $A(\gamma)^{-1}$ from the left, we obtain

$$\mathbf{p}_*\,(i) = w_i A(\gamma)^{-1}\mathbb{1} \geqq A(\gamma)^{-1} A\,(\gamma)\,\mathbf{p} = \mathbf{p}.$$

This proves the theorem.

Inequality (1) means that the real wage level for workers in country i is higher under international values than that possible in a closed economy. It is strictly higher when $\mathbf{p}^*$ is not proportional to $\mathbf{p}$. It is important to note that the conclusion of Theorem 4.2 applies only to workers who continue to be employed; these gains from trade do not apply to workers who are dismissed on account of the opening of international trade, or to entrepreneurs who are similarly obliged to close their business. Neoclassical economics usually assumes that full employment is achieved soon, if not immediately, and ignores such losses from trade. However, as Theorem 4.3 shows, it is possible that unemployment will continue if no measures are taken.

Theorem 4.3 (Existence of Unemployment)
Let $\mathcal{E}$ be an RS economy with A, J, and $\mathbf{q}$. Suppose there exists at least one pair of countries for which the minimal price vectors are not proportional to each other. Let the positive vector $\mathbf{x}(i)$ be the net product of a self-replacing state of the closed economies and $\mathbf{x} = \sum_{\mathrm{i}} \mathbf{x}(i)$ be the sum of those vectors. Finally, suppose that an international value ($\mathbf{w}$, $\mathbf{p}$) and an activity vector $\mathbf{y}$ satisfy for a suitable $\mathbf{t}$ the following four conditions:

(a) $\mathbf{y}A = \mathbf{d} \leqq \mathbf{x}$,
(b) $\mathbf{y}J = \mathbf{t} \leqq \mathbf{q}$,
(c) $J\mathbf{w} \geqq A\ \mathbf{p}$, and.
(d) $\langle \mathbf{t}, \mathbf{w} \rangle = \langle \mathbf{d}, \mathbf{p} \rangle$.

The system $\mathbf{y}$, $\mathbf{d}$, $\mathbf{w}$, and $\mathbf{p}$ forms a self-replacing state, and all operating techniques are competitive. In this self-replacing state, at least one country suffers from unemployment. □

As we have assumed, there are two countries in which minimal price vectors are not proportional. Then, there must be at least one country i where price vector $\mathbf{p}$ is not proportional to its minimal price vector. This means that

$$\mathbf{p} \leqq \mathbf{p}_* (i) \text{ and } \mathbf{p} \neq \mathbf{p}_* (i) . \tag{4}$$

Theorem 4.3 follows from a simple calculation:

$$\langle \mathbf{t}, \mathbf{w} \rangle = \langle \mathbf{d}, \mathbf{p} \rangle \leqq \langle \mathbf{x}, \mathbf{p} \rangle = \langle \sum\nolimits_i \mathbf{x}(i), \mathbf{p} \rangle$$
$$< \sum\nolimits_i \langle \mathbf{x}(i), \mathbf{p}_* (i) \rangle \leqq \sum\nolimits_{\mathrm{i}} \mathbf{q}(\mathrm{i})\, w_{\mathrm{i}} = \langle \mathbf{q}, \mathbf{w} \rangle.$$

The first equality holds from (d), the second inequality from (a), the third equality by definition, the fourth from the first part of (4) for positive $\mathbf{x}(i)$, the fifth by the fact that $\mathbf{x}(i)$ can be purchased by the wage of all workers, and the seventh by definition. The fourth inequality holds strictly, because $\mathbf{p} \neq \mathbf{p}^*(i)$ for some i. In conclusion, we obtain a strict inequality

$$\langle \mathbf{t}, \mathbf{w} \rangle < \langle \mathbf{q}, \mathbf{w} \rangle .$$

This means that the weighted sum of all countries' employment with weights w_{i} is smaller than the weighted sum of the world's labor force, and there is at least one country where some workers are unemployed. Q.E.D.

Note that in the formulation of Theorem 4.3, vector $\mathbf{t}$ in condition (b) need not be assumed to be less than vector $\mathbf{q}$. This means that unemployment is inevitable, even if workers move across country borders. Note also that the value $\mathbf{v} = (\mathbf{w}, \mathbf{p})$

is used in this proof only as weights of aggregation; it need not be the actual wages and prices of an economy.

If we combine Theorems 3.4 and 4.3, we can say the following. Theorem 3.4 tells us that there is a self-replacing state with an international value in which full employment is attained. However, Theorem 4.3 also says that with the same wage price system, unemployment necessarily occurs if world demand remains the same as before trade. Neoclassical economists have the custom of assuming that price adjustment will always suffice for full employment, but Theorem 4.3 says this is not so.

We know that Mill and his followers assumed that each commodity has an elasticity value of -1. The same assumption is adopted by Dorbusch et al. (1977). This is equivalent to assuming Cobb–Douglas demand functions. If demand changes in this way, each country's demand for competitive goods increases in such a way that total employment remains unchanged. However, if this assumption fails to hold, some countries' demand for their competitive products will become overly large, and the total necessary employment exceeds their total labor force. In this case also, the production and employment of a country are restricted to below or equal to the country's labor force, and this implies that other countries must suffer when there is a lack of effective demand.

5 Questions Related to Markup Rates

In Sects. 3 and 4, we supposed that the markup rate was 0, but this is tantamount to gross negligence. I consciously did this to avoid unnecessary complication, as a first construction. Now let us observe explicitly the effects of positive markup rates.

Markup is a common practice that is widely employed by small and large firms in determining product or service prices. If the cost is known, the sale price is the original cost plus the markup amount. This markup amount is usually calculated by a percentage markup. If the markup rate is m and the unit cost is c, then the price is set by the formula

$$p = (1 + m)\, c.$$

In actual practice, many complications intervene. What is the unit cost, and how can we calculate it? If we are concerned with a single product, we can directly estimate the unit cost by observing the accounting dataset. If we are concerned with many items—as is the case with supermarkets or multiproduct producers—it may be out of the question to calculate the exact cost for each item. Most often, costs are classified as variable (proportional) costs or fixed costs. The unit cost is the sum of all proportional costs per unit of product.[25] The total cost is the sum of the fixed costs and the unit cost times the number of units of production. One of the practical

[25]In the process of designing a car for a new market, for example, the unit cost is not a fixed constant; it is variable. It is a target to be determined, in order to produce a satisfactorily good sales

objectives of markup rates is to cover the fixed costs by the gross margin, i.e., the markup times the units of production. However, it is necessary to note that if fixed costs exist, the markup rate is not equal to the profit rate. Indeed, the total profit depends on the volume sold. If the product is sold by quantity y at the fixed markup price (i.e., at $p = (1 + m)\,c$), the gross profit is

$$m\,c\,y.$$

If the total fixed cost is F, the operating profit may be negative if

$$y < \mathrm{F}/\,(m\,c)\,.$$

The equality is attained at the breakeven point. If the sales volume exceeds this point, the operating profit is positive and increases if y further increases.

Markup rates are often determined by convention.[26] There is a certain standard level of markup rates for each industry, and it may differ by country. However, that level of markup rate is not arbitrarily determined. Roughly speaking, markup rates are determined by the state of market competition for each product;[27] this partially justifies the well-known *pricing-to-market* practice. If the price is set, the firm sells as much product as the purchase offers require. Then, with some auxiliary assumptions, a markup rate is calculated as the best policy for a firm. The key point is that one should assume that the market share of the firm's own product among competitors' products is a function of the relative prices of those products.[28] When the firm adopts this markup rate, the firm achieves maximal profit. In this sense, it is the market that determines the markup rate. However, it is also necessary to note that market competition is not independent of markup rates. As I indicate in Shiozawa (2016), if the markup rate determined by the market is overly low, the firm cannot cover the depreciation of its fixed capital. In such a case, some firms will be obliged to exit from the market, and the competition within the product pool will change. This partially explains why the markup rate differs by industry. See Shiozawa (2016; 2014 Appendix) for a fuller explanation.

---

price by way of markup pricing. This practice is well known as *target costing*. However, we are mainly concerned with the pricing practice when the manufacturing process is already in operation.

[26] On the flip side of this convention, a standard level of sales volume is often assumed. If, at this volume, profit remains negative, the firm is obliged to exit from the market and the state of competition will change. Through such long-term adjustment, the markup rate determined by the market most often produces a positive profit at the supposed standard level of sales volume. See also the next paragraph.

[27] The market's competitive condition may change in tandem with changes to currency exchange rates, as we see in Sect. 6.

[28] The markup rate is often explained in terms of kinked demand curves. However, this does not provide a good explanation of the markup rate, because a markup price is determined without clear reference to the sales volume. The markup price is an offer price that expresses the supplier's attitude that it is ready to sell any amount of product at the fixed set price (within a certain commonsense range, of course).

Demand changes for a variety of reasons, if the prices are all fixed. For example, we may enumerate weather, temperature, special events, and topics on the television or Internet. The concept of demand function is not well based. The market share function may not be exactly known, but if the market's competitive condition does not change much, the firm can derive the most profitable markup rate through trial and error.

If markup rates are positive, we need to modify the input coefficient matrix. Let us suppose that a markup rate $m(i,j)$ is fixed for each pair of country i and product j. Suppose that when a production technique h is competitive, the market price must be equal to the markup price. This means that the equation

$$\{1 + m(i,j)\} \cdot \{w(c(h))\, a_0(h) + \langle \mathbf{a}(h), \mathbf{p} \rangle\} = p_{\mathrm{g(h)}} \tag{5}$$

must hold. Here $\mathbf{a}_0(h)$ and $\mathbf{a}(h)$ expresses the input coefficient vector respectively for labor input and material inputs and differs from $\mathbf{a}(h)$ in Sect. 3. Remember that the expression $\mathbf{a}(h)$ in Sect. 3 is taken to be the net output of the production that requires a unit of labor input. The input vector $\mathbf{a}(h)$ must be transformed to express the net output. This can be done when we transform Eq. (5) into the following form:

$$w(c(h)) \cdot \mathrm{a}_0(h) = (1/\{1 + m(i,j)\}) \cdot p_{\mathrm{g(h)}} - \langle \mathbf{a}(h), \mathbf{p} \rangle.$$

The right-hand side is the net output coefficient vector which corresponds to virtual production with labor input $\mathbf{a}_0(h)$. Let us now define a new net output coefficient vector $\mathbf{a}_{\mathrm{e}}(h)$ by the vector:

$$(1/a_0(h))\,(1/\{1 + m(i,j)\})\, \mathbf{e}(g(h)) - \mathbf{a}(h).$$

If we interpret this vector $\mathbf{a}_{\mathrm{e}}(h)$ as a coefficient vector that represents a virtual production technique, we derive a new system of production techniques. The matrix A_e comprising vectors $\mathbf{a}_{\mathrm{e}}(h)$ satisfies the same value relations as (iii) of Theorem 3.4—that is,

$$J\mathbf{w} \geqq A_e\mathbf{p}.$$

We can now define the equivalent RS economy. An RS economy with A_{e}, J, and $\mathbf{q}$ is called an equivalent RS economy. With the equivalent RS economy, all stories developed in Sects. 3 and 4 hold without modification, as far as value relations are concerned. However, notions and analyses concerning quantities require more cautious treatments. For example, $\mathbf{d}$ in (i) and (iv) of Theorem 3.4 needs to be reinterpreted. If you replace A with A_{e}, you can have the same relations (i)–(iv) for the equivalent RS economy $\mathcal{E}_e$, but $\mathbf{d}$ (or perhaps $\mathbf{d}_{\mathrm{e}}$) should be interpreted as a different bundle of goods that appears in the original expression of Theorem 3.4. Let me explain.

For simplicity of discussion, let us suppose that all markup rates are the same and equal to m (where $m > 0$). Generally speaking, the production possibility set of an equivalent economy $\mathcal{E}_e = \{A_e, J, \mathbf{q}\}$ for $m > 0$ "shrinks" in comparison to the original $\mathcal{E} = \{A, J, \mathbf{q}\}$ for $m = 0$. The commodity vector $\mathbf{d}$ is a world final demand in $\mathcal{E}$. It is a sufficient demand that gives full employment through condition (ii) in Theorem 3.4. If we compare two production possibility sets of $\mathcal{E}$ and $\mathcal{E}_e$, we find that that of $\mathcal{E}_e$ is smaller than that of $\mathcal{E}$. Then, vector $\mathbf{d}$ is outside the production possibility set P_e of $\mathcal{E}_e$. Does this mean that, in an equivalent economy, the world demand for full employment could be smaller than that demanded in the original economy? How does this singular situation happen?

All these peculiarities occur on account of the ambiguity of two concepts—namely, final demand and the production possibility set. We normally assume that these concepts have well-defined meanings without any explicit reference to growth rate or other factors. In fact, we naturally think that these two concepts correspond to some objective entities and have invariant meanings that do not change, whether we think of a self-replacing state or of a proportionally growing path. However, this is a misunderstanding: these concepts implicitly depend on the economic state we imagine. To be more precise, let us suppose that we are interested in investigating a proportionally growing path. Let the growth rate be g. For simplicity, consider a closed economy of a single country. Let $\mathbf{a}_0$ be the labor input coefficient vector, A the material input coefficient matrix, and I the output coefficient matrix. Again, for simplicity, we assume that in our virtual economy there is only one production technique for each product. Then, A is a square matrix and we can assume I is an identity matrix. Let us assume we have a series of productions:

$$\mathbf{y}(0), \mathbf{y}(1), \ldots, \mathbf{y}(\mathrm{t}), \ldots.$$

If this series grows proportionally at constant rate g, the following equations hold:

$$(1+g)\,\mathbf{y}(0) = \mathbf{y}(1), (1+g)\,\mathbf{y}(1) = \mathbf{y}(2), \ldots, (1+g)\,\mathbf{y}(\mathrm{t}) = \mathbf{y}(\mathrm{t}+1), ... \quad (6)$$

In this series, what are the net products? To produce $\mathbf{y}(1), \mathbf{y}(2) \ldots, \mathbf{y}(\mathrm{t}+1), \ldots,$ we need material inputs in addition to labor inputs. These are

$$\mathbf{y}(1)A, \mathbf{y}(2)A, \ldots, \mathbf{y}(\mathrm{t}+1)\,A, \ldots.$$

A natural definition of net output for this series would be the following:

$$\mathbf{y}(0) - \mathbf{y}(1)A, \mathbf{y}(1) - \mathbf{y}(2)A, \ldots, \mathbf{y}(\mathrm{t}) - \mathbf{y}(\mathrm{t}+1)\,A, ... \quad (7)$$

These are what we can extract from the economy with constant growth. If we substitute equalities (6) into (7), we derive

$$\mathbf{y}(0)\ \{I - (1+g)\,A\}, \mathbf{y}(1)\,\{I - (1+g)\,A\}, \ldots, \mathbf{y}(\mathrm{t})\,\{I - (1+g)\,A\}, ... \quad (8)$$

It would be natural to define (8) as the net output for each period. In this formula, it is evident that the concept of "net output" depends on the growth rate.

Now let us return to the international trade economy. If the net output depends on the growth rate, the concept of the production possibility set depends on the growth rate of the underlying economy. When the uniform markup rate m and the underlying growth rate g are equal, there is no problem with simply examining the equivalent economy. The value obtained for the equivalent economy provides a value (i.e., systems of wages and prices) that determines competitive production techniques with a markup rate m. The demand vector $\mathbf{d}$ gives a bundle of commodities that can grow at the constant rate g. However, if $m \neq g$, we need an appropriate conversion for the final demand vector.

Let us consider the case where $g < m$. (The opposite case, $g > m$, is a bit more complicated, because firms cannot accumulate sufficient funds for growth from their internal reserves, and so must acquire money from workers' savings.) We can consider two equivalent economies—namely, $\mathcal{E}_e(m_0)$ and $\mathcal{E}_e(m_1)$, where $m_0 = m$ and $m_1 = g$. Each of the two economies $\mathcal{E}_e(m_0)$ and $\mathcal{E}_e(m_1)$ has a set of regular values and a set of corresponding systems of competitive techniques. It may happen that two systems of the sets of competitive techniques differ from each other; let them be $S(m_0)$ and $S(m_1)$. The production possibility set $P(m_1)$ associated with $\mathcal{E}_e(m_1)$ with growth rate $g = m_1$ is of course convex. However, if competitive techniques are chosen by the system $\mathcal{E}_e(m_0)$, the production possibility set $P(m_1, S(m_0))$ with growth rate $g = m_1$ by means of the competitive technique $S(m_0)$ set is included in the possibility set $P(m_1)$, and it may not even be convex. Takamasu (1986) provides an example of a concave production possibility frontier in the case of a closed economy when land intervenes as a production constraint. A similar situation occurs in the case of international trade, even if there are no constraints other than labor. See examples in Oka (2017, Chap. 3 of this book). He explains this fact with a set of different concepts. There is no inconvenience in this, because we are treating different trajectories with the same growth rate, and there is no logical need for those net outputs to form a convex set.

Further analyses are needed for quantity relations, but these are the foci of future research.

6 Problems with Exchange Rates

The foreign exchange market poses a delicate problem with respect to the new theory of international values. All that the new theory can say is found in Theorem 3.4, which implies that there is at least one international value vector that makes it possible to realize full employment for all countries by virtue of their competitive production techniques. In general, we have an infinite number of admissible values and a finite number of regular values for a given RS economy. When a world final demand is given, there is the strong possibility of a specific regular value being chosen. However, unlike the GE theories, the new theory does not contend that the actual international values (the actual system of wages and prices) will converge to one such regular value (thought of as a vector) and full employment will be attained. On the contrary, the new theory emphasizes the difficulties inherent in finding one such system of international values. However, if any system of international values bears some practical meaning, we need to deal with problems on how to interpret daily fluctuations in exchange rates—or rather, on how to harmonize highly volatile exchange rates—and the requirement of price stability, the latter of which the new theory presupposes as a criterion for firms in choosing more competitive production techniques.

The greatest challenge is that foreign exchange rates shift drastically and extensively. High volatility in itself is not a big problem, if the moving average shifts slowly; if this be the case, we can interpret the moving average as reflecting a slow but long-term change in economic conditions, mainly as effects of differential technological progress. However, an actual exchange rate sometimes jumps up or down over a few days and often stays at around the new rate level for many months. For example, during first the three quarters of 2012, the JPY–USD rate was at around JPY80 = USD1: then in Q4 of 2013, the rate fell to around JPY100 = USD1 and it stayed at around that level for approximately 17 months thereafter; then, in Q4 of 2014, the rate fell to around JPY120 = USD1. During the first change, the JPY fell 20%; during the second change, it fell another 17%. During that 2-year period, the JPY depreciated to two-thirds its original value. The JPY120 = USD1 exchange rate remained stable for about 1 year (i.e., throughout 2015); from the start of 2016, the trend was reversed and the JPY was appreciated, motivated by various developments, including the British referendum to leave the European Union (EU).

What was more curious about this depreciation was that Japanese exports in real terms did not increase much. Many explanations are possible, but these are beyond the scope of this paper. Our problem lies in the fact that this depreciation led to a change in relative wage rates (i.e., −33% from the Japanese side and 50% from the US side). In the new theory of international values, wage rates are expressed by a given international currency, either real or imaginary. A change in exchange rate of −33% (50%) signifies the depreciation (appreciation) of the wage rate, from the Japanese (US) perspective. Large changes to such extents will necessarily have substantial effects on the competitiveness of production techniques, on both sides. In reality, there was no large change in Japanese exports. Here arise two problems. The first problem is the speed of firms' reaction; the second is the relationship between exchange rates and the relative wages, as assumed in the new theory.

As for the first problem, firms need to take into consideration various factors and conditions. Even if a large exchange rate fluctuation were to occur, it may not be wise to react to the new situation too promptly: exchange rates are extensively volatile and it is possible for them to "swing back" relatively promptly in the opposite direction. A change in exchange rate also changes the competitive conditions of export or import markets; if those resulting changes are considerable, firms would be obliged to reconsider their markup rates. They need to observe how their competitors are behaving during extensive changes in the competitive conditions. They also need to think about preserving client loyalty, be it of consumers or of industrial purchasers.

In the case of the recent JPY depreciation (2012–2015), many consumer goods makers did not increase their product prices for about 1 year, in spite of a year-long price increase in imported materials. This reaction may have been conditioned by the long-term deflation (or rather, the price stabilization) of the Japanese economy. On the other hand, this reaction was made possible by the fact that many firms held large amounts of internal reserves. Exporters did not change their selling price in the importing currency, despite the fact that doing so would have ultimately reduced product prices and extended market share. Had the depreciation occurred in the 1960s, Japanese firms would have behaved very differently. These brief observations indicate that when large exchange rate changes occur, firms do not react promptly; rather, they "wait and see" for about 1 year.

The reactions of firms to exchange rate changes are being studied extensively, under the topic of "exchange rate pass-through to prices." Gopinath and Rigobon (2008) report that the duration of the median price in the pricing currency is 10.6 months for US imports and 12.8 months for US exports. Nakamura and Zerom (2009) report that the coffee industry changes its wholesale price 1.3 times per year and its retail price 1.5 times per year. Lewis (2016) found that pass-through is strongly nonlinear with exchange rates: the pass-through of larger bilateral exchange rate movements (i.e., more than 5%) is around four times larger than that of smaller changes. Frankel et al. (2012) report that pass-through rates and delays may change with time and by development stage.[29]

The second problem is much more difficult to answer. High volatility in the exchange rate market itself is not astonishing: it is something we see in many financial markets. The problem we face is as follows: do the relative wage rates supposed in the new theory have something to do with the level of exchange rates?

Evidently, exchange rates are always fluctuating and show no apparent tendency to converge. What kind of long-term characteristics does an exchange rate exhibit? In the decade following 1996, a closely related question was discussed under the rubric of the purchasing power parity (PPP) puzzle (Rogoff 1996). The question that was discussed at that time was not the validity of the classical PPP hypothesis, which assumes the equality of price indices (when they are converted by exchange rate to a single currency expression). It is obvious that this hypothesis will not hold if the price index includes nontradable goods and services. Because of the Balassa–Samuelson effect, the consumer price index of a high-income country tends to be higher than that of a low-income country. Even among comparable high-income countries, absolute levels of price indices have no clear tendency to converge to equality. Therefore, the question discussed after 1996 concerned the convergence of relative PPP indices.

The PPP puzzle—or the "Rogoff puzzle," as some economists have dubbed it—is a complicated question that requires deep knowledge of time series statistics (Taylor 2001; Yabu 2007). If I introduce some conclusions from the long debates, the central

[29]I do not enter into details of the measurements. We can argue, for example, whether or not a log-linear regression is adequate, if we consider the additive character of costs. Here, I simply report the raw numbers, only to derive a rough estimation of the price-adjusting behavior of firms.

question is the evaluation of the half-life of the exchange rate time series. At the time of Taylor (2001), it was believed that the half-life was "of the order of 5 years at best, and infinity at worst" (Taylor 2001, p. 473 in Abstract). The discussion continued on the nature of half-life estimation, and a new consensus emerged until 2005: the half-life of the relative PPP divergence from the limiting average was estimated to be between 3 and 5 years (Yabu 2007). It was also made clear why the unit root test behaved so poorly for about 20 years after the shift to the floating exchange rate system. Approximately 100 years of data were needed to determine whether or not the series has a unit root, if one worked in quarterly time series. Thirty to forty years after shifting to a floating system, the estimation became sufficiently accurate to discern a 4-year half-life series and the unit root series.

If we believe the recent estimation, the convergence speed of the relative PPP rate is very slow and requires 3–5 years until an accidental deviation from the trend halves. This fact, together with the delayed price response to cost changes on the scale of 1 year, teaches us the time span of the new theory of international values. This accords with recent observations in Japan. During the period of JPY depreciation, firms endured a 25–50% increase in the cost of imported goods, without raising their product prices for about a year. We have to think that the adjustment of wage levels works in this time scale.

A third question arises in relation to the Rogoff puzzle—namely, the validity of the law of one price. This question more directly relates to the new theory, because the law of one price serves as the basis of the new theory of international values (especially when the transportation cost is negligible). Studies on the law of one price should leverage not aggregated price indices, but product-wise price movement across countries. To obtain a good estimate of the effectiveness of the law, studies would be better done in a situation where trade barriers are minimal. The EU's great experiment with a unified currency provides us with a good occasion to evaluate the actual effectiveness of the law.[30] We can hope that these empirical studies will give us the opportunity to refine the theory of international values.

Division III: Extensions and Generalizations

In this division (Sects. 7, 8, 9, 10 and 11), we treat the question of how to interpret or generalize the theory and its components, to adapt them to situations not included in its typical formulations. The questions that relate to markup rates were already discussed in Sect. 5. Our first point of discussion here is how to interpret production techniques.

7 Primary Resources

The question of primary resources is not a problem of extension or generalization; rather, it is simply a question of interpretation.

[30] Although there is no specialized examination on cross-country price differences, Dhyne et al. (2006) and Vermeulen et al. (2012) are the first attempts.

In the formulation in Sect. 3, each country has its own set of production techniques. Thus, if we take the case of agriculture, the production technique that produces wheat may be different across countries, given differences in climate. In a temperate zone, we may cultivate wheat easily, whereas in arctic or tropical zones, wheat cultivation may not produce good results. These differences of efficiency that stem from climate can be easily incorporated into differences in production techniques. Our basic assumption on production techniques is that inputs are proportional to the scale of output. As far as these proportional relationships hold, there is no problem in reflecting climate conditions and other environmental effects in production techniques.

The classical question of decreasing returns arises when cultivation extends to less-fertile or poorly irrigated lands. It is a question of rents. This study does not address this question; it requires a proper theory of value that differs from domestic or international theories of value.[31] Christian Bidard, in his series of papers (Bidard 2010, 2011, 2014), studies this question energetically. For our theory of international values, the different levels of production efficiency do not pose a problem, because they can be treated as different production techniques. If no limits in the scale of production are effective, the question of which piece of land we choose is the question of the choice of production techniques.

In the same vein, underground resources pose no problem, inasmuch as extraction can be continued at the same level of efficiency and at the required volume per unit of time. If the firm has an ample demand but the production volume is bounded by the limit of underground resources, we need to appeal to the theory of rents.

The existence of primary resources does not in itself necessarily imply that production with primary resources needs to be treated as per the theory of rents. For example, Sweden produces high-quality iron core. As the potential capacity of extraction is gigantic and the actual demand for this quality of iron core is far smaller than the capacity, we can treat iron core extraction as an ordinary production process. In this sense, even in a case of production that presupposes the existence of primary resources, we can treat them as normal production techniques, and the new theory of international values will hold without explicit consideration for the amount of primary resource reserves. Just as a dissipative structure self-regulates its energy flows, the activity levels of an economy are normally limited by the internal logics of the economy itself (e.g., effective demand and profitability, inter alia).[32] Quite frequently, production capacity exceeds the required production volume; thus, the theory of international values holds in quite a broad array of situations, without appealing to the theory of rents.

[31]The classical theory of value probably comprises five fields: (1) domestic theory of value, (2) international theory of values, (3) theory of rents (land and exhaustible natural resources), (4) theory of wages (inside a country), and (5) price theory of financial markets. The first three fields have relatively firm theories of value, while the last two require completely new approaches.

[32]See, for example, Shiozawa (1996).

8 Durable or Fixed Capital Goods

Capital goods such as machines and installations can be formally treated as the production of old machines and installations, together with the main product. The existence of durable capital goods thus means that we need to assume that two or more products are produced at the same time, via a single production technique. In other words, the existence of these durable capital goods violates condition (f) in Sect. 3.

When two or more products are produced (i.e., if the net products are positive for more than two goods) in a single production technique, such a production technique is called *joint production*. John Stuart Mill, in a chapter titled "Some Peculiar Cases of Value" (Mill 1848, III. 16.), discusses joint production under the term of "joint cost of production," wherein two commodities are produced by the same "operation." His favorite example is the production of coke and coal-gas from coal (i.e., the carbonization of raw coal). In such a case, we can determine the cost of neither coke nor coal-gas. Mill argues that the cost-of-production theory of value fails in this case, and so "we must revert to a law of value anterior to cost of production, and more fundamental, the law of demand and supply" (Mill, III.16.5).[33] Von Neumann (1944) introduced the idea of treating durable capital goods as a matter of joint production; Morishima (1973) highly praised this event, calling it the "von Neumann revolution." However, a von Neumann–Morishima type of joint production treatment is too general and does not permit detailed analysis (save for balanced growth and a limited number of other cases). It is necessary to introduce some good properties that are sufficiently general and easily tractable. One such property is the widely observed custom to assume constant efficiency during the life span of the machines and that they will be freely destroyed at the end of the life span. This is the solution given in Chapter 10 of Sraffa (1960). A more general treatment that includes the choice of production technique is found in Shiozawa (1975); in this case, if a markup rate is given for production, the value contribution of the machine is calculable, and the joint production problem can be easily reduced to a simple production case. The same is possible for the international trade case.

9 Transportation and Transaction Costs and Nontraded Goods

In recent years, transportation and communication costs have decreased substantially. This, together with reductions in transportation time, has materially changed the face of the world economy and of international trade. To analyze the effects of

[33]Chapter 7 (Shiozawa 2017) examines the consequences of this "reversion" to the history of economics.

reductions in transportation and transaction costs, we need to reformulate the space of commodities in such a way that we can distinguish the same products situated in different countries. Doing so is the motivation of this section. Herein, when we say "transportation cost," we include transaction costs that relate to the transport of goods.

When we introduce transportation costs, we must distinguish the location of products. Any goods are labeled with the pair (j, i), where j is the product index and i is the country index. We refer to these as location-specified goods. Then, any commodity can be labeled with a pair of indices (j, i). Note that in this expression, the product index j precedes the country index i. When two commodities have the same product index j with two different country indices, we are considering the same good, but in different countries. In a transportation economy, therefore, there are $N \times M$ different commodities, where N is the number of products (abstract of locations) and M is the number of different countries.

Transportation is a production technique that produces product (j, i_2) with an input that comprises the same quantity (or more) of product (j, i_1). In other words, transportation is the activity by which product j situated in country i_1 is changed into product j situated in country i_2. Other inputs represent the labor and materials (e.g., packaging materials, fuels, and transportation equipment) necessary for transportation. To conserve the fundamental properties of the production technique as discussed in Sect. 3, the transportation must be simple (condition (f)). This means that the output must consist of only one product, save for transportation equipment, and that this product can be treated as a durable capital good with constant efficiency. Transportation as a production technique must satisfy another property—which is to say, labor used for production must comprise the labor of a single country (condition (h)). Therefore, the employment of a crew of mixed nationality is excluded. It would be normal to assume that material inputs are also located in the same country as the labor, but this condition is not essential to constructing the theory. In our setting, a good in country i_1 can be transported to country i_2 by a crew member of country i_3. In this case, this transportation is a production activity of country i_3.

This treatment of transportation costs may seem complicated; such complication is inevitable, if the theorems in Sect. 3 are to be extended automatically. However, this treatment has some merits of its own. Transportation costs are most often formulated as an *iceberg model*, which treats transportation cost as an evaporation of a part of the transported goods. With this modeling, there is no need to distinguish products by virtue of their location. This simplicity is the main reason why the iceberg model has been widely accepted, despite its apparent irrelevance to reality. The iceberg is a convenient parable in modeling transportation costs, but it carries some deficiencies as a model. One example is the vanishing of the Alchian–Allen effect. As Hummels and Skiba (2004, p. 1400) point out, the "iceberg hypothesis is neither correct nor innocuous."

Transportation equipment requires special treatment, because according to the standard interpretation, transportation changes a ship in the country of expedition to a ship in the country of destination. This is another form of joint production.

However, what is essential is the cost of employing transportation equipment—something that can be calculated by the standard method (if we assume constant efficiency during the life span as it is mentioned in Sect. 4).

We suppose there is at least one system of production techniques (including transportation techniques) that is productive, in the sense that any positive vector of commodities is producible with a net consumption of labor from various countries. By operating transportation techniques, no products are increased. Then, if the transportation economy is productive, the underlying RS economy will also be productive.

The competitiveness of a production technique is defined in a way similar to that seen in Sect. 3. Let h be a production technique that produces commodity (j, i) in country i. It is competitive when it satisfies the following two conditions with regard to international value $\mathbf{v} = (\mathbf{w}, \mathbf{p})$:

$$\text{(i)}\ a_0\left(c\left(h\right)\right)w_{c(h)} + \left\langle \mathbf{a}\left(h\right), \mathbf{p}\right\rangle \leqq a_0\left(c\left(h'\right)\right)w_{c(h')} + \left\langle \mathbf{a}\left(h'\right), \mathbf{p}\right\rangle$$

for all production techniques h' that produce product j in country i;

$$\text{(ii)}\ a_0\left(c\left(h\right)\right)w_{c(h)} + \left\langle \mathbf{a}\left(h\right), \mathbf{p}\right\rangle \leqq p_{(j,i)} + \left\{a_0\left(c\left(t\right)w_{c(t)} + \left\langle \mathbf{a}\left(t\right), \mathbf{p}\right\rangle\right\}\right.$$

for all transportation techniques t that transport product j from country i' to country i. Here, the value $\mathbf{v}$ is a pairing of wage vector $\mathbf{w}$ and price vector $\mathbf{p}$, but $\mathbf{p}$ represents prices for all $M \times N$ commodities (j, i). Note also that $\mathbf{a}_0(h)$ and others represent here material *input* coefficients, while $\mathbf{a}(t)$ represents material *net output* coefficients in Sect. 3. As the fundamental theorem (Theorem 3.4) holds in this case too, all commodities in any country have at least one production technique that produces it competitively (i.e., either (i) or (ii) holds with equality).

The first condition means that h is competitive among production techniques that produce j in the same country i. The second condition means that production cost by h is lower than the cost of producing product j in the other country i' and bringing it to country i. In the latter case, the transportation cost should be added to the production costs in country i'. Competitiveness of a transportation technique is defined in the same vein.

When transportation is costly, each country has a greater number of competitive techniques than when the transportation cost is negligible, because condition (ii) is more relaxed. If the transportation cost decreases uniformly, each product will be competitively produced in a smaller number of countries, and the remaining countries will begin to import the product. This explains why the general decrease in transportation cost increases both specialization and the total volume of international trade. For the same reason, a production process may be divided among many different processes in different countries. This phenomenon will be examined in more detail in Section 13, under the title of "Fragmentation and Unbundling."[34]

In arguments regarding the Balassa–Samuelson effect, the distinction between tradable and nontradable goods is important. However, there are no intrinsic properties that distinguish tradable and nontradable goods. It is the transportation cost that makes some goods tradable and some others nontradable. Indeed, if the transportation costs for a good between countries are always above the minimal difference of production costs, then such a good will not be traded. On the other hand, if the transportation costs (including transaction costs and tariffs) are negligible, the good will be traded between countries if there are small differences in cost. Therefore, if the transportation costs decrease to a small proportion of the original costs, many formerly nontradable goods become tradable and would, in effect, be traded.

Some services require face-to-face communications or proximity, and simultaneity of production and consumption. It is difficult to trade such services across countries. Even in these cases, the condition that distinguishes tradable goods from nontradable ones is the transportation cost. For example, food preparation by an especially talented chef can be exported if the demander of the service is royalty or a billionaire, and he or she is willing to pay the travel cost and the wage of the chef's time.

10 Tariffs

Examining the effects of import and export tariffs has been one of the major subjects of international trade theory. If the tariffs are proportional to the value of the imported goods (or exported goods), there is no new problem for the new theory with introducing tariffs. They can be treated as a kind of additional markup rate. In fact, in defining the competitiveness of a production technique, it is sufficient to modify the above condition (ii) into the following form:

$$\text{(ii}'\text{)}\quad a_0\left(c\left(h\right)\right)w_{c(h)} + \left\langle \mathbf{a}\left(h\right), \mathbf{p}\right\rangle \leqq \left(1+\tau\right)\left\{p_{(j,i)} + a_0\left(c\left(t\right)\right)w_{c(t)} + \left\langle \mathbf{a}\left(t\right), \mathbf{p}\right\rangle\right\}$$

Note that the markup rates required by firms are incorporated into the input coefficients by means of equivalent vectors (see Sect. 5).

[34]Using the formulation given here, Escaith and Miroudot (2016, Subsection 2.2) examine the implications of transportation cost.

11 International Migration of Labor

Some people think that wages are unequal among countries because labor does not migrate freely among countries. Although we frequently read this kind of explanation in textbooks, this understanding is not precise.

Suppose that a nonnegligible portion of the labor force of country *B* migrates to country *A*. We can consider several situations. One possible situation is that the new labor force is quickly assimilated into the labor force of country *A*. In this case, the only effects of migration are an increase in the size of the labor force of country *A* and a decrease in that of the labor force of country *B*. If the competitive patterns of countries *A* and *B* do not change on account of this migration, there would be no change in international values and the wages of both countries would remain constant. If the wage ratio between countries *A* and *B* changes on account of the migration, one of two mechanisms must be at work—namely, (1) country *B* suffers from a lack of labor, is obliged to increase the wage rate, and abandons some parts of its competitive industries, or (2) country *A*, on account of increased unemployment, must reduce its wage rate and acquire new competitive industries. However, migration is usually not so pronounced as to change trade patterns.

Of course, we can imagine a situation where migrated workers exhibit less productivity than the workers who continue to live in country *A*. However, in our assumption, the labor force is assumed to be uniform, and so we cannot treat this case. A theory of wage differentials among industries and among categories of workers needs to be constructed. (See footnote 31 regarding the five fields of value theory.) Such theory would complement the theory of international values. As it is mentioned in Sect. 3, if the relative wage rate of each category of workers is determined for institutional or other reasons, the theory of international values will work as in the case of a uniform labor force assumption.

Related to the topic of this section, it would be useful to note that monetary transfers for any reasons will not change the international values, if the world's total demand does not change on account of this transfer and transportation costs can be neglected. For example, migrated workers in country *B* may transfer some part of their wage to their family in country *A*; this may change the balance of payments, but it is possible that the volume and composition of the consumption will not change on account of this transfer. In such a case, international values would not be affected by this transfer. A change of consumption location may change the volume and direction of trade, but the competitive pattern and international values will not change if world demand remains unchanged.

The major reasons behind large wage differentials among nations are to be found in the technological differences (i.e., the sets of production techniques that each country possesses). Labor productivity constitutes one of technological differences, but it is conditioned by the working customs and incentive systems. Labor force migration changes these basic conditions, and it is not easy to determine how labor productivity changes after migration occurs.

Division IV: Applications

This division (Sects. 12, 13, 14 and 15) provides four examples of applications. The first two show how the new theory can be used to analyze dynamic changes as a coevolution in international values and sets of production techniques. Section 14 argues policy implications for economic development and makes cleat the differences of implications of the new theory with those of traditional theories. Section 15 shows that the new theory of international values has the advantage of being well matched to international IO analysis.

12 Flying Geese

Vernon's product cycle theory and Akamatsu's flying geese theory are famous as midrange theories of industrial development in the field of international trade. Akamatsu and Vernon share a similar viewpoint, but they examined the same mechanism from opposite sides. Akamatsu set his observational eye on Japan's "catching-up" process, while Vernon observed the transfer of technology and production from advanced countries to less-advanced countries. When Akamatsu started his research in the 1930s, Japan was still a "backward country," at least in the minds of Japanese scholars.

In recent discussions on East Asian economic development, it has been customary to mention Akamatsu's flying geese pattern. Many of those discussions focus on the question of whether or not the flying geese pattern, as seen in Asian countries, has changed (Boyer et al. 2012, Conclusion). However, few studies point out that this "flying geese pattern" is what Akamatsu referred to as "the third type" (Akamatsu 1962, p. 17). The original fundamental pattern of the "flying geese formation" was to explain why Japan first imported cotton thread (mainly for warp use) from abroad, then started to produce it for internal consumption (for making textiles), and finally came to export it. It was observed that many commodities traced the same pattern, and Akamatsu wanted to explain why these patterns are common. Akamatsu's logic was based on a difficult kind of Hegelian dialectics. In any case, it is not difficult to explain by way of the new theory the basic mechanism underlying the fundamental flying geese pattern.

The new theory of international values explains how wage disparity typically emerges between countries. This is one crucial difference between factor proportion theory (FPT)—which assumes the factor price equalization theorem as a standard case—and the new theory of international values. While pure theory cannot tell us how wide this disparity can be, simple observations of the real world tell us that the wage of a worker in an advanced country can be between five- and 30-fold than is seen in less-developed countries. Although China is catching up to Japan very rapidly in this respect, there still remains a large wage differential between the two countries (i.e., about fivefold).

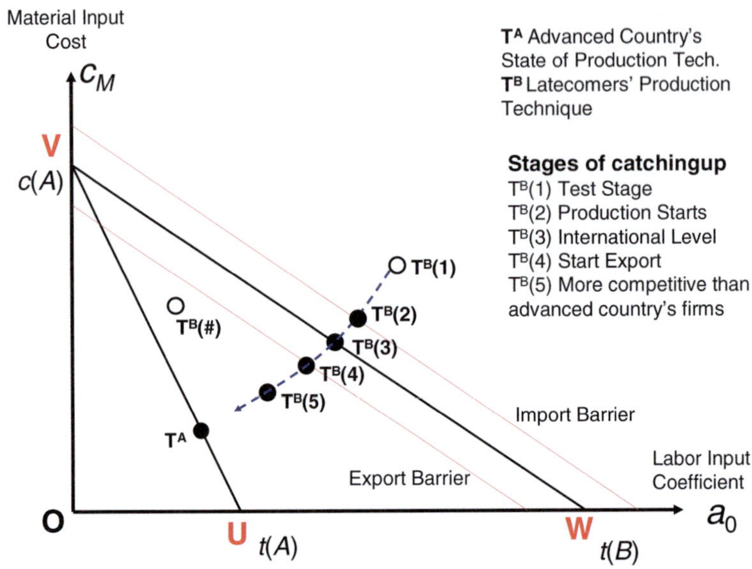

Fig. 2 The flying geese: technology path of latecomers

The basic logic behind transitioning from importation to exportation is illustrated in Fig. 2. In the following analysis, we assume that the prices of goods and the wage rates of two countries remain constant. In other words, we assume that international values except the product price concerned remain constant. Readers can weaken this assumption in various ways, once they know how to work with this assumption. The crucial assumption is the great disparity of wage rates between two countries *A* and *B*.

One point in the figure represents a state of production technique for a product. The production technique of the product is given by the input vector $(a_0, a_1, \ldots, a_N)$. The unit labor cost is $w \cdot a_0$, where w is the wage rate of either *A* or *B*. The unit material cost c_M is $a_1 p_1 + a_2 p_2 + \cdots + a_N p_N$. Then, we can express the production techniques as a point (a_0, c_M) on a plane.

In this representation, the horizontal axis is measured in real terms (e.g., units of work hours), and the vertical axis is measured in money terms.[35] In other words, the abscissa represents a labor input coefficient a_0 in real terms, and the ordinate expresses the total material input cost c_M. Because the two countries have different wage rates, the same work day will incur different labor costs. Let $a_0(A)$ and $a_0(B)$

[35] We can use mixed axes in money and real terms, because prices and wages do not change during our argument by assumption. We take the vertical axis in money terms in order to express the multidimensional point as a single real value. As for the horizontal axis, we have to take axis in real terms to express the change in production technique while the wage rates of the two countries are different.

be labor input coefficients for countries *A* and *B*, respectively. If *w*(*A*) and *w*(*B*) are the respective wage rates of the two countries, then the unit wage costs of the two countries will become equal when

$$w\,(A)\,a_0\,(A) = w\,(B)\,a_0\,(B) \text{ or } a_0\,(B) = \{w\,(A)\,/w\,(B)\} \cdot a_0\,(A)\,.$$

In these circumstances, the two countries have the same unit wage cost when $a_0(B)$ is *w*(*A*)/*w*(*B*) times as large as $a_0(A)$. Suppose (the more-advanced) country *A* has a higher wage than does (the less-advanced) country *B*. For example, if w_A is three times higher than w_B, then point $a_0(B)$ can be three times larger than $a_0(A)$.

We assume that the production technique of country *A* remains invariable. Let *c*(*A*) be the total unit cost, $c_M(A)$ the unit material cost, and $c_0(A)$ the labor input coefficient for country *A*. The two bold lines UV and WV are drawn as follows. First, plot the coordinate ($a_0(A)$, $c_M(A)$) and mark it as T^A; this represents the production technique for country *A*. By assumption, this point remains immobile. Point V is plotted at coordinate (0, *c*(*A*)). The line UV passes through point T^A and point V. All points on this line express production techniques that incur the same total unit cost for country *A*. The line of the same unit cost for country *B* is expressed by the bold line VW, which passes the point V(0, *c*(*A*)) but has a different slope. Points on the abscissa U(*t*(*A*), 0) and W(*t*(*B*), 0) have the same unit labor cost. Then, abscissa *t*(*B*) is three times larger than abscissa *t*(*A*) if w_A is three times larger than w_B.

In the following, we assume that the production technique of country *A* does not vary, but that the production technique of country *B* changes through time. (We assume that this change of the production technique does not affect the international value except the product price.) In reality, the input coefficients of *A* may also change. Readers can easily adjust the story to this case. Because point T^A moves in this case, the story becomes one of a kind of "chasing."

Now let us return to the simple case in which the production technique T^A stays invariant. We follow what happens when country *B* learns and improves its production technique. Five such points are plotted in Fig. 2. At first, the production technique is at the point $T^B(1)$, and it lies outside the triangle OWV. This means that the total unit cost is greater than that of country *A*. At this stage, country *B* cannot produce the product competitively, because the same product can be imported more cheaply than it can be produced in country *B*. If country *B* arrives at the stage where it can produce at point $T^B(2)$, production in *B* becomes competitive, if the government imposes a certain tariff. If production starts, country *B* accumulates know-how by *learning by doing* (i.e., by learning by producing) and improves its production technique (Arrow 1962; Rosenberg 1982, Chap. 6). When it arrives at point $T^B(3)$, country *B* can rival country *A* in terms of production cost, provided that the wage rate in *B* is one-third that of *A*. If the production technique reaches point $T^B(4)$, the product becomes competitive, even if the other country imposes a tariff. The width of the barrier lines depends on the rate of the tariff for the product. Finally, when the production technique in country *B* reaches point $T^B(5)$, production in country *A* will no longer be competitive. In this eventual case, the specialization pattern changes and international values may change accordingly. This may not happen easily, because country *A* improves its production technique, and this runs counter to our assumption.

The story behind the "flying geese" goes like this. Imagine a country like Japan not far in time from the Meiji revolution. Some people in Japan come to acknowledge that many convenient goods are used in advanced countries (e.g., the United States and some European countries), and so they start to import them as a part of new lifestyle. Some business owners try to produce the same products, but their relative lack of experience and a technology gap prevent them from producing them competitively, compared to the imported products. It is not a shortage of capital that prevents them from being competitive producers. If their prospectus is good and people believe they will be successful, future entrepreneurs could raise enough capital funds to purchase essential machines, installations, and materials. This is the trial phase or test stage of product nationalization. In Fig. 2, the state of input coefficients is indicated by the small circle $T^B(1)$.

Figure 2 shows the different stages of technology development for a firm in country *B*. When the state of the production technique lies at $T^B(1)$, the production cost for firms in *B* is much higher than that for firms in *A*, and entrepreneurs cannot compete with the imported products. However, they do not remain where they are; through trial and error, they arrive at a new stage where the input coefficients have been sufficiently reduced and their production cost becomes comparable to that of the advanced country. The exact cost incurred by country *B* can be somewhat higher than that of *A*, but the producers of country B may be protected by import duties, transport costs, and transaction costs. A parallel line above the second bold line indicates the import barrier. If *B*'s state of production technique comes down to this line, commercial production can start. Point $T^B(2)$ denotes this stage.

Once production starts, "learning by producing" starts. The input coefficients continue to decrease to $T^B(3)$, where country *B*'s production cost really becomes comparable to (and competitive with) that of country *A*. The parallel line below the second bold line indicates the export barrier; if "learning by producing" continues further, the coefficients decrease further and arrive to point $T^B(4)$, where country *B* can start to export the product competitively. Country *B* can still continue to reduce the input coefficients, to arrive eventually at point $T^B(5)$. At this point, producers in country *A* would be obliged to reduce their production cost, if they are to remain competitive with firms in *B*. Even at this stage, the producers in country *B* are still technologically "backward," and their production efficiency (measured by the input coefficients) is lower than that of the producers in country *A*.[36]

[36] Jane Jacobs (1969, Chap. 2) tells a similar story for Japanese bicycle manufacturing.

No one knows the limit of rationalization (i.e., the lower bound of input coefficients), but the producers in country *B* have an advantage, since they know that they can still go further: country *A* has achieved better productivity, and so they eventually can, too. This is one of the advantages that latecomers have.

Gerschenkron (1962) has highlighted several merits of "backwardness." Being privy to the existence of advanced products and technologies is another important advantage that helps "backward" countries to "catch up" to advanced countries. Akamatsu's fundamental or first pattern of flying geese shows the mechanism by which such "catching up" is achieved. Note that the flying geese pattern presupposes the importation of raw materials. In the case of the cotton industry, Japan imported raw cotton. In later stages, it exported cotton thread and cotton cloth made with this imported cotton. This pattern of trade has been called *kakō bōeki* in Japanese, and it has been an important concept in trade and industry policy discussions. Until recently, there has been no established English term for this concept, although some use the term "processing trade."[37]

If the "catching-up" process occurs in many other industries, it may pull up the wage rate of the "catching-up" country. "Chasing" will then occur between the productivity increase and the wage hike.

13 Fragmentation and Unbundling

Production processes—or parts thereof—were and are being transferred from advanced or high-wage countries to low-wage countries. This occurs for a variety of reasons and in various forms. Terms applied recently to these phenomena include "outsourcing," "offshoring," "fragmentation," "processing trade," "trade in tasks" (or task trade), and "vertical specialization." These transformations are not isolated or sporadic; they constitute a uniform and pervasive movement. We are observing a tremendous shift in production sites in the globalized world. Baldwin (2006, 2014) refers to this recent movement as the *second great unbundling* (SGU). The basic logic of unbundling is similar to flying geese, and large wage rate disparities lie at the center of this movement. On the flip side of the great unbundling is the reduced cost of transport and communication. The difference between flying geese and fragmentation, then, lies in the degree of unbundling in the production process. Flying geese suppose an overall production process, from the input of raw materials

[37]Many countries—including those within the EU—stipulate "processing trade" as a special trade regime, wherein some parts of imported intermediates and exported finished products can be traded duty-free. Processing trade represents almost one-half of China's recent exports. *Kakō bōeki* (加工貿易) is not synonymous with such a specific legal regime; rather, it refers to the full business flow, from raw material importation to product exportation, with no reference to duties. The promotion of *kakō bōeki* was a national credo in Meiji Japan. See also the last part of Sect. 18.

to the output of final products. Fragmentation, on the other hand, divides this process into two or more processes.[38]

The logic inherent in fragmentation is illustrated in Fig. 3. The coordinates in Fig. 3 have the same meanings as those in Fig. 2. The starting point of the construction is point **T**, which represents the state of production technique of a firm in the high-wage country *A*.[39] The abscissa and ordinate represent labor and material input coefficients, respectively. (To be more precise, labor is in real terms and material input is in money terms, if we want to express on a two-dimensional plane.) Suppose this process (vector **OT**) can be divided into the sum of two parts **OA** and **OC**. **OA** is the part that requires high-level technology or includes know-how that the firm wants to keep secret. **OC** is the part of the production process that the firm wants to transfer from country *A* to the low-wage country *B*. This transfer may induce a loss of efficiency, given the low production experience level and additional costs, such as the transportation costs for intermediate products, communication cost between the main office in *A* and the factory in *B*, and so on. To know the admissible range of loss, we construct two lines as follows. Draw a line through point **C**, which represents points with the same production cost when country *A*'s wage rate is applied. Let the line intercept at points **U**($c_0(A)$, 0) and **V**(0, c(**C**)), where *c*(**C**) is the total unit cost in country *A* for process **C**. Line **VW** is the set of points where production in *B* has the same total cost as the process **C** when it takes place in country *A*. Then, point **W** on the abscissa has the coordinate (c_0(**B**), 0). Production in country *A* on line **UV** and production in country *B* on line **VW** will have the same cost.

When production **OT** is divided in country *A* into the sum of **OA** and **OC**, there is no loss or gain. However, if the process part **OC** is transferred to country *B*, we can reduce at least the wage cost; this must compensate for the additional cost incurred by unbundling. Suppose the process part **OC** is realized by the state of the production technique **OB**. We assume **OB** includes the loss of efficiency and the additional cost incurred by unbundling. By construction, the total cost of the fragmented process is lower than the original integrated production in country *A*, inasmuch as point **B** remains in the interior of triangle **OVW**. A situation similar to that seen in the case of the flying geese pattern occurs. Because of the low wage rate of country *B*, the production technique **OB** can move in a wider range of efficiency states. In this case also, the low wage is the major advantage vis-à-vis cost competition. If transportation and transaction costs were reduced, **B** would be close to **C** and the chance of achieving a cost reduction by virtue of unbundling becomes higher.

It is easy to see that the above logic of fragmentation can be applied to almost all production processes. This explains the universal character of fragmentation, when

[38]Grossman and Rossi–Hansberg (2008) emphasize the necessity of paradigm change. As an RS economy includes input or intermediate goods, we need no new particular formulations.

[39]In this section, production techniques and points on the plane are denoted by bold characters, to distinguish them from country names (which are denoted by italic characters).

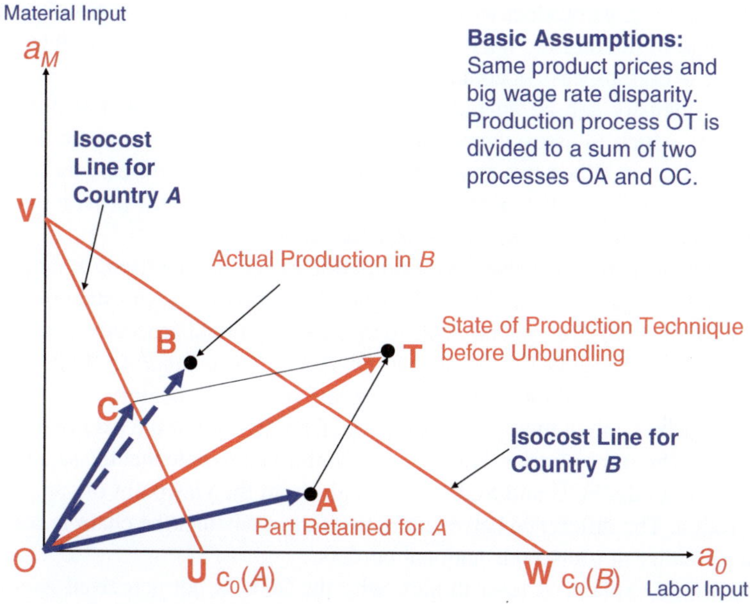

Fig. 3 Fragmentation of production process

there is a large difference in wage rates and both transportation and transaction costs are reduced.

The logic of flying geese "catching up" and fragmentation teaches us how the main message of FPT (factor proportion theory) is flawed. The HO theorem states that labor-intensive products tend to be exported from labor-abundant countries. However, if we look at Figs. 2 and 3, we see that factor intensity does not matter much. Normally, the catching-up country will have labor-intensive exports, but this is a matter of chance and probability. It may happen (as in Fig. 2) that a firm of country *B* has production technique $T^B(\#)$ with a lower unit cost, but with a higher capital intensity, than firms in country *A*.[40] As previously explained, any production technique at a point in the interior of triangle OWV has a unit cost that is lower than those in firms in country *A*. In the case of Fig. 3, the original production process was divided into two tasks, such that country *A* retained labor-intensive **OA,** and less labor-intensive **OC** was outsourced and realized as **OB**. The logic of unbundling is not based on the factor intensities of divided processes or tasks, but on the strategic decision-making of the firm. If there is no reason to retain the **OA** part in country *A*, the whole process **OT** can be transferred to country *B*. In fact, the firm of country *A*

[40]Capital and labor intensities are usually measured by the ratio of capital and labor costs to the total unit cost. Capital cost may indicate the cost of fixed capital (i.e., depreciation cost), material input cost, and the sum of the two. The slope of the line **OT** represents labor and capital, but we cannot compare them visually, as wage rates are different between the two countries.

faces the risk that its product will be produced as a whole by some firm in country *B*. The logic of unbundling itself presupposes that there is a reason to retain some part of the overall process in country *A*.

Factor intensity can be an indicator of general tendency, but it is not a good criterion by which to assess competitiveness; unit cost is a much more direct and decisive indicator. When two indicators contradict, it is the unit cost that prevails. FPT does not refer to the potential cost advantage; it merely provides a rough criterion of a not well-defined comparative advantage.

The global supply or value chain attracts managerial attention, as it includes new aspects. Crossing national borders is not the same as crossing state or regional borders. We need to gain special skills to manage additional procedures and control problems. Baldwin (2006) distinguished the *first great unbundling* (FGU) and the SGU. The first unbundling occurred in the late nineteenth century, at the country level; the second unbundling started to spread from the end of the twentieth century and occurred across countries. However, it is important to note that the same logic is at work in both the FGU and SGU—especially from the viewpoint of supply chain optimization. The difference between the two lies in whether the chain is contained within a country or transcends national borders.

This would explain, at least in part, why the SGU is not perceived as a major industrial revolution, in spite of its enormous economic consequences. Unbundling is a common technique that can be used in any industry. The production process for any given product differs in depth and width from those of another. A process may involve multiple operational stages, and it can require a large number of parts and materials. Additionally, a process can be divided infinitely many ways. Production engineers have always sought to pinpoint the optimal organization for production processes, and both the information and communication technology revolution and trade liberalization have widened the range of options open to them. As a result, top managers are now required to consider worldwide logistics. As for procurement policy, they need to adopt global optimal procurement strategies; this has been the cause of ever-increasing input trade. RS trade theory provides a general theory that incorporates all these features, because it supposes that managers will adopt the global optimal procurement policy.

14 Trade and International Wage Inequality

One of most important features of the new theory of international values is that it helps explain wage differences among countries. This contrasts with many classical and Marxian explanations, for example, which assume a minimal level of wages that makes the reproduction of the labor force socially and culturally possible. The new theory reverses the causality of things: in this new theory, it is the change in the general level of wages that pulls up or pushes down the "life level" of common workers. The new theory implies that wage differences normally occur and are maintained as far as country differences vis-à-vis technological levels continue; this is also in sharp contrast with FPT, which predicts as a standard case the equality of wages (i.e., the factor price equalization theorem). The new theory clarifies how differentiated wages emerge among countries.

The message inherent in this new theory is straightforward: it is the set of production techniques that determines the (real) wage of a country. Markups may intervene, but the margins of variance are not overly large. Production techniques can differ enormously, and this is what makes a rich country rich and a poor country poor.

Let us illustrate how the principle works, by way of a simple example. Suppose there are two countries. We ignore transportation and other costs and assume that the prices of products are uniquely determined in the two countries and the markup rates of both countries are the same for each industry. Suppose for simplicity that the production techniques of a product have the same material input coefficients for both countries.[41] Labor efficiency can differ substantially. Imagine two production teams comprising the same number of persons who work with the same machines and equipment and produce a complex machine—say, a car. The time needed by an experienced team to produce a car can be one-fifth that needed by a less-experienced team. In this case, the labor efficiency of the experienced team is five times higher than that of the less-experienced team. In terms of the labor input coefficient, this means that the coefficient of one team is one-fifth that of the other. This kind of difference can happen between countries. If the labor efficiency of all teams in country A is five times higher than that of the corresponding teams in country B, then the real wage rate of country A will be five times higher than that of country B.

In a more realistic situation, Jones's assumption does not apply, and the labor efficiency ratios between two countries are not uniform among industries. Even in such a complex case, the new theory of international values affirms that the wages between countries are determined by the same principle. It is the productivity of a country *as a whole* that determines the wage rate level of a country.

If this be the case, there are two means of increasing the real wage level: we can (1) improve the set of production techniques or (2) reduce the markup rates. Let us ignore the second possibility for the moment and concentrate on the first.

All efforts to make a production technique more productive contribute to improvements in a country's real wage rate; however, we often come to a trade-off situation. If we want to reduce the input coefficient of one input, then the coefficient of some other inputs may increase; this is the case with the choice of techniques. In general, we may have several production techniques that produce the same product, and if we are given the wage rate and prices, we can determine which of the techniques will incur the lowest cost. This is a simple question of calculation. When all firms choose anew and update their product prices, the prices will change. In a closed economy of one country, the minimal price theorem (Lemma 4.1) implies that a system of prices and a system of production techniques exist, such that the full

[41]This is the crucial assumption that Jones (1961) made in his study and is equivalent to assume our RII, but whether it holds or not is not essential to the new theory.

cost of a production technique belonging to the latter system incurs the minimal cost and is equal to the price of the product. In the case of international trade also, there exists a pair of an international value vector $\mathbf{v} = (\mathbf{w}, \mathbf{p})$ and a set of production techniques Γ, such that (1) the full cost of a production technique of Γ is equal to the price of the product, (2) the full cost of the production techniques not belonging to Γ exceeds (or is equal to) the price of the product, and (3) Γ forms a spanning tree as a competitive type.[42] A finite number of such pairs may exist, but for each of them, a steady-state economy exists. Even if final demand were to change, wage rates and prices would remain constant as long as there is no constraint by virtue of a shortage of labor force in one or more countries. This weak version of the minimal price theorem indicates that each firm's judgment in terms of production costs leads the whole system of production techniques Γ to a more efficient state. Thus, the market system has a general tendency to lead the production system into greater efficiency.

It is now evident that the prime mover in improving a country's real wage rate is firms' efforts to reduce the production costs of their products. Do not confuse this reduction of production costs with cost cuts realized by reducing the wage rate. The first cost reduction is realized at constant values (including the wage rate); this is made possible only by changing the input coefficients—in other words, only by improving the production technique. It is also important to note that the real wage rate is dependent on concerted improvements in all industries.

There are two possible paths that lead to an increased real wage rate. The first is a (relative) reduction in the price of consumer goods. A price reduction in a consumer good contributes to the improvement, by reducing prices relative to the wage rate. The second, but sometimes more important, path involves the price reduction of input goods. If a good is a material that is used widely in production, the effects of a price reduction of such a good will propagate through the input cost reduction of many products and contribute to an increase in the real wage level.

The possibility of the second path suggests why industrialization is generally preferred to improvements in agricultural products and raw materials. In the case of agricultural products, as they are proximate to final consumption within the supply chain, the effects of production improvements are confined to the first path. In the case of raw materials, the effects of the improvements will contribute to a general cost reduction. However, if the domestic industry is not well developed, the effects of a cost reduction of raw materials will contribute only to reduced production costs among foreign industries. If a country succeeds in constructing a cycle of basic input goods, the increased productivity of an industry will contribute to improved productivity among other industries. Then, a causal accumulative cycle will emerge and give impetus for wider economic development.

Improvements in production processes may contribute also to a rapid increase in the size of the market (or *débouché* in French) for products. In particular, if the cost reduction of formerly imported goods makes the production of these goods

[42] See footnote 21 attached to Definition 3.8.

competitive in the world market, this may contribute to an increase in both the domestic and world market for those goods. The increase in production volume generally contributes to cost improvements, mainly through learning by producing, but also on account of increasing returns to scale. This may produce a virtuous circle between cost improvements and the growth of production volume. This virtuous circle may spread throughout the industrial IO relation network.

The productivity of a production technique mainly depends on the production process found inside the firm. This process is most often the result of team work involving workers and production site managers. On the other hand, however, if we consider transportation and transaction costs, social infrastructure and institutions also influence the productivity of a country. In fact, each time that intermediate goods move from one production site to another, transportation costs are incurred and add to the total cost. The use of information and communication networks helps reduce information and communication costs and increase the chance of successful trade and better deals. This "business chance" creation is not often counted as value creation, but it does play a crucial role in firm growth and, by consequence, the economic growth of a country. It is evident that a sound network of roads and ports helps increase a country's total productivity.

Institutions are also important factors that help reduce both visible and invisible costs. If production and transfer of commodities require permits and approvals from administrative authorities, the transaction costs of these procedures will be substantial. The presence of good social and economic institutions also contributes to a pulldown in the markup rates, because they will generally help increase competition within local markets. Anti-monopoly legislation and its effectiveness evidently help promote competitive markets and reduce the general level of markup rates. The existence of fair and just government mechanisms is also important, because corruption and bribery pose enormous costs to private firms. For example, a study suggests that in some corrupt countries, illegal payoffs can increase the cost of public works projects by as much as 30–50% (Rose-Ackerman 1996, p. 4).

Large wage discrepancies between developed and developing countries arise from differences in their respective sets of production techniques, but they also reflect both material and institutional social infrastructure. The best policy by which to increase the real wage rate is to enhance the productivity of production techniques. As wage workers occupy the majority of working people in any capitalist economy, the policy by which to enhance the real wage rate is also the policy by which to increase a country's per capita income.[43] This suggests the effectiveness of the new theory of international values.

The difference between this new theory and other trade theories is apparent. In fact, the policies that the new theory indicates are very different from those indicated by the standard HOS theory or other FPTs, including HOV theory. FPTs focus on the capital–labor ratio, but not much on production techniques. In the worst case,

[43]The aforementioned topics are familiar to those who have worked in the economic development of developing countries. We examine the related topics in Sect. 18.

they simply assume that production techniques are the same across all countries. It is quite natural for them not to inquire into how to increase productivity or reduce production costs; their main concern, after all, is the capital–labor ratio. The new theory makes no mention of this ratio; in the new theory, the capital–labor ratio is simply irrelevant to improving the real wage rate. As we saw in Sect. 13 on Fragmentation, it is not the capital–labor ratio that should guide business decision-making; rather, if a product (specification) is given, it is the unit production cost that should determine which process is better.

Although the new theory of international values is superior to FPTs, it still has a weak point as a theory of wages. In Sect. 3, we assumed that the labor force is homogeneous across any given country; as this assumption is imperative to the theory construction, it cannot deal with problems of wage inequality inside a country. This is because the classical value theory does not yet have any good theory vis-à-vis the labor market.[44] The means of bypassing this lack of theory is to assume a constant wage ratio among different kinds of work forces; then, we can reduce different kinds of labor to a single labor force by giving weights to each labor by wages.

Here, we also see a sharp contrast between the new theory and FPT. FPT rightly assumes two kinds of labor—for example, unskilled and skilled labor. With the aid of the Stolper–Samuelson theorem, the standard theory can argue whether increased inequality of wages in some developed countries—including the United States—was brought about by deeper trade liberalization. This may be a strong point of FPT.

The two most recent decades have seen the emergence of plenty of empirical and theoretical research on increased wage inequality. (See, e.g., Kurokawa (2014).) Many of those studies argue whether this increase in wage inequality relates to trade liberalization and trade with low-wage countries. Whether or not they argue for the influence of trade, they all refer to the Stolper–Samuelson theorem, because this is the only established theory to permit the argument that international trade affects wage differences. However, I feel there is a problem here. All argue about wage differences, but focus uniquely on internal wage disparity. This may be an important and polemical question. What about wage inequalities among nations? These are as important as internal wage inequality. However, I could find only a few exceptions in the literature; two manifest exceptions are Ruffin (2009) and Waugh (2010), the first of whom discusses how globalization affects income inequality, both between countries and within countries. What does this state of economics stand for? This strongly indicates the absence, among mainstream economics, of a trade theory that can analyze international wage inequality, and it proves at least the strength and uniqueness of the new theory, without asking about its truthfulness.

[44] See footnote 31 in Sect. 7.

15 International Input–Output Tables

In the two most recent decades, world trade in intermediate goods has increased in volume and in its proportion of all trade. The increasing global fragmentation of production processes or trade in tasks has engendered among international economists and policy-makers a keen interest in the state and structure of global value chains. This interest has produced (and is producing) various stimulating studies. Investigations have required a new approach to measuring trade, such as trade in value-added (TiVA). This measurement became possible through the use of IIOTs, and they stimulated in its turn compilations of worldwide IOTs.

The first attempt to compile a multinational IOT was Leontief and his collaborators' United Nations' World Input–Output Model 1976.[45] In 1981, the Institute of Developing Economies (now IDE–JETRO) succeeded in compiling ASEAN International Input–Output Table 1975 and from 1992 started to publish Asia IOTs that covered 5-year data intervals starting from 1985. There are now a number of initiatives that compile large-scale global IO tables, including the Global Trade Analysis Project (GTAP), the Intercountry Input–Output Tables (ICIOT) of the Organisation for Economic Co-operation and Development (OECD) and the World Trade Organization (WTO), the World Input–Output Tables (WIOT) of the World Input–Output Database (WIOD), and multiregional IO tables including ADB-MRIO of the Asian Development Bank in addition to the Asian International Input–Output Tables (AIIOTs) of the IDE–JETRO.[46] The Eora multi-region IO database, a project funded by the Australian Research Council, and the EXIOBASE, a consortium of European universities and research institutes, are being compiled with the aim to contribute more specifically to environmental problems and policies. They differ in terms of purpose, coverage, period, and openness.[47] In the following, when I refer to "international IOT," I take WIOT as the standard model, because it is the most open and richly documented (Timmer et al. 2015; Dietzenbacher et al. 2013; Timmer 2012).

It is almost evident that the new theory of international values is closely related to international IOTs. Let us remember the treatment in which transportation costs are positive (Sect. 9): in this case, it is necessary to distinguish goods by their location (i.e., the country in which the product was located). In an M-country, N-good economy, we must distinguish $M \times N$ commodities. Production in a country consists of labor input and the material inputs of goods that come from various countries. An input coefficient vector is then a set comprising an input coefficient of the labor of the country of production and $M \times N$ coefficients, each representing an input of good j from country i. If we are allowed to suppose that each good or

[45]The table covers 15 regions and 45 sectors of economic activity. See Cole (1977), pp. 20–21.

[46]Jones et al. (2014) compare the first three of these global IOTs.

[47]GTAP is more oriented toward building applied general equilibrium models, while ICIOT and WIOD provide data based on official public data and are therefore consistent with them. OECD and WTO now release the ICIOT, as well as TiVA data that preceded the publication of the ICIOT.

product corresponds to an industry, we have exactly $M \times N$ vectors that express material inputs for any production technique. In the same way, the WIOT has an endogenous sector of equal size—namely, a square matrix of $M \times N$ rows and $M \times N$ columns.

We observe here some minor divergences between the new theory and the IOTs. First, in IOTs, "input" refers to the value of inputted goods as evaluated by prices (i.e., the sum of corresponding quantities times price). In the theory of values, "input" stands for some material quantity expressed in some physical unit. Second, in the IOTs, input vectors are expressed by column vectors, while in the new theory, we customarily express input coefficients through a row vector. The second difference is only a question of convention: we could start our discussion by making the input coefficient vectors into columns. We prefer them to be rows only because in the cost comparison, it is more natural to compare the cost and the price of each product on a row-by-row basis. As a consequence of this difference in conventions, material inputs and value-added within an industry are expressed by columns in IOTs. A row vector in an IOT expresses how a product is divided into different uses. In Sect. 3, the final demand vector was expressed by a row of different goods, and no difference of use was discussed there. To avoid such inconveniences, it is sufficient to take transposed expressions for all vectors and matrices in the formulation of an RS economy.

Transportation requires more cautious treatment. In Sect. 9, we argued that transportation can be interpreted as a production technique; as such, it must satisfy the requirement that labor used for transportation must comprise labor from a single country (condition (h)). In an IOT—either national or international—transportation is tabulated through a special treatment. Let us imagine the transport of product P in China to Japan, by a crew of Thai workers. According to the new theory, as a production technique, this is production in Thailand that produces commodity P in Japan by inputting product P in China. In IIOTs, this is not interpreted as a production of commodity P in Japan, but as a combination of transportation service production in Thailand and a use by Japan of product P made in China. Thus, a single operation of transportation is divided into a pairing of use and transportation services. This special treatment is made to sidestep the fact that all products pass through the transportation industry. If we tabulate in this way, we will lose sight of the material relationship between inputs and outputs—in which case, the main merit of IOTs will be lost. A similar treatment is applied to the work of wholesale agents.

A commodity (i,j) requires a special interpretation.[48] In IIOTs, a commodity with an index (i,j) does not designate the product j that is **located in** country i; rather, it refers to product j **made in** country i. Thus, a commodity in IIOTs is marked by country and industry *of origin*, and not *of location*. When the product is used in country k, transportation is required, but this activity is tabulated as a production of transportation service (possibly by a third country) and used by the country of

[48] In Sect. 9, we used the notation (j, i) to indicate the product j which is located in country i. In IIOTs, entries are arranged by the lexicographical order for pairs (i, j).

consumption k. This interpretation accords with the convention that all trade flows are expressed in free on board prices.

Some other conceptual differences between IIOTs and the new theory of international values derive from the fact that an IOT has a fixed accounting period. For example, investment in fixed capital, as well as its depreciation, is not tabulated as inputs and outputs in the endogenous sectors; rather, investments are compiled in a column of final use, and depreciation is included among rows of value-added. The new theory of international values has a more flexible viewpoint and can argue—as I did in Sect. 8—that the formation and cost imputation of fixed capital goods are something like outputs and inputs. These are actual questions to be treated theoretically, but they cannot conveniently be treated on a year-by-year statistical basis. There are methods of imputation, but they may introduce arbitrariness into figures, and it is customary to avoid the use of such imputation.

In spite of some of these conceptual differences, the new theory has the advantage of being well matched with the accounting framework of the IIOTs. The IIOTs make it possible for the new theory to derive an empirical basis through statistical analysis. IIOTs will serve as the best possible "experiments" for the new theory, to assess whether it can be employed usefully in analyzing and understanding an actual economy. On the other hand, IIOTs can leverage the new theory as background theory; the new theory may contribute to a deeper knowledge of the facts that an IIOT represents.

A key consideration relates to the stationary nature of the input coefficients of an IIOT. In the case of national IOTs—in which we can plausibly assume labor homogeneity—we have a minimal price theorem (Lemma 4.1). In this case, the theorem proves that there is a set of production techniques that gives the minimal price, provided the wage rate w is given. When this theorem holds, there are no known production techniques that can replace (or undersell) actually competitive techniques. Thus, the theorem assures us of price stability and no input substitution. If we neglect those cases in which two production techniques incur the same cost, we can safely suppose that input coefficients remain constant, even if demand composition were to change. This theorem gives us the theoretical basis to distinguish the change in input coefficients from the change in the volumes of inputs. If the input coefficients remain constant, we can estimate various quantities and values based on the constant coefficient hypothesis, when the production volume changes. This is worth noting, because many concepts—such as TiVA—are implicitly based on this assumption.

In the case of the international economy, we have no simple version of the minimal price theorem; instead, we have Theorem 3.4. Interpretation of this theorem is more subtle than that of the minimal price theorem, but the theorem assures us of the existence of a system of competitive production techniques. If production is possible through this system and within the given labor forces, there is no need for international values or production techniques to change when the production volume changes, provided that production remains within the capacities of capital installment and labor. No such assertion is provided by FPTs such as HOV theory.

A transportation flow may illustrate the present question. We can consider as an ideal limit of globalization an economy in which all transportation is free. In such an idealized (but of course unrealistic) M-country, N-good economy, the most probable state has $M + N - 1$ competitive production techniques. We can embed this economy in the IIOT framework. If production and final demand are determined, trade flows can have various solutions, as noted in Remark 3.6. In this idealized but fictitious economy, trade flows may change freely within a certain transportation polytope. Within this degree of freedom, even the trade pattern may change. The existence of (or the possibility of compiling) an IIOT does not automatically signify the constancy of its input coefficients. Such supposition requires a theoretical background. If the transportation cost is positive, the transportation will be more stable, because the cost differs according to the route and method of transport.

The WIOT, as well as other IIOTs, distinguishes commodities by country and industry of origin; for this reason, it has $M \times N$ entries as inputs and outputs. The endogenous part of the table is a square matrix of $M \times N$ rows and columns. By dividing each row by the total production of the row, we can derive the input coefficient matrix A. Let I be identity matrix of order $M \times N$. Then, by the famous Hawkins–Simon theorem, the matrix $I - A$ is nonnegatively invertible.[49] The inverse matrix is $(I - A)^{-1}$ and is called the Leontief inverse matrix.

The existence of the Leontief inverse matrix is a great merit in treating commodities in terms of country and product of origin. It provides us with a simple means of calculating the total outputs that produce a given net output vector. Using this tool, we can define various important concepts.

Let $\mathbf{f}$ be the final-use vector. We do not distinguish varieties of use class. All uses of a commodity are summed to a single total. Let $\mathbf{y}$ be the total activity vector that produces vector $\mathbf{f}$.[50] Then, we have an equality

$$\mathbf{f} = \mathbf{y} - A\mathbf{y} = (I - A)\,\mathbf{y}.$$

Multiplying $(I - A)^{-1}$ from the left, we get

$$\mathbf{y} = (I - A)^{-1}\mathbf{f}.$$

Let $\mathbf{u}$ be the value-added row vector that corresponds to unit production. Then,

$$\langle \mathbf{u}, \mathbf{y} \rangle = \sum\nolimits_{(i,k)} u\,(i,j)\,y\,(i,j)$$

[49]This follows from the fact that each column has positive value added (in other words, the value-added by labor row, e.g., has all positive entries). Final use column may be 0 for all entries if the commodity is purely an intermediate good type.

[50]Here, $\mathbf{f}$ and $\mathbf{y}$ are column vectors to make them consistent with the IOT expression of matrix A. Consequently, matrix A corresponds to the transposed of the input coefficient matrix in the new theory.

gives the total value-added that is needed to produce net output **f**. This total can be divided into value-added in different countries. If $VA(k)$ is the part of country k,

$$\langle \mathbf{u}, \mathbf{y} \rangle = \sum\nolimits_{(i,j)} u(i,j)\, y(i,j) =$$

$$\sum\nolimits_{i} \sum\nolimits_{k} u(i,j)\, y(i,j) = VA(\mathbf{f}, 1) + VA(\mathbf{f}, 2) + \cdots + VA(\mathbf{f}, M).$$

On the other hand, let **p** be the price vector (row vector). Then,

$$\mathbf{u} + \mathbf{p}A = \mathbf{p},$$

because each production technique must be competitive. We can write this as follows:

$$\mathbf{p} = \mathbf{u}(I\text{–}A)^{-1}.$$

This means that

$$\langle \mathbf{u}, \mathbf{y} \rangle = \langle \mathbf{u}, (I\text{–}A)^{-1}\mathbf{f} \rangle = \langle \mathbf{u}(I\text{–}A)^{-1}, \mathbf{f} \rangle = \langle \mathbf{p}, \mathbf{f} \rangle.$$

Thus, the total value $TV(\mathbf{f})$ of vector **f** can be divided into each country's total value-added, or

$$TV(\mathbf{f}) = VA(\mathbf{f}, 1) + VA(\mathbf{f}, 2) + \ldots + VA(\mathbf{f}, M). \tag{9}$$

If **f** is the export from country i, then (9) expresses how the total export value comprises each country's value-added. Each term of the right member of (9) is nonnegative, because each of **f**, $(I - A)^{-1}$, and **u** is nonnegative. Country i's value-added divided by the total export is called the *value-added export ratio* (Timmer et al. 2015). It is expressed by

$$VA(\mathbf{f}, i) / TV(\mathbf{f}).$$

Evidently the value-added export ratio is less than 1 (except any case where all export products are made without using any imported inputs).

If **f** expresses the total production of a product (e.g., automotives), (9) expresses each country's income in the world production of the product. *Value-added share* in world production is given by dividing each term of the right member by the total production (i.e., the left member).

An import of a country may include some portion of the value-added of the importing country. When $\mathbf{f}_{\mathrm{M}}$ is the import vector of country i and $\mathbf{f}_{\mathrm{X}}$ is the export vector, let $TV(\mathbf{f}_M, i)$ and $TV(\mathbf{f}_X, i)$ be the part of value-added for country *i* included in the import and the export, respectively. The net export value-added of the country is

$$TV\,(\mathbf{f}_X, i) - TV\,(\mathbf{f}_M, i)\,.$$

Other useful concepts can be derived in a similar way.

When value-added can be classified into several different categories—such as low-skilled workers, medium-skilled workers, high-skilled workers, and capital services (comprising profit and depreciation)—the value-added in the global value chain can be divided into the sum of value-added for each category. Indeed, if we let $\mathbf{u}$(l), $\mathbf{u}$(m), $\mathbf{u}$(h), and $\mathbf{u}$(c) be the corresponding value-added for the unit production of commodity (i,j), we have

$$\mathbf{u} = \mathbf{u}\,(\mathrm{l}) + \mathbf{u}\,(\mathrm{m}) + \mathbf{u}\,(\mathrm{h}) + \mathbf{u}\,(\mathrm{c})\,.$$

It is evident that we can define each country's contribution in each category to the total production. We can interpret this decomposition as value-added in trade, which is in turn conserved for each category. What is interpreted in HOV theory as the factor content of a commodity has a good chance of misinterpreting these relations. This theory often assumes that the factor proportions of all countries can determine international trade flows. However, it is evident that they have no such causalities, if we observe that the number of commodities is far larger than the number of different production factors.

Division V: Some Implications to Other Fields

The impact of the new theory is not limited to international trade theory. In Sects. 16 and 18, we discuss implications to two fields of economics theory. The new theory was developed on the basis of a new vision of how the market economy works. Although it is new and extremely different from the dominant equilibrium framework, its underlying thinking does go back to classical economics. Section 16 argues the rational core of classical value theory and the common base between the two theories. Section 18 is in part a continuation of Sect. 14. The question of international wage inequality has always been at the core of development economics, but the traditional theory of international trade has not provided a suitable theory for discussing it. On this question, the new theory has a strong message, and development economics must have something to inspire it. Section 17 takes up international political economy—a new science that emerged on account of the inability of traditional trade theory to analyze the cause of trade conflict, among other things. The new theory may provide an economic basis for this science, which has principally been considered a political science, in spite of its name.

16 Classical Theory of Value

The classical theory of value has a close relationship with the new theory of international values.[51] Many theories of value coexist in the classical political economy period. In fact, a single economist could concurrently subscribe to two different versions of value: Adam Smith, for example, embraced both the labor theory of value and the labor command theory of value. Other political economists, such as Jean B. Say, have their own versions of the utility theory of value. The law of demand and supply existed before Smith and persisted throughout the classical era of economics. Ricardo was often interpreted as preaching the labor theory of value, but his core theory was what we can now call the *cost-of-production theory of value* (Takenaga 2004). Hereafter, when I use the expression "classical theory of value," I am referring to Ricardo's cost-of-production theory of value.

The classical theory of value and the neoclassical theory of value stand in opposition. The most important transition from the classical political economy to neoclassical economics is, as Hicks (1976) puts it, a change in focus from production to exchange. Value theory was at the center of this change. The classical theory of value posits that the value of a commodity is determined by the production conditions, while the neoclassical theory of value posits that that value is determined by demand and supply, and demand by psychological factors. The shift from classical to neoclassical economics marked a shift from the objective to subjective theory of values.

The classical theory of value had many good points, but the neoclassical theory of value supplanted it. Why was this so? What were the defects of the classical theory? In my opinion, the most important defect was that it lacked a theory of international values. As I argued in Sect. 2 of this chapter and will argue in Shiozawa (2017, Chap. 7 of this book), Mill—intending to solve the problem that Ricardo left unsolved—opened the way to the reversion to the law of demand and supply (i.e., the old common wisdom). However, there is a good chance that the classical theory of value will be revived (Shiozawa 2016).

Under the influence of Keynesian thinking, twentieth-century economics discovered the fixed-price economy, markup pricing, quantity adjustment, and IOTs. By constructing a new theory of international values, the intellectual power balance has changed. For approximately 150 years, classical economics was on the defensive; it is now time for a counterattack. After the Lehman Brothers collapse, many people—economists and noneconomists alike—started to doubt whether economics is running in the right direction. Students are asking for a more pluralistic education, and a substantial number of established economists are now reconsidering economics.[52] The new theory of international values may play an important part in this rethinking of economics.

[51] As I have written a separate paper on this theme (Shiozawa 2016), I will be brief in this section.

[52] The *International Student Initiative for Pluralism in Economics* is a network of students in more than 30 countries, who are clustered into various groups with names like *Rethinking Economics*.

17 International Political Economy

The international political economy (IPE) treats a wide variety of topics concerning international relations. I here confine myself to discussion concerning international trade conflicts. The IPE picks up problems that we can refer to as "trade conflicts." It is natural for the IPE to be a part of political science, because political science is always concerned with conflicts; however, the IPE is also a part of economics. Problems arise from the fact that standard international economics does not in principle speak to trade conflicts.

Economists in the neoclassical tradition, or those who think within the general equilibrium framework, deny the existence of trade conflicts—and if they do admit to them, they say that they are only transitory events and that "all goes well in the end." Paul Krugman, one of most prestigious trade theorists and a famous polemicist with the *New York Times*, once wrote a paper entitled "The Illusion of Conflict in International Trade" (Krugman 1996, Chap. 5). This essentially means that the IPE has no support from international economics; we could even go so far as to say that the IPE emerged precisely *because* no economists have debated or analyzed trade conflicts. However, with the emergence of the new theory of international values, the intellectual situation can change substantially. The new theory is on a strand of economics different from neoclassical economics: it is a theory that positively affirms unemployment, for example. (Theorem 4.3 is one example.) The new theory of international values may provide the IPE with a powerful economic tool.

To illustrate the aforementioned contention, let us cite Oatley (2004), a book compiled by a successful textbook writer in the field of the IPE. That text intends to give students the opportunity to reflect on the crucial questions pertaining to the IPE. Topics on the IPE are always polemic, because it focuses on issues in which opinions are divided widely and deeply. The book presents two opposing stands, and it asks students to reflect on this divergence. A particular chapter—which pits Robert E. Scott against Douglas A. Irwin—is illustrative.[53] The former argues that trade deficits indicate job loss; the latter counters, saying that there has been no job loss on account of free-trade agreements. This debate reveals the strength and weakness of the IPE. Mainstream economics textbooks do not treat this problem in a symmetrical way. There are occasional mentions of popular opinions, but the text is rather dismissive, given its emphasis that those opinions are misconceived.

They aim to "demystify, diversify, and invigorate economics." *Institute for New Economic Thinking* is an open forum that comprises more professional economists.

[53]The chapter is entitled "Trade and Jobs in the United States" (Part II, Chap. 1). A short introduction by Oatley is followed by Robert E. Scott's paper "Fast Track to Lost Jobs: Trade Deficits and Manufacturing Decline Are the Legacies of NAFTA and the WTO" and Douglas A. Irwin's "The Employment Rationale for Trade Protection."

In Oatley (2004), Irwin contends that

> the overall impact of trade on the number of jobs in an economy is best approximated as zero. Total employment is not a function of international trade, but the number of people in the labor force. (Oatley 2004, p. 27)

Here, Irwin simply repeats the claim he made in his book *Free Trade under Fire* (2002, p. 115). All his arguments are based on the GE framework, in which there is no unemployment. One of the fundamental assumptions inherent in the GE model is the efficient use of all resources, including that of the labor force. This is what Irwin supposes and also what he deduces as a conclusion; clearly, this is a case of petitio principii.

Scott's arguments and analyses follow the IPE tradition and are based on either real or supposed conflicts of interest. It is less dogmatic and reflects the existing psychology of society. These characteristics constitute the strength of the IPE: it is in closer proximity to the actual conflicts that exist. Paradoxically, they also reveal the weakness of the IPE, because although it correctly raises the appropriate questions, it cannot delve into economics arguments and provide criticism. The IPE needs to produce or otherwise provide a theory that serves as an alternative to the existing economic explanations. The new theory of international values may provide the IPE with such a framework.

18 Development Economics

Development economics emerged after World War II, and it underwent substantial changes in terms of its leading ideas (Lindauer and Pritchett 2002). We can detect three distinct generations of thinking in development economics.

In the 1950s and 1960s, the era of the first generation, the "Big Idea," was to attain economic independence. In line with this thinking, the state should play a leading role in undertaking accumulations and industrializations. Import substitution was the focus; foreign direct investment was to be avoided. There was a big swing between the first and second generations; a neoclassical counterrevolution took place in the 1970s, and many aspects of the "Big Idea" were reversed.

The policies most frequently advocated in the 1980s and 1990s were collectively dubbed "the Washington Consensus." The general orientation was to "let the market go." State interventions were interpreted as the main obstacle to development, and investment emphasis changed from public to private ones. Trade and foreign direct investment were welcomed. Deregulations were recommended, and it was thought that the market economy should be reinforced. Exports became the strategic target of development policies; however, the liberalization of trade and finance brought about a series of financial crises. Economic performance in the 1990s differed substantially from country to country, and the true effectiveness of the second set of "Big Ideas" was unclear. Both the East Asian Miracle and China introduced much more confusion than clarity about what two generations of development theories had produced.

Krugman (1992) calls for a counter-counterrevolution and argues that the high-development theory of the first generation of development theory appears to be more sensible, if we take into account the new development of theories that incorporate increasing returns to scale. However, the situation was not as straightforward as he had imagined. Stiglitz (1992) argues in his comment to Krugman (1992) that Krugman's vison is too narrow and ignores equally important factors. Rodrik (1998) shows, based on a 1992 cross-country study, that the usual "rules of thumb" regarding what makes for good policy (e.g., uniformity, transparency, nonselectivity) are quite useless in predicting which policy regimes will perform better in practice. Lindauer and Pritchett (2002) talk about the "end of big ideas."

The economic success of East Asia, Southeast Asia, China, and India revealed that unexpected processes were under way in these countries and areas, and this helped in overcoming a classical dilemma—namely, causally circular conditionals of economic development. Previously, industrialization required that a whole set of industries be developed. When such a nexus of products and techniques was lacking, it was difficult for an industry to develop in isolation. In the case of the aforementioned Asian countries and regions, foreign trade made it possible to separate one industry from others. As mentioned, this phenomenon is known as the SGU (Baldwin 2006, 2014).

Why did this phenomenon emerge at the end of the twentieth century? In essence, it was a result of drastic reductions in transportation and communication costs and increased speed in these two areas.[54] The formerly "bundled" manufacturing process was unbundled and divided into chains of fragmented processes, whereupon a portion of the chain was transferred to countries that paid lower wages. Here, however, a peculiar problem occurs. We lacked a general theory of international trade in which input goods are traded. This deficiency was noticed as early as the late 1950s, but the theory of input trade had not been developed, mainly on account of the mathematical difficulty inherent in formulating price theories.

In spite of this critical shortcoming, trade theory continued to play an important role in formulating industrial and trade policies. This "state of the art" produced a series of erroneous policies, and it became one of the reasons why the first and second generations of development policies failed.

Although various pieces of evidence refute FPT (or the HOS and HOV theories), and few people consider them economically relevant, economists continue to recommend policies that rely on factor proportion arguments. Even those economists who are critical of mainstream economics sometimes argue along this line of thinking.

Take as an example high-technology industries in India. Is it not good to develop these industries as possible export industries, given that India is still a labor-abundant country? India is a very large country with the largest army of skilled engineers and a proportionally small capital–labor ratio. If we follow what the FPT recommends, it would be advisable to concentrate on industries that have smaller capital–labor ratios. The new theory of international values points to another

[54] Section 12 addresses the logic that underlies unbundling.

possibility: it would be wise to develop any industry that can produce a product (of a given quality) at a competitive cost, given the actual wage disparity between India and other, more-developed countries. For example, a cutting-edge industry that features a relatively high rate of capital and requires a substantial number of skilled engineers could be highly competitive, if the labor cost of skilled engineers were one-third that seen in the United States. In such a case, FPT would make a completely wrong recommendation. The new theory of international values, as a theory that analyzes firm-level competiveness, would provide more plausible policy recommendations than could the FPT, which considers only country-level factor differences.

At one time, the lack of an appropriate theory of trade pushed development economics in the wrong direction. A typical case was the dependency theory. People who embraced this theory worried that the terms of trade for developing countries were worsening, and they recommended import substitution industrialization policy.[55] Emmanuel's (1969) theory of unequal exchange provided a reason for their orientations. Dependency theorists argued that the high wages of developed countries worsened the terms of trade for less-developed countries. It is evident that they were aligning themselves with John Stuart Mill: for Mill, the terms of trade are not determined by production relations, but by the law of demand and supply. Dependency theorists thought that these terms of trade were ultimately determined by power relations between developed and developing countries.

The new theory does not think in the same way. If we assume a predetermined volume and composition of the world demand, wage disparity is substantially determined by differences in the technologies that each country possesses. This is not to claim that institutions or knowledge do not matter; they are indeed important factors that determine the present set of production techniques. There are many other factors that influence the state of technology in each country. For example, the infrastructure of a society helps reduce transport costs and make production techniques more efficient. Good ports, roads, and railways reduce real transport costs and contribute to making almost all production techniques more efficient. The important thing to know here is that wage disparities are determined by virtue of the sets of production techniques; we cannot change them through trade policy. The terms of trade reflect wage disparities between countries—not vice versa, as Emmanuel (1969) imagines. To ameliorate wage disparities, we need to improve the production techniques in low-wage countries. This is a lesson drawn directly from the new theory of international values.

A low wage rate in itself is a bad thing, but it can serve as a powerful arm for exportation; full and effective use of this arm makes it possible to promote export-oriented industrialization. Even in such a case, initial production experience

[55] I do not deny that import substitution industrialization policy had some plausibility. Given an economic structure that is completely dependent on the former colonial powers, it was necessary to formulate a more independent and internally self-supporting and circulating economy. For that purpose, some measures that exclude foreign commodities were justifiable. Here, I am questioning their implicit theory of international values.

is crucial. Japan, the "four little tigers" (i.e., Hong Kong, Taiwan, South Korea, and Singapore), China, India, and the Southeast Asian countries accumulated this social capability in very different ways. In the case of prewar Japan, it was the flying geese pattern. South Korea followed a path similar to Japan's. Taiwan accumulated experience through contract manufacturing. The Southeast Asian countries gained it by undertaking manufacturing initiated by foreign direct investment. China and India kept their respective economies rather closed for an extended period; they have since worked to grow their potential, and in the 1990s, they opened their countries. In all cases, processing trade in the wide sense (in the Japanese sense of *kakō bōeki*, see footnote 37) has been a key concept in trade policy. It is astonishing that trade theory lacked this concept until the 1990s. One easy explanation for this absence is the influence of trade theory: input trade (or the trade of intermediate goods) was excluded from the trade theory, and this may have retarded the recognition of input trade and the strategic importance of processing trade.

Development economics requires a sound trade theory. The new theory of international values may act as the new theory it needs, as it is a unique general theory that can treat input trade. Exceptions may include those trade theories based on the GE framework, but we may contend (with good reason) that GET does not serve as a good framework for development economics.

Acknowledgments I owe much to discussions I have had with members of the Workshop on the Theory of International Values. I express deep thanks to Mitsuharu Ito, who advised me to organize the group. Hubert Escaith, Sébastien Miroudot, and Toshihiro Oka each helped me greatly, by offering detailed comments on the first draft of this paper.

Appendix: Proof of the Fundamental Theorem

This appendix gives a proof of Theorem 3.4. Although it is desirable to give a comprehensive proof, it would occupy too much space to start from the very beginning of the theory of linear inequality. I cite two theorems from the theory of linear inequality (or real linear algebra) and a theorem from convex polytope theory. Based on these classical results, I will prove the fundamental theorem.

Let me first cite these three theorems. In the following, $A = (a_{ij})$ denotes a real valued $M \times N$ matrix. Vector $\mathbf{x}$ is an M-row vector and $\mathbf{p}$ an N-column vector. Vector $\mathbf{b}$, which appears in Theorem 2, is an N-row vector.

Theorem 1 (Tucker's Key Theorem)
The equation

$$\mathbf{x}A = \mathbf{0}, \quad \mathbf{x} \geqq \mathbf{0}$$

and the inequality

$$A\mathbf{p} \geqq \mathbf{0}$$

have a couple of solutions **x** *and* **p** *such that*

$$\mathbf{x}A + \mathbf{p}^t > \mathbf{0}.$$

Here $\mathbf{p}^t$ *denotes the transposed line vector of* **p**.

Theorem 1 was first given by Tucker (1956, p.5) as a lemma and is now often called Tucker's Key Theorem, because it provides many (or almost all) important theorems in the theory of linear inequalities. You can find this theorem in various textbooks on the theory of linear inequality, game theory, theory of optimization, and theory of finance. Among more recent papers, it is easy to access Giorgi (2014).

Theorem 2 (Farkas)
Equation

$$\mathbf{x}A = \mathbf{b}$$

has a nonnegative solution if and only if

$$\langle \mathbf{b}, \mathbf{p} \rangle \geqq 0 \quad \textit{for} \quad \forall \mathbf{p} : A\mathbf{p} \geqq \mathbf{0}.$$

Here $\langle \mathbf{b}, \mathbf{p} \rangle$ means the scalar product of two (row and column) vectors of the same number of entries, which is to say,

$$\langle \mathbf{b}, \mathbf{p} \rangle = \sum_{i=1}^{N} b_i p_i.$$

This theorem is equivalent to Theorem 2.6 in Gale (1960, p.44) and Theorem (Minkowski-Farkas lemma) in Nikaido (1961, p.128). Minkowski conjectured this theorem in his attempt to establish a theorem on the mutual duality between a polyhedral convex cone and its dual. Farkas proved Theorem 2. This is often called the Minkowski-Farkas Theorem as Nikaido (1961) does.

The third theorem is more rarely found in a formal proposition. It may be intuitively too apparent to be called a theorem. We may deem it a kind of definition.

Theorem 3 (Unique Normal Direction)
Let P *be a full dimensional convex polytope in* $\mathbb{R}^N$ *and* F *its facet (i.e., a face of codimension 1 or dimension* N–1*). Then, there is an* N*-column vector* **p** *and a real number* c *such that*

$$\{\mathbf{x} | \langle \mathbf{x}, \mathbf{p} \rangle = c\} \cap P = F.$$

All vectors **p** *that define facet* F *are proportional with each other.*

The next theorem follows easily from Theorem 1. This is one of various examples that we can deduce from Theorem 1.

Theorem 4 (Positive Solution for an Inequality)
There exists a vector **p** *that satisfies*

$$A\mathbf{p} \leqq \mathbf{0} \quad \text{and} \quad \mathbf{p} > \mathbf{0}, \tag{A.1}$$

if and only if inequality

$$\mathbf{x}A \geqq \text{but} \neq \mathbf{0} \quad \text{and} \quad \mathbf{x} \geqq \mathbf{0} \tag{A.2}$$

has no solution **x**.

This theorem, which appears in Nikaido (1961, pp.157–158) in the form of a lemma, is rarely found in textbooks, but plays a crucial role in the theory of international values as we will see in the proof of Theorem 7. A similar theorem is more famous.

Theorem 5 (Stiemke, Positive Solution for an Equality)
Equality

$$A\mathbf{p} = \mathbf{0} \quad \text{and} \quad \mathbf{p} > \mathbf{0}$$

has a solution **p** *if and only if inequality*

$$\mathbf{x}A \leqq \text{but} \neq \mathbf{0}$$

has no solution **x**.

We can find this theorem in Gale (1960, p.49) as Corollary 2 to Theorem and in Nikaido (1961, p.116 and p.121) as Theorems 1 and 2 in Chap. 3.

Theorems 4 and 5 are too similar to memorize without confusion, but the two theorems are independent in the sense that there is no direct easy proof of one theorem from another.[56] Theorem 4 has a weaker condition and a weaker conclusion than Theorem 5. The conditions of the former in the “if and only if” clause is restricted to positive vectors, whereas those of the latter concern all real vectors. Another similar but distinct theorem appears in Gale (1960, p.49) as Theorem 2.10.

We can easily derive Theorem 4 from Theorem 1. Indeed, equation

$$A\mathbf{p} = \mathbf{0}, \quad \mathbf{p} \geqq \mathbf{0}$$

[56] Both can be derived from Theorem 1. Curiously enough, we can derive Theorem 1 from Theorems 4 and 5 by an elementary argument.

and the inequality

$$\mathbf{x}A \geqq \mathbf{0}$$

have a couple of solutions **x** and **p** that satisfies condition

$$\mathbf{x}A + \mathbf{p}^t > \mathbf{0}. \tag{A.3}$$

If the sufficient condition with the inequalities (A.2) of Theorem 4 holds, there is no nonnegative **x** that satisfies $\mathbf{x}A \geqq$ and $\neq \mathbf{0}$. This means $\mathbf{x}A = \mathbf{0}$. Then condition (A.3) signifies that $\mathbf{p} > \mathbf{0}$. Thus, the necessary conditions follow. Conversely, if there exists a positive solution that satisfies inequality (A.1),

$$\langle \mathbf{x}A, \mathbf{p} \rangle = \langle \mathbf{x}, A\mathbf{p} \rangle \leqq \mathbf{0}.$$

Because $\mathbf{x}A$ is nonnegative and **p** is positive for all its entries, this means that $\mathbf{x}A = \mathbf{0}$. This is what we have to prove.

Now let us proceed to the proof of Theorem 3.4, the fundamental theorem in the theory of international values. We proceed in three steps. First, we find a necessary and sufficient condition that a point belongs to the production possibility set. Then, we find a necessary and sufficient condition that a point belongs to the maximal frontier of the production possibility set. The uniqueness follows as a corollary of Theorem 3 with the aid of Theorem 2. The following is essentially a reproduction of Section 5 of Shiozawa (2007).

Let $\{A, J, \mathbf{q}\}$ be an RS economy and P be its production possibility set. A and J are, respectively, $T \times N$ and $T \times M$ matrices. Here, T is the number of production techniques, N the number of goods, and M the number of countries of the economy.

Theorem 6 (A Point in the Production Possibility Set)
A point **x** *in* $\mathbb{R}^N$ *belongs to the production possibility set* P *if and only if the inequality*

$$\langle \mathbf{x}, \mathbf{p} \rangle \leqq \langle \mathbf{q}, \mathbf{w} \rangle$$

holds for any couple of M- and N-column vectors **w**, **p** *that satisfy the condition.*

$$\mathbf{w} \geqq \mathbf{0} \text{ and } J\mathbf{w} \geqq A\mathbf{p}.$$

Proof of Theorem 6
If **x** is a point of P, there exists a nonnegative activity vector $\mathbf{s} = (s_\tau)$ such that

$$\mathbf{x} = \mathbf{s}A \quad \text{and} \quad \mathbf{s}J \leqq \mathbf{q}.$$

The last condition is equivalent to the existence of a nonnegative vector **t** such that

$$\mathbf{x} = \mathbf{s}A \quad \text{and} \quad \mathbf{s}J + \mathbf{t} = \mathbf{q}. \tag{A.4}$$

Take an $(M + T) \times (M + N)$ matrix

$$\begin{bmatrix} E & O \\ J & -A \end{bmatrix},$$

where E is the $M \times M$ identity matrix and O is an $M \times N$ matrix whose entries are all 0. Condition (A.4) can be expressed as

$$(\mathbf{t}, \mathbf{s}) \begin{bmatrix} E & O \\ J & -A \end{bmatrix} = (\mathbf{q}, -\mathbf{x}) . \tag{A.5}$$

By Theorem 2, there exists a nonnegative vector $(\mathbf{t}, \mathbf{s})$ that satisfies (A.5) if and only if the inequality

$$\langle (\mathbf{q}, -\mathbf{x}) , (\mathbf{w}, \mathbf{p}) \rangle \geqq 0 \quad \text{or} \quad \langle \mathbf{x}, \mathbf{p} \rangle \leqq \langle \mathbf{q}, \mathbf{w} \rangle$$

holds for all vectors $\mathbf{w}$, $\mathbf{p}$ that satisfy

$$\begin{bmatrix} E & O \\ J & -A \end{bmatrix} \begin{bmatrix} \mathbf{w} \\ \mathbf{p} \end{bmatrix} \geqq \mathbf{0}.$$

In another expression, a point $\mathbf{x}$ is a point of P if and only if the following relations are satisfied:

$$\langle \mathbf{x}, \mathbf{p} \rangle \leqq \langle \mathbf{q}, \mathbf{w} \rangle$$

for all $\mathbf{w}$ and $\mathbf{p}$ that satisfy

$$\mathbf{w} \geqq \mathbf{0} \text{ and } J\mathbf{w} \geqq A\text{p}.$$

□

Theorem 7 (A Point on the Maximal Frontier)
A point $\mathbf{x}$ *of* P *is on the maximal frontier* F *if and only if there exists a couple of positive vectors* $\mathbf{w}$, $\mathbf{p}$ *such that*

$$J\mathbf{w} \geqq A\boldsymbol{p} \text{ and } \langle \mathbf{x}, \mathbf{p} \rangle = \langle \mathbf{q}, \mathbf{w} \rangle.$$

To prove Theorem 7, we need the following lemma:

Lemma 8 (Necessary Condition to be a Maximal Point)
If $\mathbf{x}$ *is a maximal point of P, there exists a nonnegative activity vector* $\mathbf{s}$ *such that*

$$\mathbf{x} = \mathbf{s}A \quad \text{and} \quad \mathbf{s}J \leqq \mathbf{q}.$$

Proof of Lemma 8
Suppose that there exists a nonnegative vector $\mathbf{s}$ that satisfies

$$\mathbf{x} = \mathbf{s}A \quad \text{and} \quad \mathbf{s}J \leqq \text{and} \neq \mathbf{q}.$$

Let $\mathbf{r} = (r_j) = \mathbf{s}J$. Then, for a country index i, we have $r_i < q_i$. By assumption of RS economy, country i has a productive system of production techniques. There exists a nonnegative activity vector $\mathbf{t} = (t_\tau)$ such that

$$\mathbf{t}A > 0 \text{ and } \mathbf{t}J = \mathbf{r}.$$

Here, $t_\tau = 0$ for all τ except for some production techniques belonging to country i. Then, taking a positive ϵ sufficiently small, we get

$$\mathbf{y} = (\mathbf{s} + \epsilon\mathbf{t})\,A > \mathbf{x} \quad \text{and} \quad (\mathbf{s} + \epsilon\mathbf{t})\,J \leqq \mathbf{q}.$$

This means that vector $\mathbf{y}$ is a point of P, but this contradicts the assumption that $\mathbf{x}$ is maximal in P. This proves the lemma. □

Proof of Theorem 7
First, suppose that $\mathbf{x}$ is a maximal point of P. Then, inequality

$$(\mathbf{s}, 1)\begin{bmatrix} -J & A \\ \mathbf{q} & -\mathbf{x} \end{bmatrix} \geqq \text{and} \neq \mathbf{0} \tag{A.6}$$

has no nonnegative solution $\mathbf{s}$. Let us prove this fact by contradiction. Suppose that there exists a nonnegative $\mathbf{s}$. Then, depending on where we have the strict inequality, at least one of two cases holds:

(a) $\mathbf{s}A \geqq$ and $\neq \mathbf{x}$ and $\mathbf{s}J \leqq \mathbf{q}$.
(b) $\mathbf{s}J \leqq$ and $\neq \mathbf{q}$ and $\mathbf{s}A \leqq \mathbf{x}$.

Case (a) contradicts that $\mathbf{x}$ is a maximal point in P, because condition (a) means that $\mathbf{y} = \mathbf{s}A$ is a point in P and $\mathbf{x}$ cannot be a maximal point. Case (b) also contradicts the fact that $\mathbf{x}$ is a maximal point in P by virtue of Lemma 8. Therefore, inequality (A.6) has no nonnegative solution $\mathbf{s}$. Theorem 4 tells that equation

$$\begin{bmatrix} -J & A \\ \mathbf{q} & -\mathbf{x} \end{bmatrix} \begin{bmatrix} \mathbf{w} \\ \mathbf{p} \end{bmatrix} \leqq \mathbf{0}$$

has a positive solution $(\mathbf{w}, \mathbf{p})$. This means that there exists a positive $\mathbf{w}$, $\mathbf{p}$ that satisfy

$$J\mathbf{w} \geqq A\,\mathbf{p} \quad \text{and} \quad \langle \mathbf{q}, \mathbf{w} \rangle \leqq \langle \mathbf{x}, \mathbf{p} \rangle .$$

On the other hand, as $\mathbf{x}$ is a point of P, we have, from Theorem 6, $\langle \mathbf{x}, \mathbf{p} \rangle \leqq \langle \mathbf{q}, \mathbf{w} \rangle$ for any positive vectors $\mathbf{w}$ and $\mathbf{p}$ that satisfies $J\mathbf{w} \geqq A\mathbf{p}$. Combining these two results, we have $\langle \mathbf{x}, \mathbf{p} \rangle = \langle \mathbf{q}, \mathbf{w} \rangle$.

Conversely, let us assume for a point $\mathbf{x}$ of P that there exists two positive vectors $\mathbf{w}$ and $\mathbf{p}$ such that

$$J\mathbf{w} \geqq A\mathbf{p} \quad \text{and} \quad \langle \mathbf{x}, \mathbf{p} \rangle \leqq \langle \mathbf{q}, \mathbf{w} \rangle .$$

Then, $\mathbf{x}$ must be a maximal point of P. We prove this fact by contradiction. Suppose that $\mathbf{x}$ is not a maximal point. There exists a point $\mathbf{y} \in P$ such that

$$\mathbf{y} \geqq \mathbf{x} \quad \text{and} \quad \mathbf{y} \neq \mathbf{x}.$$

Then

$$\langle \mathbf{y}, \mathbf{p} \rangle > \langle \mathbf{x}, \mathbf{p} \rangle = \langle \mathbf{q}, \mathbf{w} \rangle .$$

Theorem 6 implies that $\mathbf{y}$ cannot be a point of P. This is a contradiction.This stands that point $\mathbf{x}$ is a point of the maximal frontier. □

With these preparations, we can now prove the fundamental theorem (Theorem 3.4). Take a point $\mathbf{d}$ of the maximal frontier; then, Theorem 7 implies that there are two positive vectors $\mathbf{w}$ and $\mathbf{p}$ that satisfy

$$\mathbf{s}A = \mathbf{d}, \quad \mathbf{s}J = \mathbf{q}, \quad J\mathbf{w} \geqq A\mathbf{p} \text{ and } \langle \mathbf{q}, \mathbf{w} \rangle = \langle \mathbf{d}, \mathbf{p} \rangle .$$

If a point $\mathbf{d}$ is in a regular domain, or in the relative interior of a facet of codimension 1, the above vectors give a boundary half-space

$$\langle \mathbf{x}, \mathbf{p} \rangle = \langle \mathbf{q}, \mathbf{w} \rangle$$

and by Theorem 3 the direction of vector $\mathbf{p}$ is uniquely determined. It remains to prove that the international vector $\langle \mathbf{w}, \mathbf{p} \rangle$ itself is unique up to a scalar multiplication. We will show this in the next proposition.

Proposition 9 (Uniqueness)
In an RS economy $\{A, J, \mathbf{q}\}$, *let* $\mathbf{x}$ *be a point of a regular domain D of the maximal frontier. There exists a pair of positive vectors* $(\mathbf{w}, \mathbf{p})$ *that satisfies*

$$J\mathbf{w} \geqq A\mathbf{p} \quad \text{and} \quad \langle \mathbf{x}, \mathbf{p} \rangle = \langle \mathbf{q}, \mathbf{w} \rangle .$$

Let $\mathbf{y}$ *be any point in the domain D and any pair* $(\mathbf{u}, \mathbf{p})$ *satisfies*

$$J\mathbf{u} \geqq A\mathbf{p} \quad \text{and} \quad \langle \mathbf{y}, \mathbf{p} \rangle = \langle \mathbf{q}, \mathbf{w} \rangle ,$$

then $\mathbf{u} = \mathbf{w}$.

The existence part of the proposition is a repetition of Theorem 7. If point $\mathbf{x}$ is in a regular domain, the equation

$$\langle \mathbf{y}, \mathbf{p} \rangle = \langle \mathbf{q}, \mathbf{w} \rangle .$$

defines a supporting hyperplane and is unique up to a scalar multiplication. Then, the uniqueness of price vectors $\mathbf{p}$ is evident. The question is whether $(\mathbf{w}, \mathbf{p})$ is unique up to a scalar multiplication.

Suppose there exists another pair of international vectors $\mathbf{u}$, $\mathbf{p}$ that satisfies the same equation as the proposition:

$$J\mathbf{u} \geqq A\mathbf{p} \quad \text{and} \quad \langle \mathbf{y}, \mathbf{p} \rangle = \langle \mathbf{q}, \mathbf{u} \rangle .$$

This equation holds for any point $\mathbf{y}$ in the regular domain. As $\mathbf{y}$ is a point of production possibility set P, it is expressed by an activity vector $\mathbf{t}$ as $\mathbf{y} = \mathbf{t}A$. Let $\mathbf{r} = \mathbf{t}J$. As $\mathbf{y}$ is maximal in P, for any country i, $r_i > 0$, where $\mathbf{e}(i)$ is the unit vector whose nonvanishing entry is at the ith position. This means that there exists a production technique $\tau = (\mathbf{e}(i), \mathbf{a}(\tau))$ of country i and satisfies the equation

$$u_i = \langle \mathbf{u}, \mathbf{e}(i) \rangle = \langle \mathbf{a}(\tau), \mathbf{p} \rangle . \tag{A.7}$$

On the other hand, the production technique τ must be competitive with regard to international value $(\mathbf{w}, \mathbf{p})$. In fact, if τ is not competitive, we have $\langle \mathbf{a}(\tau), \mathbf{p} \rangle < w_i$. Then, as $\mathbf{t}$ has a positive entry $t(\tau)$, we get

$$\langle \mathbf{y}, \mathbf{p} \rangle = \langle \mathbf{t}A, \mathbf{p} \rangle = \langle \mathbf{t}, A\mathbf{p} \rangle < \langle \mathbf{t}, J\mathbf{w} \rangle = \langle \mathbf{t}J, \mathbf{w} \rangle \leqq \langle \mathbf{q}, \mathbf{w} \rangle .$$

This contradicts the assertion that $\mathbf{y}$ is maximal (Theorem 7). So, the production technique τ is competitive. This means

$$w_i = \langle \mathbf{w}, \mathbf{e}(i) \rangle = \langle \mathbf{a}(\tau), \mathbf{p} \rangle . \tag{A.8}$$

Combining equations (A.7) and (A.8), $u_i = w_i \ \forall \ i$ or $\mathbf{u} = \mathbf{w}$. This proves the proposition. □

References

Akamatsu, K. (1962). A historical pattern of economic growth in developing countries. *The Developing Economies, 1*(s1), 3–25.

Arrow, K. J. (1951) Alternative proof of the substitution theorem for Leontief models in the general case. In Koopmans (1951a) pp 155–164.

Arrow, K. (1962) The economic implications of learning by doing. *Review of Economic Studies, 29*(3), 155–173.

Arrow, K. J., & Debreu, G. (1954). Existence of an equilibrium for a competitive economy. *Econometrica, 22*(3), 265–290.

Baldwin, R. (2006) Chapter 1: Globalization: The great unbundling(s). In *Globalization challenges to Europe*, Secretariat of the Economic Council, Finnish Prime Minister's Office. ISBN 952–6531-15-X.

Baldwin, R. (2014) Chapter 5: Trade and industrialization after globalization's second unbundling: How building and joining a supply chain are different and why it matters. in R.C. Feenstra and A.M. Taylor (_Eds.) Globalization in an age of crisis: multilateral economic cooperation in the twenty-first century, University of Chicago Press, London pp. 165–212.

Bidard, C. (2010). The dynamics of intensive cultivation. *Cambridge Journal of Economics, 34*(6), 1097–1104.

Bidard, C. (2011), *The ricardian dynamics revisited* (Working Paper), EconomiX.

Bidard, Ch. (2014) *The ricardian rent theory: An overview* (Centro Sraffa Working Papers No.8).

Boyer, R., H. Uemura, and I Isogai (Eds.) (2012) Diversity and transformations of Asian capitalisms, Rougledge, New York.

Chipman, J. S. (1965). A survey of the theory of international trade: Part 1, The classical theory. *Econometrica***33**(3): 477–519. Part 2, The neoclassical theory. *Econometrica***33**(4): 685–760. 477–519. Part 3, The modern theory. *Econometrica, 34*(1), 18–76.

Cole, S. (1977) *Global models and the international economic order: A paper for the United Nations Institute for Training and Research Project on the Future* (New edn). Oxford: UNITAR/Pergamon.Press.

Dhyne, E., Álvarez, L. J., Le Bihan, H., Veronese, G., Dias, D., Hoffmann, J., Jonker, N., Lünnemann, P., Rumler, F., & Vilmunen, J. (2006). Price setting in the Euro area/some stylized facts from individual consumer price data. *Journal of Economic Perspective, 20*(2), 171–192.

Dietzenbacher, E., Los, B., Stehrer, R., Timmer, M., & de Vries, G. J. (2013). The construction of world input–output tables in the WIOD project. *Economic Systems Research, 25*, 71–98.

Dorbusch, R., Fischer, S., & Samuelson, P. (1977). Comparative advantage, trade, and payments in ricardian model with a continuum of goods. *American Economic Review, 67*(5), 823–839.

Emmanuel, A. (1969). *L'échange inégale*. François Mapérot.

Ethier, W. J. (1999). Profile: Jones and trade theory. *Review of International Economics, 7*(4), 764–768.

Escaith, H., & Inomata, S. (2013). Chapter 5: Geometry of global value chains in East Asia: The role of industrial networks and trade policies. In D. K. Elms & P. Low (Eds.), *Global value chains in a changing world* (pp. 135–157). Geneva: WTO Publications.

Escaith, H., & S. Miroudot. (2016). *Industry-level competitiveness and inefficiency spillovers in global value chains*. A paper presented at the 24th international input-output conference 4–8 July 2016, Seoul.

Frankel, J., Parsley, D., & Wei, S.-J. (2012). Slow passthrough around the world: A new import for developing countries? *Open Economies Review, 23*(2), 213–251.

Gale, D. (1960). *The theory of linear economic models*. New York: McGrow-Hill Book Company.

Georgescu-Roegen, M. (1951). Some properties of a generalized Leontief model. In Koopmans (1951a) pp. 165–173.

Gerschenkron, A. (1962). *Economic backwardness in historical perspective*. Cambridge: Belknap Press of Harvard University Press.

Giorgi, G. (2014, October) *Again on the Farkas and the Tucker key theorem proved easily* (DEM Working Paper Series) (University of Pavia) #94. ftp://ftp.repec.org/opt/ReDIF/RePEc/pav/demwpp/DEMWP0094.pdf

Gopinath, G., & Rigobon, R. (2008). Sticky border. *Quarterly Journal of Economics, 123*, 531–575.

Graham, F. D. (1948). *The theory of international values*. Princeton: Princeton University Press.

Grossman, G., & Rossi-Hansberg, E. (2008). The rise of offshoring: It's not wine for cloth anymore. *American Economic Review, 98*, 1978–1997.

Haberler, G. (1933) *Die internationale Handel*, Berlin. English Translation: *The Theory of International Trade, with Its Applications to Commercial Policy*. London: William Hodge and Company, Ltd., 1936.

Hicks, J. R. (1976). Revolutions' in economics. In S. Latsis (Ed.), *Method and appraisal in economics*. Cambridge: Cambridge University Press.

Higashida, K. (2005) Chūkanzai to Kokusai Seisan Tokka Patān/Tasūzai Tasūkoku Moderu (in Japanese: Intermediate Goods and International Production Specialization/Multi-good, Multi-country Model) in Ishikawa and Furusawa (Eds.) Kokusai Bōeki Riron no Tenkai, Tokyo: Bunshindo.

Hummels, D., & Skiba, A. (2004). Shipping the good apples out? An empirical confirmation of the Alchian-Allen conjecture. *Journal of Political Economy, 112*(6), 1384–1402.

Irwin, D. A. (2002) *Free trade under fire* (2nd edn) Princeton: Princeton University Press. Paperback (3rd edn.) in 2009.

Jacobs, J. (1969). *The economy of cities*. New York: Random House.

Jones, R. (1961). Comparative advantage and the theory of tariffs: A multi-country, multi-commodity model. *Review of Economic Studies, 28*(3), 161–175.

Jones, L., Li Xin, and CH Degain (2014) *The similarities and differences among three major Inter-country input-output database and their implications for trade in value-added estimates* (Office of Economics Working Paper, No. 2014–128), U.S. International Trade Commission.

Koopmans, T. C. (ed.). (1951a) Activity analysis. *Cowles commission for research in economics* (Monograph 13). http://cowles.yale.edu/sites/default/files/files/pub/mon/m13-all.pdf

Koopmans, T.C. (1951b) Alternative proof of the substitution theorem for Leontief models in the case of three industries. In Koopmans (1951a), pp. 147–154.

Krugman, P. (1979). Increasing returns, monopolistic competition, and international trade. *Journal of International Economics, 9*, 469–479.

Krugman, P. (1980). Scale economies, product differentiation, and the pattern of trade. *American Economic Review, 70*, 950–959.

Krugman, P. (1992) Towards a counter-counterrevolution in development theory [With comments by J. Stiglitz, L. Jayawardena and Floor Discussion] World Bank Annual Conference on Development Economics 1992. Washington, DC: World Bank, pp. 15–61.

Krugman, P. (1996) *Pop internationalism*. Cambridge, MA: The MIT Press. Paperback ed. 1997.

Kurokawa, Y. (2014). A survey of trade and wage inequality: Anomalies, resolutions and new trends. *Journal of Economic Surveys, 28*(1), 169–193.

Lewis, J. (2016). *What can big data tell us about the passthrough of big exchange rate changes? Bank of England Staff Working Paper No.579*. London: Bank of England.

Lindauer, D.L., and L. Pritchett (2002) What's the big idea? The third generation of policies for economic growth. *Economia3*(1) 1–39. With Comments by Dany Rodrick and R. S. Eckaus.

McKenzie, L. (1954). Specialization and efficiency in world production. *Review of Economic Studies, 21*(3), 165–180.

Melitz, M. J. (2003). The impact of trade on intra-industry reallocations and aggregate industry productivity. *Econometrica, 71*(6), 1695–1725.

Mill, J. S. (1848). *Principles of political economy with some of their applications to social philosophy.* Citations from Library of Economics and Liberty. http://www.econlib.org/library/Mill/mlP.html

Morishima, M. (1973). *Marx's economics: A dual theory of value and growth.* Cambridge: Cambridge University Press.

Nakamura, E., & Zerom, D. (2009). Accounting for incomplete pass-through. *Review of Economic Studies, 77*(3), 1192–1230.

Nikaido, F. (1961). *Linear mathematics for economics* (in Japanese: *Keizai no Tame no Senkei Sūgaku*). Tokyo: Baifūkan.

Oately, T. (2004) *The global economy: Contemporary debates* (2nd edn.). New York: Longman, 2011.

Ohlin, B. (1933) *Interregional and international trade*, Cambridge, MA: Harvard University Press. Swedish edition 1931.

Oka, T. (2017). The relation between value and demand in the new theory of international values. In Y. Shiozawa, T. Oka, & T. Tabuchi (Eds.) *A new construction of Ricardian theory of international values: Analytical and historical Approach* (Chapter 3). Tokyo: Springer.

Ricardo, D. (1817–21 [1951]) On the principles of political economy and taxation. In P. Sraffa (Ed.), *Volume 1 of the works and correspondence of David Ricardo*. Cambridge: Cambridge University Press.

Rodrik, D. (1998) Globalization, social conflict and economic growth. In *Conferencia de Raúl Prebisch. Ginebra. Versión revisada (en inglés) disponible en.*http://www.ksg.harvard.edu/rodrik/global.pdf

Rogoff, K. (1996). The purchasing power parity puzzle. *Journal of Economic Literature, 34*, 647–668.

Rosenberg, N. (1982). *Inside the black box: Technology and economics*. Cambridge: Cambridge University Press. Chap. 7 is titled "Learning by using."

Rose-Ackerman, S. (1996). The political economy of corruption—Causes and consequences. Viewpoint (World Bank) Note No.74, 1–4. https://openknowledge.worldbank.org/handle/10986/11629

Ruffin, R. J. (2009). Globalization and income inequality. *Trade and Development Review, 2*(2), 56–69.

Samuelson, P. A. (1948). International trade and the equalisation of factor prices. *Economic Journal, 58*(230), 163–184.

Samuelson, P. A. (1949). International factor-price equalisation once again. *Economic Journal, 59*(234), 181–197.

Samuelson, Paul A. 1951. Abstract of a theorem concerning substitutability in open Leontief models. In Koopmans (1951a) 142–146.

Samuelson, P. A. (1953). Prices of factors and goods in general equilibrium. *Review of Economic Studies, 21*(1), 1–20.

Samuelson, P. (1964). Theoretical notes on trade problems. *Review of Economic Studies, 46*(2), 145–154.

Samuelson, P. (2001). A Ricardo-Sraffa paradigm comparing gains from trade in inputs and finished goods. *Journal of Economic Literature, 39*(4), 1204–1212.

Shiozawa, Y. (1975) *Durable capital goods and their valuation* (Kyoto Institute Of Economic Research Discussion Paper Series No.91) Available from ResearchGate.

Shiozawa, Y. (1985). International trade and Choice of Techniques (in Japanese: Kokusai Bōeki to Gijutsu Sentaku). *Keizaigaku Zasshi (Osaka City University), 85*(6), 44–61.

Shiozawa, Y. (1996) Economy as dissipative structure. A presentation made in Keihanna Prigogine conference, organized and sponsored by Keihanna Co. and Sankei Newspaper, May 28 1996. Available from ResearchGate.

Shiozawa, Y. (2007). A new construction of ricardian trade theory: A many-country, many-commodity case with intermediate goods and choice of techniques. *Evolutionary and Institutional Economics Review, 3*(2), 141–187.

Shiozawa, Y. (2014). *Rikādo Bōeki Mondai no Saishū Kaiketsu* (In Japanese: A Final Solution of the Ricardo Problem on International Values), Iwanami.

Shiozawa, Y. (2015). International trade theory and exotic algebra. *Evolutionary and Institutional Economics Review, 12*, 177–212.

Shiozawa, Y. (2016). The revival of classical theory of values. In Yokokawa and others (Eds.). *The Rejuvenation of political economy* (Chap. 8, pp. 151–172). London/New York: Routledge.

Shiozawa, Y. (2017). An origin of the neoclassical revolution: Mill's "reversion" and its consequences. In Y. Shiozawa, T. Oka, & T. Tabuchi (Eds.), *A new construction of Ricardian theory of international values: Analytical and historical approach* (Chapter 7). Tokyo: Springer

Sraffa, P. (1960). *Production of commodities by means of commodities*. Cambridge: Cambridge University Press.

Stiglitz, J. (1992). Comment on "Krugman's towards a counter-counterrevolution in development theory." In World Bank annual conference on development economics 1992. Washington, DC: World Bank, 1992, pp 39–49.

Takamasu, A. (1986). On the production possibility frontier in a Sraffa-Leontief economy. *The Manchester School, 54*(2), 202–207.

Takenaga, S. (2004) *The structure of the theory of value of David Ricardo*. A paper read at the ESHET 8th annual conference at Ca'Foscari University in Treviso, Italy, 28 Feb. 2004.

Taylor, A. M. (2001). Potential pitfalls for the purchasing-power-parity puzzle? Sampling and specification biases in mean-reversion tests of law of one price. *Econometrica, 69*(2), 473–498.

Timmer, M. P.. 2012. The World Input-Output Database (WIOD): Contents, sources and methods Working Paper Number: 10 (Edited by Marcel P. Timmer, with contributions of the WIOD Consortium Members).

Timmer, M. P., Dietzenbacher, E., Los, B., Stehrer, R., & de Vries, G. J. (2015). An illustrated user guide to the world input–output database: The case of global automotive production. *Review of International Economics, 23*, 575–605.

Tucker, A. W. (1956). Chapter 1: Dual systems of homogeneous linear relations. In H. W. Kuhn & A. W. Tucker (Eds.), *Linear inequalities and related systems, Ann. of Math. Studies No. 38* (pp. 3–18). Princeton: Princeton University Press.

Vermeulen, P., Dias, D. A., Dossche, M., Gautier, E., Hernando, I., Sabbatini, R., & Stahl, H. (2012). Price setting in the Euro Area: Some stylized facts from individual producer price data. *Journal of Money, Credit and Banking, 44*(8), 1631–1650.

Viner, J. 1937. *Studies in the theory of international trade*. London: Allen and Unwin. Reprinted by Augustus M Kelley in 1975.

von Neumann, J. (1944). A model of general economic equilibrium. *Review of Economic Studies 13*(1) 1–9. English translation of a paper of 1938 in German. Reprinted in Readings in Price Theory.

Waugh, M. E. (2010). International trade and income differences. *American Economic Review, 100*(5), 2093–2124.

Yabu, T. (2007). Kōbairyokuheika Pazuru No Kaimei: Jikeiretus apurōch No Shiten Kara (in Japanese: Elucidation of PPP Puzzle: From the View Point of Time Series Approach). *Kinyū Kenkyū (Institute for Monetary and Economic Studies, Bank of Japan), 26*(4), 75–106.

Part II
Theoretical Topics in and about the New Theory

The New Theory of International Values in the Context of the Ricardo-Sraffian Theory of Value and Distribution

Tosihiro Oka

Abstract In this article we introduce readers to the new theory of international values by placing it in the context of the Ricardo-Sraffian theory of value and distribution. Ricardo's theory is described as that in which exchangeable values of commodities are regulated by the quantities of labour bestowed in their production, on which he established his theory on the distribution of the produce of the earth. Contemporary classical theory, founded by Sraffa, is described as preserving Ricardo's perspective of the value independent of distribution and of demand by replacing the labour theory with the production-cost theory. After noting that Ricardo left the question of determination of values of the commodities traded internationally, it is shown that J. S. Mill argued that the law of demand and supply determines them, which conflicts with the classical perspective. We then demonstrate how the new theory of international values solves the question in line with the classical vision. Lastly, the similarity between this theory and Sraffa's treatment of multiple products is indicated.

Keywords J. S. Mill • Sraffa on multiple products • Classical theory of value and distribution

1 Introduction

In this article we will introduce readers to the new theory of international values. The classical theory of value is characterised as determination of relative prices by production costs, which is contrasted with the neoclassical theory in which relative prices are determined as equilibria between demand and supply. In the case of a closed economy with simple production, the minimum price theorem (often referred to as the non-substitution theorem) guarantees that prices will be determined by production costs. International trade and multiple production (including existence

T. Oka (✉)
Faculty of Economics, Fukui Prefectural University, Fukui, Japan
e-mail: oka@fpu.ac.jp

Y. Shiozawa et al. (eds.), *A New Construction of Ricardian Theory of International Values*, Evolutionary Economics and Social Complexity Science 7,
DOI 10.1007/978-981-10-0191-8_2

of other production factors than labour) are the two causes that undermine the dominance of the cost side in price determination. We will show how the new theory retains the classical nature of value theory when international trade exists.

First, I will describe Ricardo's theory of value and distribution and his vision on the dynamics of distribution. Second, I will show Sraffa's way to retain Ricardo's vision of distribution that is independent of value, without the labour theory of value, on which Ricardo depended. It is explained that the classical theory of value is reformulated as the production-cost theory of value. Third, it is shown that the production-cost theory of value seems to become invalid when foreign trade is introduced, because Ricardo left the determination of international values unsolved and J. S. Mill presented a solution using a law of supply and demand, which is in opposition to the production-cost theory of value. By using Mill's numerical example, I will show that Mill's solution is not valid and international values can be determined by cost-side factors according to the new theory. Lastly, it is indicated that international trade is, in a sense, regarded as a case of the same nature as that in which multiple products exist.

2 Ricardo's Theory of Value and Distribution

David Ricardo, quoting Adam Smith, argued that utility is necessary for a commodity to have exchangeable value, but that utility is not the measure of exchangeable value. He then identified two sources of commodities' exchangeable value; one is scarcity and the other is quantity of labour required to obtain them (Ricardo 1951a, pp. 12–13). He insisted, however, that commodities that derive their exchangeable value from scarcity are limited to those whose supply cannot be increased by labour and that the greatest part of goods that are the object of desire are produced by labour, and their exchangeable value is determined solely by the quantity of labour bestowed on their production.

Thus, the theory of value of the classical school is characterised as the labour theory of value; relative prices are in proportion to the quantity of labour bestowed on their production. The quantity of labour includes the labour used directly in production and the labour embodied in the goods used in production, such as raw materials, fuels, chemicals, machines and tools. This theory can be expressed well by the equation

$$v_j = v_1 a_{1j} + v_2 a_{2j} + \cdots v_n a_{nj} + l_j, \tag{1}$$

where v_j represents the value of a good j, l_j the quantity of labour directly used in the production of one unit of j, a_{ij} the quantity of an input $i (i = 1, 2, \cdots, n)$ required to produce one unit of j and v_i the value of the good $i (i = 1, 2, \cdots, n)$.

On the basis of this theory of value, Ricardo established a proposition on the law of the distribution of products: profits will fall only when wages rise. Wage is the price of labour, and it is regulated by the same law that regulates all prices of commodities: relative prices are in proportion to the quantities of labour bestowed on their production; thus, wage is in proportion to the 'value of labour', in other words, the amount of labour required to produce the foods, necessaries and conveniences to enable the labourers 'to subsist and to perpetuate their race without either increase or diminution' (Ricardo 1951a, p. 93). If we let ω represent the value of one unit of labour, then the equation (1) becomes

$$v_j = v_1 a_{1j} + v_2 a_{2j} + \cdots v_n a_{nj} + \omega l_j + \pi_j, \tag{2}$$

where π_j represents the profit from the production of one unit of commodity j. Comparing (1) and (2), we have

$$\pi_j = (1 - \omega) l_j. \tag{3}$$

This equation expresses straightforwardly the relation between wage and profit. It also shows that a change in ω does not have any effect on the value of the products; it affects only the distribution of the products between capitalists and labourers.

Ricardo dismissed scarcity as a source of exchangeable value for the greatest part of goods, the supply of which can be increased by labour. It is only the scarcity of land that plays a role in Ricardo's theory of value and distribution for such goods. In the progress of population, demand for foods increases, which requires land of inferior quality to be called into cultivation. Cultivation of such land will require a larger quantity of labour to produce the same amount of product, which will raise the value of the product. The increase in the value gives rise to a surplus in the cultivation of land of superior quality. The surplus will be obtained by the landowner as a rent (Ricardo 1951a, p. 70).

If $a^1_{in} (i = 1, 2, \cdots, n)$ and l^1_n denote the quantity of input i and labour, respectively, to produce one unit of the good n, say, corn, on the land of superior quality, and $a^2_{in} (i = 1, 2, \cdots, n)$ and l^2_n the quantity of input i and labour, respectively, to produce one unit of corn on the land of inferior quality, then

$$\begin{cases} v_n = v_1 a^1_{1n} + v_2 a^1_{2n} + \cdots v_n a^1_{nn} + l^1_n + \tau \\ v_n = v_1 a^2_{1n} + v_2 a^2_{2n} + \cdots v_n a^2_{nn} + l^2_n, \end{cases}$$

where τ represents the rent that accrues from the land of superior quality. When lands of further inferior quality come into cultivation, the values would have to meet the equations

$$\begin{cases} v_n = v_1 a^1_{1n} + v_2 a^1_{2n} + \cdots v_n a^1_{nn} + l^1_n + \tau^1 \\ v_n = v_1 a^2_{1n} + v_2 a^2_{2n} + \cdots v_n a^2_{nn} + l^2_n + \tau^2 \\ \vdots \\ v_n = v_1 a^{\nu-1}_{1n} + v_2 a^{\nu-1}_{2n} + \cdots v_n a^{\nu-1}_{nn} + l^{\nu-1}_n + \tau^{\nu-1} \\ v_n = v_1 a^{\nu}_{1n} + v_2 a^{\nu}_{2n} + \cdots v_n a^{\nu}_{nn} + l^{\nu}_n, \end{cases} \tag{4}$$

where a_{in}^k, l_n^k $(k = 1, 2, \cdots, \nu)$ represent the quantity of input i and labour, respectively, to produce one unit of corn on the kth superior land, and τ^k $(k = 1, 2, \cdots, \nu - 1)$ is the rent to that land. The ν-th land is one of the poorest qualities, or the marginal land, and generates no rent.

Collecting the equation (1) for n commodities, we have

$$\begin{cases} v_1 = v_1 a_{11} + v_2 a_{21} + \cdots v_n a_{n1} + l_1 \\ v_2 = v_1 a_{12} + v_2 a_{22} + \cdots v_n a_{n2} + l_2 \\ \quad \vdots \\ v_n = v_1 a_{1n} + v_2 a_{2n} + \cdots v_n a_{nn} + l_n. \end{cases} \tag{5}$$

This defines the system of values. When rents exist for scarce land, the last equation in (5) is replaced by (4).

The principal problem for Ricardo was to determine the laws that regulate the distribution of the produce of the earth, that is, how it is divided among the three classes of the community (i.e. how it is divided among rents, profits and wages) (Ricardo 1951a, p. 5). The most important part of the law is given by equation (3), which shows that profits decrease only when wages increase. As the population grows, lands of inferior quality are called into cultivation with the emergence of rents, accompanied by a rise in the value of corn, which brings about a rise in wages and a fall in profits. This is an outline of the dynamics of distribution viewed by Ricardo.

The rate of profit is the profit divided by the value of capital; using the notation in (2), the rate of profit, ρ_j, is expressed as

$$\rho_j = \frac{\pi_j}{v_1 a_{1j} + v_2 a_{2j} + \cdots v_n a_{nj}} = \frac{(1-\omega) l_j}{v_1 a_{1j} + v_2 a_{2j} + \cdots v_n a_{nj}},$$

when wages are paid after production and as

$$\rho_j' = \frac{\pi_j}{v_1 a_{1j} + v_2 a_{2j} + \cdots v_n a_{nj} + \omega l_j} = \frac{(1-\omega) l_j}{v_1 a_{1j} + v_2 a_{2j} + \cdots v_n a_{nj} + \omega l_j},$$

when wages are paid before production.

The rates of profit for different commodities are, in general, not equal to each other (i.e. $\rho_j \neq \rho_k$, $\rho_j' \neq \rho_k'$), unless the capital-labour ratios of two industries are equal:

$$\frac{l_j}{v_1 a_{1j} + v_2 a_{2j} + \cdots v_n a_{nj}} = \frac{l_k}{v_1 a_{1k} + v_2 a_{2k} + \cdots v_n a_{nk}}.$$

The higher the capital-labour ratio, the lower the rate of profit. Inequality in the rate of profit would cause shifts of capital between industries, which would lower the relative price of the products of the industry that has a higher profit rate. Such shifts of capital will continue until the profit rates in all the industries become equal.

Ricardo, thus, acknowledged that the law that relative prices of commodities are in proportion to the quantity of labour bestowed on their production must be modified when taking the difference in the capital-labour ratio into account. In his expression the cause of the modification was the difference in the degree of durability of capital or in the time which must elapse before one set of commodities can be brought to market (Ricardo 1951a, p. 34), but that is the same as the difference in the capital-labour ratio.

Ricardo's proposition on the distribution must also be influenced by this modification in the theory of value: a change in the value of labour does not affect the value of the commodity, but only affects the amount of profit. Ricardo, however, did not think this modification had a serious impact on the whole body of his theory of value and distribution, because, he thought, the effects of a change in distribution on relative prices were slight in comparison to the effects of a change in the quantity of labour required for production (Ricardo 1951a, p. 45).

3 Contemporary Classical Theory

The above-mentioned problem, the incompatibility between the labour theory of value and a uniform rate of profits, was solved in the direction of abandoning the labour theory while preserving the fundamental views Ricardo intended to advocate: first, the distribution that is independent of the values and second, the values that are independent of the demands. This solution was proposed by Sraffa, who had been engaged in editing *The Works and Correspondence of David Ricardo*. He discovered a rudimentary version of the first fundamental view in Ricardo's *An Essay on the Influence of a Low Price of Corn on the Profits of Stock* (Ricardo 1951b) and found that it was held through all the editions of the *Principles* (Sraffa 1951, pp. xxx–xxxv).

According to Sraffa, the system of prices is formulated as

$$\begin{cases} p_1 = (1+r)(p_1a_{11} + p_2a_{21} + \cdots + p_na_{n1}) + wl_1 \\ p_2 = (1+r)(p_1a_{12} + p_2a_{22} + \cdots + p_na_{n2}) + wl_2 \\ \vdots \\ p_n = (1+r)(p_1a_{1n} + p_2a_{2n} + \cdots + p_na_{nn}) + wl_n, \end{cases} \tag{6}$$

where $p_j (j = 1, 2, \cdots, n)$ is the price of commodity j and w is the wage rate.[1] Sraffa devised a composite commodity which, when used as a measure of value,

[1]This expression is different from Sraffa's in that the equations are written in terms of one unit of output.

held constant the ratio of outputs to inputs, or capital, and expressed the relation between the wage rate and the profit rate without being influenced by the change in relative prices. The composite commodity is called the 'standard commodity' and is defined as the set of commodities, $(x_1, x_2, \cdots, x_n)$, that satisfies

$$\begin{cases} x_1 = (1+R)(a_{11}x_1 + a_{12}x_2 + \cdots + a_{1n}x_n) \\ x_2 = (1+R)(a_{21}x_1 + a_{22}x_2 + \cdots + a_{2n}x_n) \\ \qquad \vdots \\ x_n = (1+R)(a_{n1}x_1 + a_{n2}x_2 + \cdots + a_{nn}x_n). \end{cases} \tag{7}$$

Assuming the total quantity of labour employed to produce $(x_1, x_2, \cdots, x_n)$ unity (i.e. $\sum_j l_j x_j = 1$), adopting this commodity as the standard of value and defining the value of the net product of this commodity produced by one unit of labour as unity (i.e. $\frac{R}{1+R}\sum_i p_i x_i = 1$), we have the equation

$$r = R(1 - w) \tag{8}$$

by comparing the sum of each equation in (6) multiplied by x_j:

$$\sum_j p_j x_j = (1+r)\sum_i \sum_j p_i a_{ij} x_j + w \sum_j l_j x_j$$

with the sum of each equation in (7) multiplied by p_i:

$$\sum_i p_i x_i = (1+R)\sum_i \sum_j p_i a_{ij} x_j.$$

Equation (8) shows that wage rate and profit rate are in a relation expressed as a straight line.

In addition to this straightforward expression, Sraffa gave a proof for the proposition that the real wage rate must be lowered in terms of any commodity as a standard when profit rate, r, increases. Suppose the commodity i is the standard of value (i.e. $p_i = 1$); if w does not fall when r rises, $p_1 a_{1i} + p_2 a_{2i} + \cdots + p_n a_{ni}$ must decline, and the price of some commodities, therefore, must decline. Suppose j is such a commodity; then $p_1 a_{1j} + p_2 a_{2j} + \cdots + p_n a_{nj}$ must decline at a greater rate than p_j does, but for the commodity whose price declines at the greatest rate, that is impossible, and consequently, w must fall (Sraffa 1960, p.47).

That values are independent of demands is self-evident in equation (6), where demands never appear. The significance of this view of values being independent of demands, however, becomes great when the choice of production techniques is considered. Let us assume labour as the standard of value (i.e. $w = 1$), in (6), and assume that a new technique or a method of production appears for, say, commodity 1 with a set of coefficients, $a'_{11}, a'_{21}, \cdots, a'_{n1}, l'_1$, which, under the present prices satisfying (6), brings about a surplus to the production of commodity 1:

$$p_1 > (1+r)(p_1 a'_{11} + p_2 a'_{21} + \cdots + p_n a'_{n1}) + l'_1.$$

The surplus, or extra profit, will make this new technique prevail in industry 1, which will, in turn, eliminate the extra profit by lowering the price of commodity 1 until it equals its cost of production including normal profit. In this state, the prices must satisfy the equations:

$$\begin{cases} p'_1 = (1+r)(p'_1 a'_{11} + p'_2 a'_{21} + \cdots + p'_n a'_{n1}) + l'_1 \\ p'_2 = (1+r)(p'_1 a_{12} + p'_2 a_{22} + \cdots + p'_n a_{n2}) + l_2 \\ \quad \vdots \\ p'_n = (1+r)(p'_1 a_{1n} + p'_2 a_{2n} + \cdots + p'_n a_{nn}) + l_n. \end{cases} \tag{9}$$

Here, p'_1 must be smaller than p_1 when $w = 1$ in (6). If commodity 1 never enters the production of other commodities (i.e. if $a_{1j} = 0$ for $j = 2, 3, \cdots, n$), their prices will not change (i.e. $p'_j = p_j$ for $j = 2, 3, \cdots, n$). In this case, we will have

$$p'_1 < (1+r)(p'_1 a_{11} + p'_2 a_{21} + \cdots + p'_n a_{n1}) + l_1,$$

because the first equation in (6) is met, $p'_i = p_i (i = 2, 3, \cdots, n)$ and $1-(1+r)a_{11} > 0$. This means the previous method for commodity 1 runs a deficit under the present prices; this method will never revive. If commodity 1 enters the production of any other commodities, say, commodity 2 (i.e. $a_{12} > 0$), then p'_2 must also be smaller than p_2. Furthermore, if commodity 2 enters the production of another commodity, its price will be lowered. The prices of the commodities, the production of which commodity 1 enters directly or indirectly, thus, should be lowered. Suppose commodity k is the one with the greatest rate of reduction in price among those commodities whose price is lowered. Because from (6)

$$p_k[1-(1+r)a_{kk}] = (1+r)(p_1 a_{1k} + \cdots + p_{k-1} a_{k-1,k} + p_{k+1} a_{k+1,k} + \cdots + p_n a_{nk}) + l_k,$$

and $p_k[1-(1+r)a_{kk}] \geqq (1+r)p_1 a_{1k}$, if the rate of reduction in p_k is larger than that in p_1, the production of commodity k will run a deficit. The rate of reduction in the price of commodity k must, therefore, not be larger than that in the price of commodity 1; as a result, the rate of reduction in p_1 is the greatest. Because from the first equation of (6)

$$p_1[1-(1+r)a_{11}] = (1+r)(p_2 a_{21} + p_3 a_{31} + \cdots + p_n a_{n1}) + l_1,$$

and p_1 will face the greatest rate of reduction, we will have

$$p'_1[1-(1+r)a_{11}] < (1+r)(p'_2a_{21}+p'_3a_{31}+\cdots+p'_na_{n1})+l_1.$$

This means the previous method for commodity 1 will run a deficit under the new set of prices; therefore, that method will never revive.

Consequently, with a given rate of profit, r, a method of production for a commodity that has the least cost of production under the present prices will be chosen. If the chosen method is different from the one that consists of the set of methods of the products on which the present prices are based, the new method will bring about a new set of prices, under which the newly chosen method will continue to be competitive; under this set of prices, producers have no incentive to return to the previous method.

Let us refer to a set of production methods, or techniques, each of which produces a certain net quantity of each of the products demanded by the society as a 'system of techniques'; a set of the coefficients of production of the commodity j

$$(a_{1j}, a_{2j}, \cdots, a_{nj}, l_j)$$

defines a method of production of j, and a set of such sets of coefficients for all the commodities

$$\begin{cases} (a_{11}, a_{21}, \cdots, a_{n1}, l_1) \\ (a_{12}, a_{22}, \cdots, a_{n2}, l_2) \\ \qquad\vdots \\ (a_{1n}, a_{2n}, \cdots, a_{nn}, l_n) \end{cases}$$

defines a system of techniques.

Through consecutive application of the above process, a system of techniques and a set of prices, each of which is the lowest in terms of wage, are determined independently of demands. Shiozawa (1981, pp. 104–109, 2007, p. 143, 2014, p. 71) called this proposition the 'minimum price theorem'.[2] The demand side has a relation to the quantities of the commodities supplied but does not concern their prices.

When the rate of profit changes, the system of techniques chosen will also change, but the direction of the change has no relation to the intensity of capital or labour; we cannot say that when the rate of profit rises, the methods of production with smaller intensity of capital become more competitive.

In the neoclassical theory of prices and distribution, a rise in the rate of profits, r, causes a change in production techniques in the direction of decreasing capital

[2]This theorem is generally known as the 'non-substitution theorem' (Dorfman et al. 1958). Pasinetti (1977) discussed the meaning of the theorem from the Sraffian viewpoint.

intensity or increasing labour intensity. This fact, when combined with discrepancies in capital (or labour) intensity among products and with product prices dependent on demands, brings about the effects of changes in demands on profit rates (or wage rates); thus, when demand increases for a product with high capital intensity and its price rises, the rate of profits will rise and the rate of wages will decline. Those relations are the key in the Heckscher-Ohlin-Samuelson theory of international trade.

In the contemporary classical theory, as formulated above, the quantity of capital cannot be defined other than as a value of the commodities which enter production, and it is dependent on the prices of the commodities, which, in turn, are dependent on the rate of profits. The rate of profits, therefore, cannot be 'explained' by such a concept as 'the marginal products of capital'. This is a characteristic of the classical theory that is in contrast to the neoclassical theory, and the famous 'capital controversy' concerned this point.[3]

When the demand for foods increases and Ricardian rents arise, the n-th equation in (6) is replaced with

$$\begin{cases} p_n = (1+r)\left(p_1 a_{1n}^1 + p_2 a_{2n}^1 + \cdots p_n a_{nn}^1\right) + w l_n^1 + \tau^1 \\ p_n = (1+r)\left(p_1 a_{1n}^2 + p_2 a_{2n}^2 + \cdots p_n a_{nn}^2\right) + w l_n^2 + \tau^2 \\ \qquad\vdots \\ p_n = (1+r)\left(p_1 a_{1n}^{\nu-1} + p_2 a_{2n}^{\nu-1} + \cdots p_n a_{nn}^{\mu-1}\right) + w l_2^{\nu-1} + \tau^{\nu-1} \\ p_n = (1+r)\left(p_1 a_{1n}^{\nu} + p_2 a_{2n}^{\nu} + \cdots p_n a_{nn}^{\nu}\right) + w l_n^{\nu}. \end{cases} \tag{10}$$

Sraffa (1960, pp. 75–76) pointed out there can be another type of rent—a rent which would emerge from the coexistence of two different methods applied to the land with a uniform quality. In the increase in demand for foods, a method of producing foods that produces more corn per acre with higher cost per unit of product may be required. Rent would have to emerge in order for that method of production to be equally competitive with another method with smaller cost. In this case,

$$\begin{cases} p_n = (1+r)(p_1 a_{1n} + p_2 a_{2n} + \cdots p_n a_{nn}) + w l_n + \tau t \\ p_n = (1+r)\left(p_1 a'_{1n} + p_2 a'_{2n} + \cdots p_n a'_{nn}\right) + w l'_n + \tau t', \end{cases} \tag{11}$$

[3]The controversy began with the question raised by Joan Robinson (1953–1954) on the concept of 'capital' in the aggregate production function, which became popular in growth theory (Solow 1957). The possibility of the 'reswitching' of techniques indicated by Sraffa (1960) became a focal point, because it implies that capital intensity has no relation to the rate of profits. The proposition posed by Levhari (1965) that reswitching does not occur when all the commodities are 'basics' turned out to be false (Levhari and Samuelson 1966). Solow (1967) tried to save the neoclassical capital theory by proposing an interpretation that the profit rate is regarded as the rate of the increase in consumption for the saving necessary for switching from one technique to another, but the interpretation was criticised by Pasinetti (1969) as nothing but an accounting identity.

where $(a_{1n}, a_{2n}, \cdots, a_{nn}, l_n, t)$ and $(a'_{1n}, a'_{2n}, \cdots, a'_{nn}, l'_n, t')$ are two sets of coefficients (t and t' are the coefficients for land input) representing two different methods applied to land of the same quality.

Sraffa included the chapter titled 'Land' in Part II of his book, which deals with 'multiple-product industries and fixed capital', and regarded the case where rents arise for the land with a uniform quality as of the same nature with the case of multiple products from a single industry, in the sense that 'at least one commodity is produced by *more than one* method'; that is, multiple processes producing the same products can coexist (Sraffa 1960, p. 78).

The contemporary classical theory of value is characterised as the cost of production theory, which retains Ricardo's vision of distribution that is independent of values without relying on the labour theory of value; it established the vision of values that are independent of demands taking choice of techniques into account.[4]

4 Foreign Trade and J.S. Mill's Solution for the Determination of International Values

In Ricardo's theory, profits decrease only when wages rise, and vice versa, and wages fall and rise according to the fall and rise in the quantity of labour required to produce the necessaries on which wages are expended (Ricardo 1951a, pp. 48–49, 126, 132). On the basis of this theory, Ricardo argued that 'the natural tendency of profits is to fall; for, in the progress of society and wealth, the additional quantity of food required is obtained by the sacrifice of more and more labour' (*op. cit.*, p. 120). Foreign trade is regarded as a factor that lowers wages through the imports of cheaper food (*op. cit.*, p. 132) and maintains the rate of profit against the tendency to decline.

Here, it is meant that the labour to produce a certain quantity of an export exchangeable for a quantity of food produced in a foreign country is less than the labour to produce the same quantity of food domestically. The absolute quantity of labour bestowed to produce food in the foreign country does not matter; the quantity of an export needed in exchange for a certain quantity of an import 'is not determined by the respective quantities of labour devoted to the production of each' (*op. cit.*, p. 135). The famous example given by Ricardo demonstrates that: to produce cloth requires the labour of 100 men for one year, and to make wine requires the labour of 120 men for the same time in England; to produce the wine in Portugal requires the labour of 80 men for one year, and to produce the cloth requires the labour of 90 men for the same time. In this situation, Portugal exports wine in exchange for cloth from England, and it is advantageous for both countries.

[4]Shiozawa (2016) proposes to rename Ricardo's theory of value the *cost of production theory of value* from the modern viewpoint.

The labour bestowed in Portugal is not exchanged for the same quantity of labour in England; if it is, neither cloth nor wine will ever be exported from England because they are too expensive.

In autarky, 100 units of wine is worth 120 units of cloth, or the relative price of wine is 6/5 in terms of cloth in England; 90 units of wine is worth 80 units of cloth, or the relative price of wine is 8/9 in terms of cloth in Portugal.[5] When the relative price of wine is greater than 6/5, cloth will not be produced in England; when it is smaller than 6/5, wine will not be produced in England. When the relative price of wine is greater than 8/9, cloth will not be produced in Portugal; when the relative price of wine is smaller than 8/9, wine will not be produced in Portugal. In order for cloth to be produced in England and for wine to be produced in Portugal, the relative price should be between 8/9 and 6/5. When the relative price of wine is 8/9, only cloth is produced in England, both wine and cloth can be produced in Portugal, and one unit of labour in Portugal is equal to $8/9 \times 100/80 = 10/9$ units of labour in England; when the relative price of wine is 6/5, only wine is produced in Portugal, both wine and cloth can be produced in England, and one unit of labour in Portugal is equal to $6/5 \times 100/80 = 12/8$ units of labour in England. When the relative price is greater than 8/9 and smaller than 6/5, England specialises in the production of cloth and Portugal in the production of wine.

Ricardo left the question at what point between the two poles the relative price is determined unresolved. John Stuart Mill addressed this problem and gave an answer that it is the law of supply and demand that determines the relative price of a commodity traded internationally.

He made a 2-country, 2-commodity example: 10 yards of broadcloth cost as much labour as 15 yards of linen in England and as much as 20 yards of linen in Germany. In this situation, ‘it would be the interest of England to import linen from Germany, and of Germany to import cloth from England’ (Mill 1909, p. 585). The relative price of linen in terms of cloth should be between 10/20 and 10/15. The demand for linen in England and for cloth in Germany depends on the relative price. Suppose it is 10/17. Under this price, if the demand for cloth in Germany is 10,000 yards and the demand for linen in England is 17,000 yards, Germany can pay for 10,000 yards of cloth by 17,000 yards of linen, and England can pay for 17,000 yards of linen with 10,000 yards of cloth; the demand and the supply of both goods coincide with each other.

If, however, the demand for linen in England is only 13,600 yards under the price of 10/17, Germany cannot pay for 10,000 yards of cloth with 13,600 yards of

[5]Here, Ricardo’s four numbers are interpreted as representing the quantities of labour needed to produce a unit of wine and of cloth in both countries. Tabuchi (2006, 2017) and Faccarello (2015) argue that Ricardo’s numbers should not be interpreted as such. They insist that those numbers are not technical coefficients, and Ricardo assumed from the outset some amount of wine produced by using 80 men a year in Portugal is exchangeable for some amount of cloth produced by using 100 men a year in England. See also Chap. 9 in this volume.

linen, which is worth only 8000 yards of cloth; the demand for linen is below its supply, and vice versa for cloth. The price of linen must be lowered in order for the demand to increase. Suppose when the price becomes 10/18, the demand for linen in England increases to 16,200 yards, and the demand for cloth in Germany decreases to 9000 yards. Now England can pay for the 16,200 yards with 9000 yards of cloth; the demand coincides with the supply for both commodities. When, on the contrary, the demand for linen in England under the price 10/17 is too large, say 20,400, the price should rise for the demand to meet its supply.

Mill says, 'it may be considered, therefore, as established, that when two countries trade together in two commodities, the exchange value of these commodities relatively to each other will adjust itself to the inclinations and circumstances of the consumers on both sides, in such manner that the quantities required by each country, of the articles which it imports from its neighbour, shall be exactly sufficient to pay for one another' (Mill 1909, p. 587). Mill, here, seems to regard the constraint of the equality between payments and receipts as being able to adjust the exchange value of the traded commodities. He, in fact, emphasises that 'the exports of each country must exactly pay for the imports; meaning now the aggregate exports and imports, not those of particular commodities taken singly'(*op. cit.*, p. 590) and named this the 'equation of international demand' or the 'law of international values' (*op. cit.*, p. 592). Mill considered this law to be 'but an extension of the more general law of Value, which we called the Equation of Supply and Demand' (*ibid.*), but the equation of supply and demand evidently refers to the equality between the quantities of particular commodities (*op. cit.*, pp. 446–448), and the law, if it is about the adjustment of the international values, should concern the equality of demand and supply of particular commodities.

Mill linked the determination of the relative value with the determination of the share of the advantage of trade; taking the above example of cloth and linen, when 10 yards of cloth is equal to 15 yards of linen, England will not get any share of the advantage of trade, while Germany will obtain the entire advantage, and when 10 yards of cloth is equivalent to 20 yards of linen, England will obtain the entire advantage, with Germany receiving no share. The distribution of the advantage is, thus, determined according to the law of supply and demand; the greater the intensity of demand of the exported commodity from the foreign country, the more share of advantage the exporting country will obtain.

Mill considered the law to be valid also when applied to cases of more than two commodities and presented the above cited numerical example with the addition of a third commodity, iron. In the example above, where 10 yards of cloth was of equal value with 15 yards of linen in England and of equal value with 20 yards of linen in Germany, Mill assumed the terms of interchange to be 10 yards of cloth for only 16 yards of linen, because the demand of England for linen is much greater than that of Germany for cloth. Then it was assumed that the quantity (called hundredweight) of iron which is of equal value with 10 yards of cloth in England will, if it is produced in Germany, cost as much labour as 18 yards of linen (*op. cit.*, p. 590). Iron, which is a product England can export, will improve the terms of interchange for England in comparison with the previous circumstances in which England exported only cloth. As a result, Mill supposed, the rate of interchange will be 10 yards of cloth for 17 yards of linen.

Suggesting that the same argument can be applied when the 4th, 5th and 6th commodities are included, Mill presented the following conclusions:

> If, therefore, it be asked what country draws to itself the greatest share of the advantage of any trade it carries on, the answer is, the country for whose productions there is in other countries the greatest demand, and a demand the most susceptible of increase from additional cheapness.... It gets its imports cheaper, the greater the intensity of the demand in foreign countries for its exports. It also gets its imports cheaper, the less the extent and intensity of its own demand for them. (*op. cit.*, p. 591)

It is this view that the new theory of international values denies. In the next section, I demonstrate how the view is denied by the new theory of international values by using Mill's numerical example.

5 The New Theory of International Values

Mill drew his conclusion by focusing on the case where the value of cloth is 10/17 in terms of linen for the cloth-linen-iron, England-Germany trade economy. This is an example of a 2-country, 3-commodity trade economy. The new theory follows the method of the contemporary classical theory; prices are proportional to their costs of production including the profits to the capital, and a commodity appears both as a product and as an input to other products. To describe Mill's example, however, it is appropriate to ignore the inputs of commodities, which enables one to ignore profits, but it is indispensable to express the quantity of labour and the wage rates.

A characteristic of the new theory, in contrast to the contemporary classical theory for the domestic economy, is that the labour in one country is a different factor of production from the labour in another country, and the wage rates are also different between those countries.

Let us take the quantity of labour required to produce one yard of linen as unity in England and in Germany. Since the labour in England and that in Germany are different things, this assumption merely expresses a choice of the unit of those kinds of labour. In addition, let us take the labour in England as the standard of value (i.e. the wage rate in England is assumed to be unity). Suppose p_c^e, p_l^e and p_i^e represent the prices of cloth, linen and iron, respectively, in England and p_c^g, p_l^g and p_i^g those in Germany; domestic prices in England in autarky should meet

$$p_c^e = \frac{3}{2}, \quad p_l^e = 1, \quad p_i^e = 15.$$

If w is the wage rate in Germany, the prices in Germany in autarky should meet

$$p_c^g = 2w, \quad p_l^g = w, \quad p_i^g = 18w.$$

When trade begins between the two countries, which commodities produced in each country become competitive depends on w:

1. When w is greater than 1, no commodity from Germany becomes competitive; if trade takes place, production is carried out only in England, and the international prices, p_c, p_l, p_i become identical to those in England:

$$p_c = \frac{3}{2}, \quad p_l = 1, \quad p_i = 15.$$

2. When $w = 1$, the domestic prices in Germany become

$$p_c^g = 2, \quad p_l^g = 1, \quad p_i^g = 18;$$

linen from Germany becomes as competitive as that from England, but the other two commodities are not competitive. All the commodities from England are competitive; so, the international prices, p_c, p_l, p_i, become identical to those in England:

$$p_c = \frac{3}{2}, \quad p_l = 1, \quad p_i = 15.$$

Germany specialises in the production of linen.
3. When $5/6 < w < 1$, the domestic prices in Germany become

$$p_c^g > \frac{5}{3}\left(> \frac{3}{2}\right), \quad p_l^g < 1, \quad p_i^g > 15;$$

so, linen from Germany becomes more competitive than that from England, but the other commodities from Germany are still not competitive. Linen in England loses competitiveness. Therefore, Germany specialises in linen, and England specialises in cloth and iron. This is Mill's case. The international prices will become as follows:

$$p_c = \frac{3}{2}, \quad \frac{5}{6} < p_l < 1, \quad p_i = 15.$$

Mill assumed $p_l = (10/17)p_c = 15/17$, which satisfies $5/6 < p_l < 1$.
4. When $w = 5/6$, the domestic prices in Germany become

$$p_c^g = \frac{5}{3}\left(> \frac{3}{2}\right), \quad p_l^g = \frac{5}{6}(< 1), \quad p_i^g = 15;$$

linen from Germany is more competitive than that from England, and iron becomes as competitive as that from England, but cloth is still not competitive. The international prices would be

$$p_c = \frac{3}{2}, \quad p_l = \frac{5}{6}, \quad p_i = 15.$$

Cloth is produced only in England, and linen only in Germany, but iron is produced in both countries.

5. When $3/4 < w < 5/6$, the domestic prices in Germany become

$$p_c^g > \frac{3}{2}, \quad p_l^g < \frac{5}{6}(< 1), \quad p_i^g < 15;$$

linen and iron from Germany become more competitive than those from England, but cloth from England is more competitive than that from Germany. International prices should meet

$$p_c = \frac{3}{2}, \quad \frac{3}{4} < p_l < \frac{5}{6}, \quad \frac{27}{2} < p_i < 15.$$

England specialises in cloth, and Germany in linen and iron.

6. When $w = 3/4$, the domestic prices in Germany would meet

$$p_c^g = \frac{3}{2}, \quad p_l^g = \frac{3}{4}(< 1), \quad p_i^g = \frac{27}{2}(< 15);$$

all the commodities from Germany become competitive; cloth is as competitive as that from England, and linen and iron are more competitive than those from England. The international prices should meet

$$p_c = \frac{3}{2}, \quad p_l = \frac{3}{4}, \quad p_i = \frac{27}{2}.$$

Under those prices, England is competitive only in cloth and specialises in it, while Germany will produce the full set of commodities.

7. When $w < 3/4$, every commodity will lose competitiveness, if produced in England, and production will be carried out only in Germany; the international prices will be identical to those in Germany:

$$p_c = 2w, \quad p_l = w, \quad p_i = 18w.$$

Table 1 shows how the competitive industries change according to the change in w. When the wage rate in Germany equals unity, only linen is competitive in Germany. As the wage rate declines, iron and then cloth is called into production in Germany, while in England linen and then iron are withdrawn.

At the wage rates of 1, 5/6 and 3/4, the prices are uniquely determined only by the technological factors, that is, by the costs of production; the law of demand and supply has no role here. When the wage rate is between those values, prices are not determined just by the technological factors, but they cannot be said to be

Table 1 Competitive industries and the wage rate

	England			Germany			International price		
w	Cloth	Linen	Iron	Cloth	Linen	Iron	Cloth	Linen	Iron
$1 < w$	+	+	+	−	−	−	3/2	1	15
$w = 1$	+	+	+	−	+	−	3/2	1	15
$5/6 < w < 1$	+	−	+	−	+	−	3/2	w	15
$w = 5/6$	+	−	+	−	+	+	3/2	5/6	15
$3/4 < w < 5/6$	+	−	−	−	+	+	3/2	w	$18w$
$w = 3/4$	+	−	−	+	+	+	3/2	3/4	27/2
$w < 3/4$	−	−	−	+	+	+	$2w$	w	$18w$

'+' means 'competitive', and '−' means 'uncompetitive'

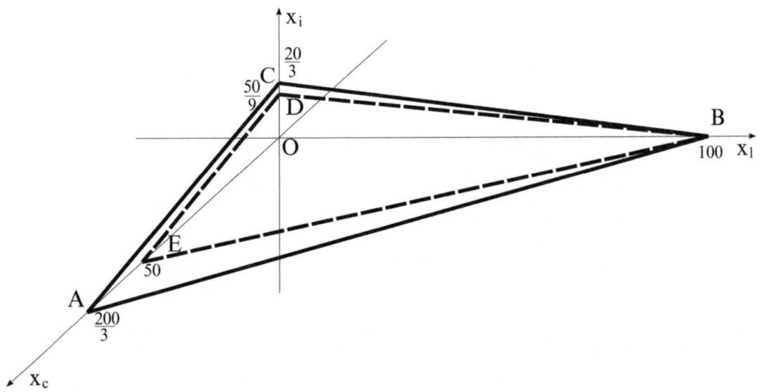

Fig. 1 Production possibility set in autarky

determined by the law of demand and supply; they are still under the constraint of technological factors, not just in the sense Mill noticed, that is, in the sense that prices have upper and lower bounds, but also in the sense that not all the prices can change independently; for example, $p_i/p_c = 10$ when $5/6 < w < 1$, and $p_i/p_l = 18$ when $3/4 < w < 5/6$.[6]

When the quantity of labour is limited, a set of coefficients for labour input gives a production possibility set. The triangle ABC in Fig. 1 represents the production possibility frontier in autarky for England when the quantity of labour is 100. The triangle EBD is that for Germany. When trade begins between them, the production possibility set is enlarged; the faces ABC, CFDA and AED in Fig. 2 are the production possibility frontier. The triangle ABC corresponds to the case where cloth, linen and iron are competitively produced in England and just linen is produced in Germany; the vector (3/2, 1, 15), which consists of the prices in this

[6]The cases with the wage rates of 1, 5/6 and 3/4 are examples of Graham's (1948) 'linkage case', whereas the cases with the wage rates between those values are examples of the 'limbo case', as explained by Sato in Chap. 10 of this volume.

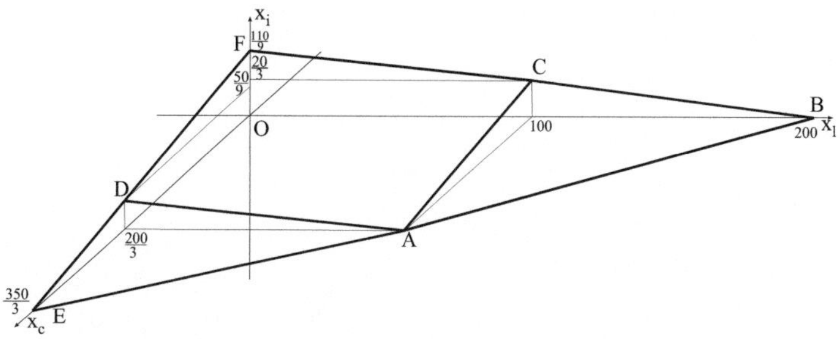

Fig. 2 Production possibility set in the trade between England and Germany

case, is normal to the triangle. The parallelogram CFDA corresponds to the case where cloth and iron are competitively produced in England and linen and iron are competitively produced in Germany; the price vector (3/2, 5/6, 15) is normal to the face. The triangle AED corresponds to the case where just cloth is produced in England and all three commodities are produced in Germany; the price vector (3/2, 3/4, 27/2) is normal to the face.

A point on one of these faces excluding its edges can be produced only by using the system of techniques consisting of the methods that become competitive under the set of prices, the vector of which is normal to the face. For example, a point on the face CFDA can be produced only by using the cloth and iron industries in England and the linen and iron industries in Germany; those industries become competitive under the price set (3/2, 5/6, 15).

Mill's case refers to a point on the ridge AC, where the price of linen has the freedom to change from 5/6 to 1; the law of demand and supply is able to influence the terms of trade.[7] The new theory reveals that those are very special points on the entire production possibility frontier. On the overwhelmingly large area, the set of prices is determined solely by technological considerations and has no freedom to change depending on supply and demand.

There is no point on the frontier at which prices can change freely to any direction and at which the law of supply and demand can be said to determine the prices, other than the points A, B, C, F, D and E, none of which exists in the positive space. This is in contrast with Ricardo's and Mill's examples of a 2-country, 2-commodity case. Figure 3 shows the production possibility frontier of Mill's cloth-linen, England-Germany example when the quantity of labour in both countries is 100. Point M is a point in the positive area at which the relative price p_c/p_l can change freely within the range [3/2,2]. The number of commodities must not be larger than the number

[7]The ridges of the production possibility frontier represent the limbo cases as explained in Chap. 10 of this volume; Mill, therefore, dealt with only a case of limbo.

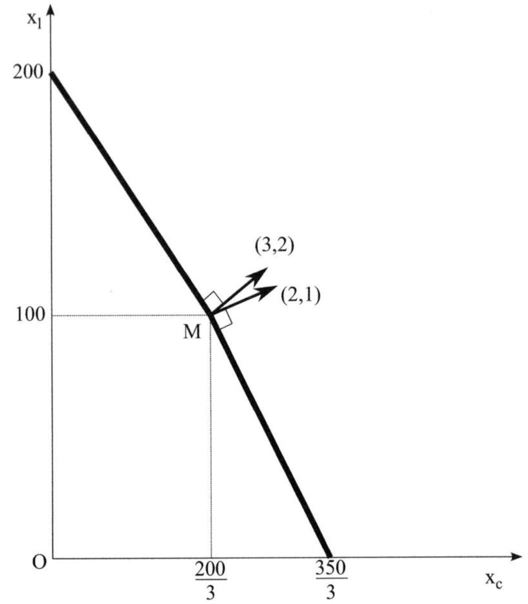

Fig. 3 Production possibility frontier of 2-country, 2-commodity trade economy

of countries in order for such a point to exist in the positive space. When the number of commodities is larger than the number of countries, such points do not exist in the positive space.

6 Introduction of Intermediate Goods and General Expression of the New Theory of International Values

When wage is paid before production, profits can arise even if there are no inputs other than labour. In this case, the above analysis is made valid by replacing the sum of wage with the sum of wage multiplied by 1 plus the rate of profit, or $(1 + r)wl$, where r is the profit rate, w is the wage rate and l is the coefficient of labour input. Here, the rate of profit in one country can be different from that in the other country; it can be varied even among industries.

When there are inputs other than labour, in the case of 2-country, 3-commodity, the prices, p_1, p_2, p_3, and the wage rate of country 2, w (assuming the wage rate of country 1 is unity), should meet

$$\begin{cases} p_1 \leqq (1+r^1)\left(p_1 a_{11}^1 + p_2 a_{21}^1 + p_3 a_{31}^1\right) + l_1^1 \\ p_2 \leqq (1+r^1)\left(p_1 a_{12}^1 + p_2 a_{22}^1 + p_3 a_{32}^1\right) + l_2^1 \\ p_3 \leqq (1+r^1)\left(p_1 a_{13}^1 + p_2 a_{23}^1 + p_3 a_{33}^1\right) + l_3^1 \\ p_1 \leqq (1+r^2)\left(p_1 a_{11}^2 + p_2 a_{21}^2 + p_3 a_{31}^2\right) + wl_1^2 \\ p_2 \leqq (1+r^2)\left(p_1 a_{12}^2 + p_2 a_{22}^2 + p_3 a_{32}^2\right) + wl_2^2 \\ p_3 \leqq (1+r^2)\left(p_1 a_{13}^2 + p_2 a_{23}^2 + p_3 a_{33}^2\right) + wl_3^2, \end{cases} \tag{12}$$

where r^k represents the rate of profit in country k and a_{ij}^k and l_j^k the quantity of commodity i and labour in the production of commodity j in country k. Four of the inequalities will be met as equations, and the rest will be met as strict inequalities when a point on one of the faces of the production possibility frontier excluding its edges should be produced. For example, when industry 1 and 3 in country 1 and industry 2 and 3 in country 2 are competitive,

$$\begin{cases} p_1 = (1+r^1)\left(p_1 a_{11}^1 + p_2 a_{21}^1 + p_3 a_{31}^1\right) + l_1^1 \\ p_3 = (1+r^1)\left(p_1 a_{13}^1 + p_2 a_{23}^1 + p_3 a_{33}^1\right) + l_3^1 \\ p_2 = (1+r^2)\left(p_1 a_{12}^2 + p_2 a_{22}^2 + p_3 a_{32}^2\right) + wl_2^2 \\ p_3 = (1+r^2)\left(p_1 a_{13}^2 + p_2 a_{23}^2 + p_3 a_{33}^2\right) + wl_3^2. \end{cases}$$

The new theory established that a set of values p_1, p_2, p_3, w exists that satisfies (12) and

$$p_1 y_1 + p_2 y_2 + p_3 y_3 = q^1 + wq^2,$$

where $y_i(1 = 1, 2, 3)$ is the net output of commodity i and $q^k(k = 1, 2)$ is the existing quantity of labour in country k that satisfy

$$\begin{cases} s_1^1 + s_1^2 = (1+r^1)\left(a_{11}^1 s_1^1 + a_{12}^1 s_2^1 + a_{13}^1 s_3^1\right) + (1+r^2)\left(a_{11}^2 s_1^2 + a_{12}^2 s_2^2 + a_{13}^2 s_3^2\right) + y_1 \\ s_2^1 + s_2^2 = (1+r^1)\left(a_{21}^1 s_1^1 + a_{22}^1 s_2^1 + a_{23}^1 s_3^1\right) + (1+r^2)\left(a_{21}^2 s_1^2 + a_{22}^2 s_2^2 + a_{23}^2 s_3^2\right) + y_2 \\ s_3^1 + s_3^2 = (1+r^1)\left(a_{31}^1 s_1^1 + a_{32}^1 s_2^1 + a_{33}^1 s_3^1\right) + (1+r^2)\left(a_{31}^2 s_1^2 + a_{32}^2 s_2^2 + a_{33}^2 s_3^2\right) + y_3 \\ q^1 = l_1^1 s_1^1 + l_2^1 s_2^1 + l_3^1 s_3^1 \\ q^2 = l_1^2 s_1^2 + l_2^2 s_2^2 + l_3^2 s_3^2, \end{cases} \tag{13}$$

where s_i^k is the level of activity for industry i in country k. This is what Theorem 4.3 in Chap. 1 means for the 2-country, 3-commodity case. Such a price vector (p_1, p_2, p_3) is normal to the face on which the point (y_1, y_2, y_3) satisfying (13) exists. This proposition has been established for more general cases with m countries and n commodities, allowing the possibility of having more than one method of producing the same commodity in a country.

The prices in autarky for country 1 should meet the same equations as (6):

$$\begin{cases} p_1 = \left(1 + r^1\right)\left(p_1 a^1_{11} + p_2 a^1_{21} + \cdots + p_n a^1_{n1}\right) + l^1_1 \\ p_2 = \left(1 + r^1\right)\left(p_1 a^1_{12} + p_2 a^1_{22} + \cdots + p_n a^1_{n2}\right) + l^1_2 \\ \qquad\vdots \\ p_n = \left(1 + r^1\right)\left(p_1 a^1_{1n} + p_2 a^1_{2n} + \cdots + p_n a^1_{nn}\right) + l^1_n, \end{cases} \tag{14}$$

under the assumption that the wage rate is in unity. An equation for the n-th commodity from country 2

$$p_n = (1 + r^2)(p_1 a^2_{1n} + p_2 a^2_{2n} + \cdots + p_n a^2_{nn}) + l^2_n$$

cannot coexist in general with the system of equations (14). If the method for the n-th commodity from country 2 is superior to that in country 1, the n-th equation in (14) will be excluded. If, however, the wage rate in country 2, w, is included, the equation

$$p_n = (1 + r^2)(p_1 a^2_{1n} + p_2 a^2_{2n} + \cdots + p_n a^2_{nn}) + w l^2_n$$

becomes compatible with the equations in (14); we can have a system of international values:

$$\begin{cases} p_1 = \left(1 + r^1\right)\left(p_1 a^1_{11} + p_2 a^1_{21} + \cdots + p_n a^1_{n1}\right) + l^1_1 \\ p_2 = \left(1 + r^1\right)\left(p_1 a^1_{12} + p_2 a^1_{22} + \cdots + p_n a^1_{n2}\right) + l^1_2 \\ \qquad\vdots \\ p_n = \left(1 + r^1\right)\left(p_1 a^1_{1n} + p_2 a^1_{2n} + \cdots + p_n a^1_{nn}\right) + l^1_n \\ p_n = \left(1 + r^2\right)\left(p_1 a^2_{1n} + p_2 a^2_{2n} + \cdots + p_n a^2_{nn}\right) + w l^2_n. \end{cases} \tag{15}$$

If the method for commodity $n-1$ in country 2 is to join, either the production of commodity n or commodity $n-1$ in country 1 must retire. Suppose the production of the n-th commodity in country 1 retires; we have a system of international values:

$$\begin{cases} p_1 = \left(1 + r^1\right)\left(p_1 a^1_{11} + p_2 a^1_{21} + \cdots + p_n a^1_{n1}\right) + l^1_1 \\ p_2 = \left(1 + r^1\right)\left(p_1 a^1_{12} + p_2 a^1_{22} + \cdots + p_n a^1_{n2}\right) + l^1_2 \\ \qquad\vdots \\ p_{n-1} = \left(1 + r^1\right)\left(p_1 a^1_{1,n-1} + p_2 a^1_{2,n-1} + \cdots + p_n a^1_{n,n-1}\right) + l^1_{n-1} \\ p_{n-1} = \left(1 + r^2\right)\left(p_1 a^2_{1,n-1} + p_2 a^2_{2,n-1} + \cdots + p_n a^2_{n,n-1}\right) + w l^2_{n-1} \\ p_n = \left(1 + r^2\right)\left(p_1 a^2_{1n} + p_2 a^2_{2n} + \cdots + p_n a^2_{nn}\right) + w l^2_n. \end{cases} \tag{16}$$

In autarky, there can be n independent equations with n price variables. In the 2-country trade economy, $n + 1$ independent equations become able to exist by including a new variable w. When a third country joins the trade economy, an additional variable w' will appear, and for another commodity, say the $(n - 2)$-th one, two equations will become able to coexist. According to Sraffa's logic about the similarity of the case with rent to the multiple-product case cited above, foreign trade can also be regarded as another case at which 'at least one commodity is produced by *more than one* method', and it is thus of the same nature as the multiple-product case.

7 Conclusions

The new theory of international values established the existence of international values that can be interpreted as being independent of demand and being mainly determined by technology. The fact that wage rates are different among countries enables the coexistence of processes in different countries of producing the same commodity and enables a system of equations sufficient to determine prices and wages just on the basis of technological information, except for a distributional variable—the rates of profit. Difference in wage rates among countries is the principal indicator of inequality in international income distribution. According to the new theory, the difference in wage rate should be understood as a result of the difference in technology among countries.

References

Dorfman, R., Samuelson, P. A., & Solow, R. M. (1958). *Linear programming and economic analysis*. New York: McGraw-Hill.

Faccarello, G. (2015). Comparative advantage. In H. D. Kurz & N. Salvadori (Eds.), *The Elgar companion to David Ricardo* (pp. 69–77). Cheltenham: Edward Elgar.

Graham, F. D. (1948). *The theory of international values*. Princeton: Princeton University Press.

Levhari, D. (1965). A nonsubstitution theorem and switching of techniques. *Quarterly Journal of Economics, 79*, 98–105.

Levhari, D., Samuelson, P. A. (1966). The non-switching theorem is false. *Quarterly Journal of Economics, 80*, 518–519.

Mill, J. S. (1909). *Principles of political economy with some of their applications to social philosophy*, New edition. London: Longmans, Green & Company.

Pasinetti, L. L. (1969). Switches of techniques and the "rate of return" in capital theory. *Economic Journal, 79*, 508–531.

Pasinetti, L. L. (1977). *Lectures on the theory of production*. New York: Columbia University Press.

Ricardo, D. (1951a). On the principles of political economy and taxation. In P. Sraffa (Ed.), *The works and correspondence of David Ricardo* (Vol. I). Cambridge: Cambridge University Press.

Ricardo, D. (1951b). An essay on the influence of a low price of corn on the profits of stock. In P. Sraffa (Ed.), *The works and correspondence of David Ricardo* (Vol. IV, pp. 3–41). Cambridge: Cambridge University Press.

Robinson, J. (1953–1954). The production function and the theory of capital. *Review of Economic Studies, 21*, 81–106.
Shiozawa, Y. (1981, in Japanese). *Foundations of mathematical economics*. Tokyo: Asakura Syoten.
Shiozawa, Y. (2007). A new construction of Ricardian trade theory—a many-country, many-commodity case with intermediate goods and choice of production techniques. *Evolutionary and Institutional Economics Review, 3*(2), 141–187.
Shiozawa, Y. (2014, in Japanese). *A final solution of Ricardo problem on international values*. Tokyo: Iwanami Syoten.
Shiozawa, Y. (2016). The revival of classical value theory. In N. Yokokawa, K. Yagi, H. Uemura, & R. Westra (Eds.), *The rejuvenation of political economy* (pp. 151–172). London: Routledge.
Solow, R. M. (1957). Technical change and aggregate production function. *Review of Economics and Statistics, 39*, 312–320.
Solow, R. M. (1967). The interest rate and transition between techniques. In C. H. Feinstein (Ed.), *Socialism, capitalism, and economic growth: Essays presented to Maurice Dobb* (pp. 30–39). Cambridge: Cambridge University Press.
Sraffa, P. (1951). Introduction. In P. Sraffa (Ed.), *The works and correspondence of David Ricardo* (Vol. I, pp. xiii–lxii). Cambridge: Cambridge University Press.
Sraffa, P. (1960). *Production of commodities by means of commodities: Prelude to a critique of economic theory*. Cambridge: Cambridge University Press.
Tabuchi, T. (2006, in Japanese). *Trade, money, and power: a critique to international economics*. Tokyo: Hôsei Daigaku Syuppankyoku.
Tabuchi, T. (2017). Yukizawa's interpretation of Ricardo's "theory of comparative costs". In S. Senga, M. Fujimoto, & T. Tabuchi (Eds.), *Ricardo and international trade*. London: Routledge.

The Relation Between Value and Demand in the New Theory of International Values

Tosihiro Oka

Abstract The principal theorem of the new theory of international values for a Ricardo-Sraffa trade economy is presented and then illustrated using a two-country, two-commodity model and a two-country, three-commodity model. It is shown that the classical vision of values as independent of demand is preserved, even when international trade takes place. In other words, values are mainly determined by costs of production or, ultimately, by technology. The values are, however, not determined uniquely, and demand plays a role in selecting a set of values from among those that are admissible under present technology and mark-up rates. Three different production possibility frontiers are introduced: R-efficient locus, physical maximal frontier and capitalistically feasible frontier. It is argued that distinguishing among these three frontiers is necessary in order to comprehend the role of demand in determining international value. Lastly, the similarity of this relation of value and demand to that of rent theory is pointed out.

Keywords International values and demand • Production possibility frontier • R-efficient locus • Capitalistically feasible frontier • Growth rate and profit rate

1 Introduction

The classical theory of value is characterized as the value determined by production costs; the relative prices are determined by technology, independent of demand, if distributive variables, a set of profit rates, are given. This is true for a closed economy, which has no international trade. The new theory of international values retains the classical characteristic in the case of open economies in which commodities for final consumption and intermediate commodities are traded. In the case of international trade, however, value is not determined uniquely, even when a set of profit rates is given, and there is a room for demand to play some role in determining which value is selected from those that are feasible. The new

T. Oka (✉)
Faculty of Economics, Fukui Prefectural University, Fukui, Japan
e-mail: oka@fpu.ac.jp

Y. Shiozawa et al. (eds.), *A New Construction of Ricardian Theory of International Values*, Evolutionary Economics and Social Complexity Science 7,
DOI 10.1007/978-981-10-0191-8_3

theory has been developed for the 'equivalent economy', as defined by Shiozawa (2014, p. 110). That is an economy where the input coefficients include profits on advance capital. In an equivalent trade economy, the relation between prices and demand is established in such a way that the price vector is normal to the facet of the production possibility set on which the demand vector exists. An equivalent economy is, however, a hypothetical economy, defined using input coefficients that include profits. Thus, the production possibility set for the economy is different from that of a real economy. The relation between demand and value should be real, and thus the relation in an equivalent economy should be restated for a real economy. This is the objective of this chapter. We first define a Ricardo-Sraffa (RS) trade economy, for which the principal theorem of the new theory of international values is presented in a general form. Then, the theorem is illustrated using a two-country, two-commodity and a two-country, three-commodity examples for an RS trade economy. Next, we distinguish among three kinds of production possibility frontier, and, using these concepts, we identify the relation between demand and value. Here, we describe how demand affects the selection of the system of techniques and then determines international values.

2 The Principal Theorem of New Theory of International Values

The new theory of international values, developed by Shiozawa (2014), has established the existence of a combination of prices and wages that enables a set of production techniques to be adopted competitively and that gives no incentive to change to other techniques. This theorem is established for the Ricardo-Sraffa (RS) trade economy, which is defined as follows.

There are M countries and N commodities; a technique τ of producing a commodity is identified by the vectors of net output and labour input coefficients:

$$\boldsymbol{a}(\tau) = (a_1^\tau, a_2^\tau, \cdots, a_N^\tau), \quad \boldsymbol{u}(\tau) = (u_1^\tau, u_2^\tau, \cdots, u_M^\tau).$$

Here, a_j^τ represents the net output of commodity $j (j = 1, 2, \cdots, N)$, and u_k^τ denotes the labour input of country $k (k = 1, 2, \cdots, M)$ for technique τ. The net output of commodity j is the gross output minus the input of commodity j, though 'input' here includes profit on advance capital. Thus, $a_j^\tau = b_j^\tau - c_j^\tau(1 + r^\tau)$, where b_j^τ is the gross output of j, c_j^τ is the physical input of j, and r^τ is the mark-up rate, or the profit rate, for technique τ.

Simple production is assumed, so there is no multiple production. Thus, every technique produces only one commodity; suppose commodity n is produced using technique τ, $a_n^\tau > 0$ and $a_j^\tau \leqq 0$ for $j \neq n$. Every technique is assumed to belong to only one country. If technique τ belongs to country m, $u_m^\tau > 0$ and $u_k^\tau = 0\,(k \neq m)$. Since any scalar multiple of a technique is assumed to be feasible, we can assume

$u_m^\tau = 1$, for normalization. Therefore, only one component of $\boldsymbol{u}(\tau)$ is positive, with a value of unity, and the values of the other components are 0.

Arranging the vectors $\boldsymbol{a}(\tau)$, $\boldsymbol{u}(\tau)$ for T techniques vertically, we have the following matrices:

$$A = \begin{bmatrix} a_1^1 & a_2^1 & \cdots & a_N^1 \\ a_1^2 & a_2^2 & \cdots & a_N^2 \\ \vdots & \vdots & & \vdots \\ a_1^T & a_2^T & \cdots & a_N^T \end{bmatrix} \quad J = \begin{bmatrix} u_1^1 & u_2^1 & \cdots & u_M^1 \\ u_1^2 & u_2^2 & \cdots & u_M^2 \\ \vdots & \vdots & & \vdots \\ u_1^T & u_2^T & \cdots & u_M^T \end{bmatrix}.$$

The vector $\boldsymbol{y} = (y_1, y_2, \cdots, y_N)$, defined as

$$\boldsymbol{y} = \boldsymbol{s}A,$$

represents the net products, where $\boldsymbol{s} = (s^1, s^2, \cdots, s^T)$ is the vector whose component, s^τ, represents the size of the operation of technique τ. The labour inputs are represented by $\boldsymbol{s}J$. When country m has q_m of labour, the labour quantities in the world are represented by $\boldsymbol{q} = (q_1, q_2, \cdots, q_M)$.

Using this notation, the production possibility set $\mathcal{P}$ is defined as

$$\mathcal{P} = \{\boldsymbol{y} \in \mathbb{R}^N \mid \boldsymbol{y} = \boldsymbol{s}A, \boldsymbol{s}J \leqq \boldsymbol{q}, \boldsymbol{s} \geqq \boldsymbol{0}, \boldsymbol{s} \in \mathbb{R}^T\}.$$

An element of $\mathcal{P}$, $\boldsymbol{y}$, is called a maximal element, when $\boldsymbol{z}$ meeting $\boldsymbol{z} \geq \boldsymbol{y}$, $\boldsymbol{z} \in \mathcal{P}$ does not exist.[1] The set of maximal elements is called the maximal boundary or the production possibility frontier (PPF). The PPF consists of a finite number of facets. The interior of such a facet is called a regular domain.[2] Note that a maximal element of $\mathcal{P}$ represents a combination of commodities that can be consumed with leaving sufficient capital for all techniques to grow at a rate of $r^\tau (\tau = 1, 2, \cdots, T)$. The hypothetical economy, with the growth rates equivalent to the profit rates, is called an 'equivalent economy' (Shiozawa 2014, p. 110).[3]

Based on this concept of an RS trade economy, the following principal theorem of the new theory of international values is established:

Theorem 1 *Provided that* $\boldsymbol{y}$ *is a maximal element of the production possibility set, there exists a vector of commodity prices,* $\boldsymbol{p} = [p_1, p_2, \cdots, p_N]'$, *and a vector of wages,* $\boldsymbol{w} = [w_1, w_2, \cdots, w_M]'$, *under which no technique obtains extra profit* ($J\boldsymbol{w} \geqq A\boldsymbol{p}$) *and the total value of the net products is equal to the total sum of wages*

[1]Here, $\boldsymbol{z} \geq \boldsymbol{y}$ means $\boldsymbol{z} \geqq \boldsymbol{y}$ and $z_i > y_i$ for at least one component i.

[2]See Definition 3.3 in Chap. 1 of this volume.

[3]When the actual rate of growth for the technique τ is different from its rate of profit, the actual production possibility set will be different from $\mathcal{P}$. This difference matters when we deal with the role of demand in determining the international value. We will return to this question later.

$(\langle y, p\rangle = \langle q, w\rangle)$,[4] *under which every country has at least one competitive technique and labour is fully employed. Conversely, if there is a set of* p *and* w *that satisfies* $\langle y, p\rangle = \langle q, w\rangle$ *and* $Jw \geqq Ap$, *then* y *is a maximal element.*[5]

A vector $(p, w)'$ is called an 'international value'. An international value that satisfies $Jw \geqq Ap$ and $\langle y, p\rangle = \langle q, w\rangle$ is called an 'admissible value', and an admissible international value corresponding to a net product vector on a regular domain is called 'regular value'.[6] When technique τ is competitive under an admissible international value $(p, w)'$, the τth component of the inequality $Jw \geqq Ap$ is satisfied with equality (i.e. $\langle u(\tau), w\rangle = \langle a(\tau), p\rangle$), and for uncompetitive techniques, a strict inequality holds. The set of competitive techniques is called a 'system of techniques'.

3 The Two-Country, Two-Commodity Model for an Equivalent Economy

A diagram illustration is a useful way to explain this theorem. Since David Ricardo (1951, p. 135), two-country, two-commodity models have been used repeatedly. The model has a risk of opening the way to the supply and demand theory of international value, but it can express the principal characteristics of the new theory.

Table 1 gives an example of the labour and commodity input coefficients of a two-country, two-commodity RS trade economy. Here, the numbers represent the quantity of inputs per unit of gross output. Assuming that the rate of profit is unity for all the techniques, the net output coefficients per unit of labour for the equivalent economy are as shown in Table 2. Let us suppose that the quantities of labour in countries A and B are unity and five, respectively. When all existing labour in both

Table 1 An example of a two-country, two-commodity RS trade economy: input coefficients

		Input coefficient		
		Labour	Commodity 1	Commodity 2
Country A	Production of commodity 1	1/10	0	1/4
	Production of commodity 2	1/50	1/20	9/20
Country B	Production of commodity 1	1/10	9/20	1/4
	Production of commodity 2	1/10	1/20	0

[4] $\langle y, p\rangle$ and $\langle q, w\rangle$ represent the scalar products of y and p and q and w, respectively.

[5] This theorem is equivalent to Theorem 3.4 in Chap. 1, the proof of which is presented in Appendix to Section 3 of Chap. 1.

[6] See Definition 3.7 in Chap. 1. Shiozawa (2014) gave different definitions; an international value that satisfies $Jw \geqq Ap$ is called 'admissible', and an admissible international value that satisfies $\langle y, p\rangle = \langle q, w\rangle$ is called 'regular' (Shiozawa 2014, p. 351).

Table 2 Net output coefficients for the two-country, two-commodity example per unit of labour

		Net output coefficient	
		Commodity 1	Commodity 2
Country A	Production of commodity 1	10	−5
	Production of commodity 2	−5	5
Country B	Production of commodity 1	1	−5
	Production of commodity 2	−1	10

Fig. 1 Production possibility set of a two-country, two-commodity model

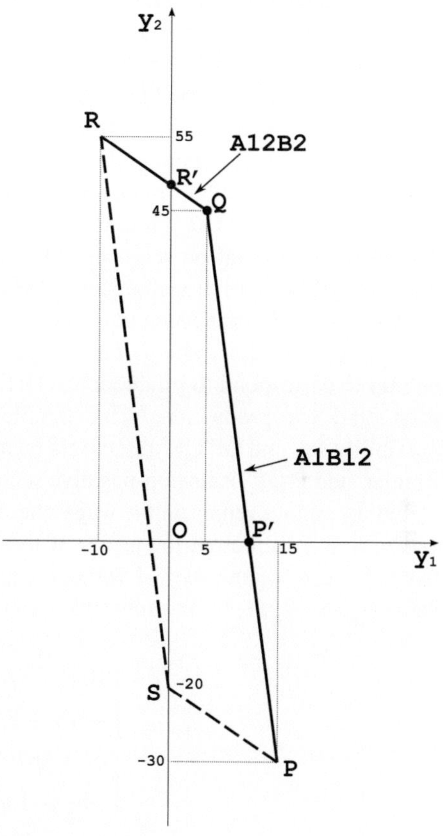

countries is directed to the production of commodity 1, the combination of the net products will be represented by the vector

$$1 \times (10, -5) + 5 \times (1, -5) = (15, -30),$$

which is also represented by point P in Fig. 1, where y_1 and y_2 represent the net products of commodities 1 and 2, respectively.

When all labour in country A is directed to the production of commodity 1, and all labour in country B to commodity 2, the net products are

$$1 \times (10 - 5) + 5 \times (-1, 10) = (5, 45),$$

represented by point Q in Fig. 1. When all labour in both countries is used to produce commodity 2, the net products will be (−10, 55), represented by point R. Point S represents the case where all labour in country A is directed to commodity 2 and all labour in country B is directed to commodity 1.

Segment PQ represents the net products that can be produced by directing all labour in country A to the production of commodity 1 and labour in country B to the production of both commodities. Similarly, segment QR represents the net products that can be produced by directing labour in country A to both commodities and all labour in country B to commodity 2. Let us call the activity that produces the net products on segment PQ 'production A1B12' and the activity that produces the net products on segment QR 'production A12B2'. Let us also call the set of techniques that produces a point on segment PQ 'system of techniques A1B12' and that on segment QR 'system of techniques A12B2'. Similarly, segment PS corresponds to production A12B1 or system of techniques A12B1, and segment RS corresponds to production A2B12 or system of techniques A2B12. Using this notation, point P can be said to correspond to production A1B1, Q to production A1B2, R to production A2B2 and S to production A2B1. Parallelogram PQRS represents the production possibility set, and OP′QR′ represents its non-negative section. PQR is the maximal frontier, and P′QR′ is its non-negative section.

Let w_A and w_B represent the wage rates in country A and country B, respectively, and p_1 and p_2 represent the prices of the first and the second commodity, respectively. We can assume $w_B = 1$ without losing generality. In order for an international value $(p_1, p_2, w_A, 1)$ to be admissible, it should meet

$$\begin{cases} 10p_1 - 5p_2 \leqq w_A \\ -5p_1 + 5p_2 \leqq w_A \\ p_1 - 5p_2 \leqq 1 \\ -p_1 + 10p_2 \leqq 1. \end{cases} \tag{1}$$

When production A1B12 becomes competitive, the first, third and fourth inequalities should be met with equality:

$$\begin{cases} 10p_1 - 5p_2 = w_A \\ p_1 - 5p_2 = 1 \\ -p_1 + 10p_2 = 1, \end{cases}$$

which implies $p_1 = 3, p_2 = 2/5$, and $w_A = 28$. Under these values, the second inequality of (1) is met with a strict inequality, which means the production of commodity 2 in country A is not competitive and will never be carried out. The vector of prices $(3, 2/5)'$ is normal to segment PQ. Any point on segment PQ, $\boldsymbol{y}$, has a value equal to the y_1 coordinate of point P′, in terms of commodity 1, when valued by the price vector $(3, 2/5)'$ (i.e. $\langle \boldsymbol{y}, \boldsymbol{p}\rangle / p_1 = 11$). Thus, $\langle \boldsymbol{y}, \boldsymbol{p}\rangle = 33$, which is equal to $\langle \boldsymbol{q}, \boldsymbol{w}\rangle = 1 \times 28 + 5 \times 1$. Therefore, this international value is admissible. Since any point on the interior of segment PQ can be produced using competitive techniques under the value $(3, 2/5, 28, 1)'$, this international value is regular.

In order for production A12B2 to be competitive, the first, second and fourth inequalities of (1) should be met with equality:

$$\begin{cases} 10p_1 - 5p_2 = w_A \\ -5p_1 + 5p_2 = w_A \\ -p_1 + 10p_2 = 1, \end{cases}$$

which implies $p_1 = 1/14, p_2 = 3/28$, and $w_A = 5/28$. Under these values, the third inequality of (1) is met with strict inequality. The production of commodity 1 in country B is not competitive. The vector of prices $(1/14, 3/28)'$ is normal to segment QR. Any point on this segment, $\boldsymbol{y}$, has a value equal to the y_2 coordinate of point R′ in terms of commodity 2, when valued by the price vector $(1/14, 3/28)'$ (i.e. $\langle \boldsymbol{y}, \boldsymbol{p}\rangle / p_2 = 145/3$). Thus, $\langle \boldsymbol{y}, \boldsymbol{p}\rangle = 145/28$, which is equal to $\langle \boldsymbol{q}, \boldsymbol{w}\rangle = 1 \times 5/28 + 5 \times 1$. Therefore, this international value is admissible and regular.

Point Q corresponds to production A1B2. In order for this production to be competitive,

$$\begin{cases} 10p_1 - 5p_2 = w_A \\ -p_1 + 10p_2 = 1 \end{cases}$$

must be met, which implies

$$p_1 = \frac{2w_A + 1}{19}, \quad p_2 = \frac{w_A + 10}{95}. \tag{2}$$

The other two techniques will not become competitive when

$$\begin{cases} -5p_1 + 5p_2 < w_A \\ p_1 - 5p_2 < 1 \end{cases}$$

are met. Combining the above result, this implies

$$\frac{5}{28} < w_A < 28.$$

The change in the value of w_A from 28 to 5/28 corresponds to a change in p_2/p_1 from 2/15 to 3/2. As long as (2) is met within this range,

$$\langle \boldsymbol{y}, \boldsymbol{p} \rangle = \langle \boldsymbol{q}, \boldsymbol{w} \rangle = w_A + 5.$$

Therefore, the international value that meets (2) is admissible, but is not regular, because points on the interior of segments PQ or QR cannot be produced with competitive techniques under this value.

Production A2B12 realizes the net products on segment RS. The value that makes production A2B12 competitive must meet the equations:

$$\begin{cases} -5p_1 + 5p_2 = w_A \\ p_1 - 5p_2 = 1 \\ -p_1 + 10p_2 = 1, \end{cases}$$

which implies $p_1 = 3, p_2 = 2/5$, and $w_A = -13$. In other words, the wage rate in country A is negative, and under this wage rate, the production of commodity 1 in country A earns extra profit (i.e. $10p_1 - 5p_2 = 28 > -13$). Therefore, the value, $(p_1, p_2, w_A, w_B) = (3, 2/5, -13, 1)$, is not admissible. Similarly, the international value that would make production A12B1 competitive is not admissible; the variables cannot all be non-negative—assuming $w_B = 1$, $(p_1, p_2, w_A, w_B) = (-2/13, -3/13, -5/13, 1)$ and assuming $w_A = 1$, $(p_1, p_2, w_A, w_B) = (2/5, 3/5, 1, -13/5)$.

Consequently, the maximal frontier of the production possibility set consists of two segments, each of which has a vector that is normal to it, and that vector represents the prices that construct a regular international value. Only for production at the vertex of the segments can there be admissible international values, the price vector of which does not have a unique slope, but the slope must lie between the slopes of the two vectors that are normal to the two segments. This is the case which Graham (1948) called 'limbo', as described in Chap. 10 by Sato in this volume.

Like Graham (1923, 1948), the new theory stresses the importance of production on segments other than their endpoints and regards the limbo case as improbable. This point will be made clearer in the two-country, three-commodity case. The two-country, three-commodity model is the minimal model needed to represent the RS trade economy where the number of countries, M, is smaller than the number of commodities, N.

4 Two-Country, Three-Commodity Model for the Equivalent Economy

Table 3 presents an example of the two-country, three-commodity case. When the profit rate is unity for all techniques (the uniform rate of profit is not a necessary assumption, just for simplification), the net output coefficients per unit of labour will be as shown in Table 4.

Table 3 An example of a two-country, three-commodity RS trade economy: input coefficients

		Input coefficient			
		Labour	Commodity 1	Commodity 2	Commodity 3
Country A	Production of commodity 1	1/10	0	1/4	0
	Production of commodity 2	1/50	1/20	9/20	0
	Production of commodity 3	1/100	3/200	1/20	0
Country B	Production of commodity 1	1/10	9/20	1/4	0
	Production of commodity 2	1/10	1/20	0	0
	Production of commodity 3	1/20	1/40	0	0

Table 4 Net output coefficient for the two-country, three-commodity example per unit of labour

		Net output coefficient		
		Commodity 1	Commodity 2	Commodity 3
Country A	Production of commodity 1	10	−5	0
	Production of commodity 2	−5	5	0
	Production of commodity 3	−3	−10	100
Country B	Production of commodity 1	1	−5	0
	Production of commodity 2	−1	10	0
	Production of commodity 3	−1	0	20

Assuming the quantity of labour in country A is 1 and that in country B is 5, the production possibility set is the nonahedron shown in Fig. 2. This diagram is drawn in the same way as in the two-country, two-commodity case. In this diagram, A_iB_j represents the net products by applying all labour in country A to the production of commodity i and all labour in country B to the production of commodity j, and the nonahedron is drawn by connecting all A_iB_js $(i, j = 1, 2, 3)$. The non-negative section is described by hexahedron OGHIJKL in Fig. 3. Triangles GHI and HKL and tetragon HIJK are the maximal frontier.

Assuming $w_B = 1$, an admissible value should meet the inequalities:

$$\begin{cases} 10p_1 - 5p_2 \leqq w_A \\ -5p_1 + 5p_2 \leqq w_A \\ -3p_1 - 10p_2 + 100p_3 \leqq w_A \\ p_1 - 5p_2 \leqq 1 \\ -p_1 + 10p_2 \leqq 1 \\ -p_1 + 20p_3 \leqq 1. \end{cases}$$

When production A123B2 is carried out competitively, the first, second, third and fifth inequalities must be met with equality, which implies

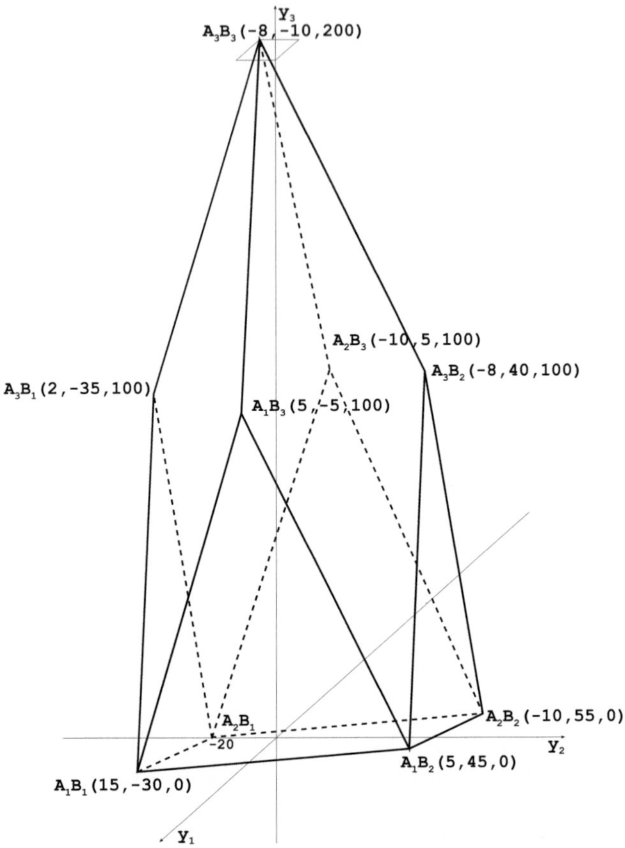

Fig. 2 Production possibility set of the two-country, three-commodity example

$$\begin{cases} p_1 = 1/14 \\ p_2 = 3/28 \\ p_3 = 41/2800 \\ w_A = 5/28. \end{cases} \tag{3}$$

Under this international value, the residual inequalities are met with strict inequality. Production A123B2 can realize any point on triangle HKL, and because the price vector $(1/14, 3/28, 41/2800)$ is normal to HKL, any point (y_1, y_2, y_3) on the plane satisfies

$$\frac{1}{14}(y_1 - 5) + \frac{3}{28}(y_2 - 45) + \frac{41}{2800}y_3 = 0,$$

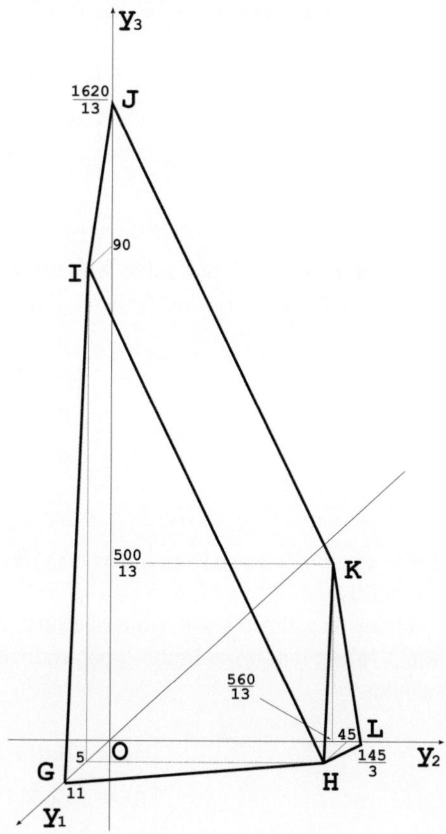

Fig. 3 Production possibility set of the two-country, three-commodity example: the non-negative section

or

$$\frac{1}{14}y_1 + \frac{3}{28}y_2 + \frac{41}{2800}y_3 = \frac{145}{28},$$

while the sum of wages is

$$1 \times \frac{5}{28} + 5 = \frac{145}{28};$$

thus, $\langle \boldsymbol{y}, \boldsymbol{p} \rangle = \langle \boldsymbol{q}, \boldsymbol{w} \rangle$. The international value (3) is, therefore, admissible. It is also regular, because any point in the interior of triangle HKL can be produced with the techniques that are competitive under this international value.

The same reasoning establishes that

$$\begin{cases} p_1 = 9/17 \\ p_2 = 13/85 \\ p_3 = 13/170 \\ w_A = 77/17 \end{cases}$$

is a regular international value which makes production A13B23 that produces a point on tetragon HIJK competitive. The international value

$$\begin{cases} p_1 = 3 \\ p_2 = 2/5 \\ p_3 = 1/5 \\ w_A = 28 \end{cases}$$

is also regular and makes production A1B123 that produces a point on triangle GHI competitive.

Other than those regular international values, there are two admissible international values that bring limbo-type productions into existence. One is the value that satisfies

$$\begin{cases} p_1 = (2w_A + 1)/19 \\ p_2 = (w_A + 10)/95 \\ p_3 = (27w_A + 23)/1900 \\ 5/28 < w_A < 77/17, \end{cases}$$

and the other is

$$\begin{cases} p_1 = (2w_A + 1)/19 \\ p_2 = (w_A + 10)/95 \\ p_3 = (w_A + 10)/190 \\ 77/17 < w_A < 28. \end{cases}$$

The former makes production A13B2 competitive, and the latter makes production A1B23 competitive. In the two-commodity case, the relative price p_1/p_2 could change freely between the upper and the lower limits for the limbo-type production to be carried out. Here, in the three-commodity case, the relative prices p_1/p_2 and p_3/p_2 are constrained as

$$p_1 = 10p_2 - 1 = \frac{200p_3 - 1}{27}$$

in the case of A13B2, and

$$p_1 = 10p_2 - 1 = 20p_3 - 1$$

in the case of A1B23. This is because these limbo cases bring about net products on the edges HK and HI, respectively, and do not bind them at any vertex of the polyhedron. There is no vertex in the positive octant, because the number of countries is less than the number of commodities. As the difference between the numbers expands, the degree of freedom in the relative prices decreases, and the degree at which demand affects prices becomes smaller.

We have 15 systems of techniques. Table 5 shows the international values that corresponds to the 15 systems of techniques. The net products that system A*ijk*B*l* can produce with the full employment of labour in both countries are shown as triangle $(A_iB_l)(A_jB_l)(A_kB_l)$ in Fig. 2, the net products of system A*i*B*jkl* as triangle $(A_iB_j)(A_iB_k)(A_iB_l)$, and the net products of system A*ij*B*kl* as tetragon $(A_iB_k)(A_iB_l)(A_jB_l)(A_jB_k)$.

The first three systems in Table 5 have regular international values, as described above. The other systems do not have admissible values. The fourth to seventh systems have positive international values, because the plane including the polygon each system can produce with full employment of labour has a positive normal vector and has a non-negative section. Their international values, however, are not admissible, because they give extra profit to some technique not belonging to the system. The 8th to 11th systems of techniques have non-negative price vectors, but the wage rate of either country becomes negative, because the plane including the

Table 5 International values in the systems of techniques

System of techniques	International value						
	p_1	p_2	p_3	w_A	w_B	Remarks	
A123B2	1/14	3/28	41/2800	5/28	1	Regular value	
A13B23	9/17	13/85	13/170	77/17	1		
A1B123	3	2/5	1/5	28	1		
A12B23	1/14	3/28	3/56	5/28	1	Positive value	Extra profit to A3
A13B12	3	2/5	41/100	28	1		Extra profit to B3
A123B3	10/31	15/31	41/620	25/31	1		Extra profit to B2
A3B123	3	2/5	1/5	7	1		Extra profit to A1
A123B1	2/13	3/13	41/1300	5/13	−1	Non-negative price vector Negative wage for either country	
A23B13	2	3/5	1/20	−7	−1		
A2B123	3	2/5	1/5	−13	1		
A23B12	3	2/5	0	−13	1		
A12B13	−2/13	−3/13	11/260	−5/13	1	Price vector with negative components	
A23B23	−3/10	7/100	7/200	37/10	1		
A13B13	2/3	−1/15	1/12	7	1		
A12B12	–					No value makes A12B12 competitive	

polygon each system can produce does not have a non-negative section. The 12th to 14th systems do not have positive normal price vectors, and the last system does not have an international value that makes all the techniques constructing the system competitive, except for the edges.

5 Discrepancy Between the Growth Rate and the Profit Rate

The matrix A denotes net outputs. 'Net output', here, means gross output minus input multiplied by one plus the profit rate; $a_j^\tau = b_j^\tau - c_j^\tau(1 + r^\tau)$. Net products $\boldsymbol{y}$ is defined using this concept, and thus, the production possibility set should also be understood in terms of the input coefficients that include profits on advance capital. Shiozawa argued, 'in this case, the production possibility set should be interpreted as the set of net surplus product in the growing economy with growth rate $1 + r$' (Shiozawa 2007, p. 146).

As is evident from the fact that the profit rate is expressed with superscript τ above, the rate can vary among techniques; thus, it can also vary among industries and among countries. Therefore, the production possibility set should be interpreted as the set of net surplus product in the growing economy with growth rates that are equal to the profit rates, which can vary among industries and among countries.

The growth rate of an industry is, however, not necessarily equal to its profit rate. With regard to the world economy as a whole, its growth rate is, in general, different from the rate of profit on capital. Let us investigate what occurs to the relation between net products and international value when growth rates are different from profit rates.

Let us consider the case where there is a unique profit rate r for all industries in the world, and the growth rate of all industries is zero. This is a special case, but the analysis can be extended to more general cases.

From the two-country, two-commodity example given in Table 1, we have the net output coefficients per unit of labour under the zero rate of growth, as shown in Table 6. The net products from full employment of labour can exist on segments TU, UV, VW and WT in Fig. 4. Allowing for underemployment, the inside of parallelogram TUVW and the area of tetragon OTUV can also be produced. Thus,

Table 6 Net output coefficient for the two-country, two-commodity example per unit of labour under a zero rate of growth

		Net output coefficient	
		Commodity 1	Commodity 2
Country A	Production of commodity 1	10	−5/2
	Production of commodity 2	−5/2	55/2
Country B	Production of commodity 1	11/2	−5/2
	Production of commodity 2	−1/2	10

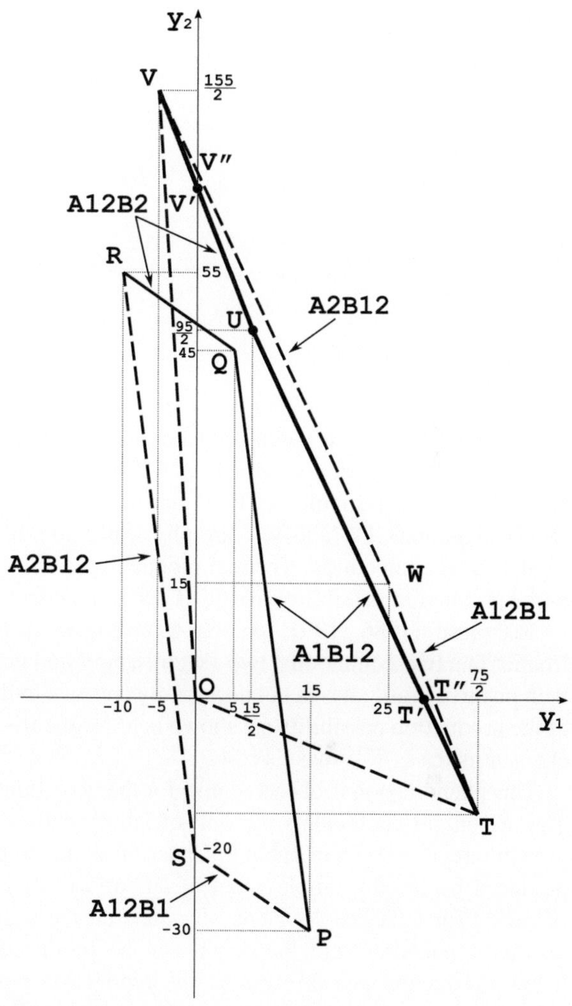

Fig. 4 Production possibility set under a real growth rate for the two-country, two-commodity model

the production possibility set is represented by tetragon OTWV, and TWV is the production possibility frontier.

This production possibility frontier, however, will not be produced under the actual profit rate, which is unity in this case. This is because in order for the points on TW to be produced, production A12B1 must be carried out, but system of techniques A12B1 does not have an admissible international value, as observed above. Similarly, the points on VW cannot be produced under the existing profit rate, because there is no admissible international value that makes A2B12 competitive. Under a profit rate of unity, only systems of techniques A1B12 and A12B2 have admissible international values.

Table 7 Net output coefficient for the two-country, three-commodity example per unit of labour under a zero rate of growth

		Net output coefficient		
		Commodity 1	Commodity 2	Commodity 3
Country A	Production of commodity 1	10	−5/2	0
	Production of commodity 2	−5/2	55/2	0
	Production of commodity 3	−3/2	−5	100
Country B	Production of commodity 1	11/2	−5/2	0
	Production of commodity 2	−1/2	10	0
	Production of commodity 3	−1/2	0	20

Therefore, under a profit rate of unity, only productions A1B12 and A12B2 can be carried out, which will produce the net products on segments TU and UV, the non-negative section of which is T′U and UV′. Hence, the actual production possibility frontier should be TUV, and the production possibility set should be OTUV. Segments TW and WV are physically possible, but will not appear under capitalistical competition. The real production possibility set OTUV is not convex, or the production possibility frontier TUV is not concave to the origin.

Takamasu (1986, 1991, pp. 63–67) showed that the production possibility frontier in a two-commodity, two-factor (labour and land) Sraffa-Leontief economy with positive profit rates is not necessarily concave to the origin. The non-convexity of the production possibility set shown here for the RS trade economy has the same characteristics as Takamasu's case.

This phenomenon is observed also for the two-country, three-commodity model. The net output coefficients per unit of labour derived from Table 3 under a zero growth rate are shown in Table 7. Under these net output coefficients, net products produced by the full employment of labour will be on the nonahedron in Fig. 5. Allowing for underemployment, all points on the segments connecting the points on this nonahedron and the origin can be produced; these points are included in the production possibility set. The production possibility frontier consists of triangle $(A_3B_1)(A_3B_2)(A_3B_3)$, triangle $(A_1B_1)(A_2B_1)(A_3B_1)$ and parallelogram $(A_2B_1)(A_2B_2)(A_3B_2)(A_3B_1)$, the non-negative section of which consists of triangle MNP, QRS and pentagon MPQST in Fig. 6.

Net products on this production possibility frontier could be produced by systems of techniques A123B1, A23B12 or A3B123. These systems of techniques, however, will never appear under the present rate of profit. As shown in Table 5, under the profit rate of unity, a negative wage rate is necessary to make A123B1 and A23B12 feasible, and the value making A3B123 feasible gives extra profit to A1. This corresponds to the fact that plane $(A_1B_1)(A_2B_1)(A_3B_1)$ and plane $(A_2B_1)(A_3B_1)(A_3B_2)(A_2B_2)$ do not have non-negative section and that triangle $(A_3B_1)(A_3B_2)(A_3B_3)$ is behind the maximal frontier in Fig. 2.

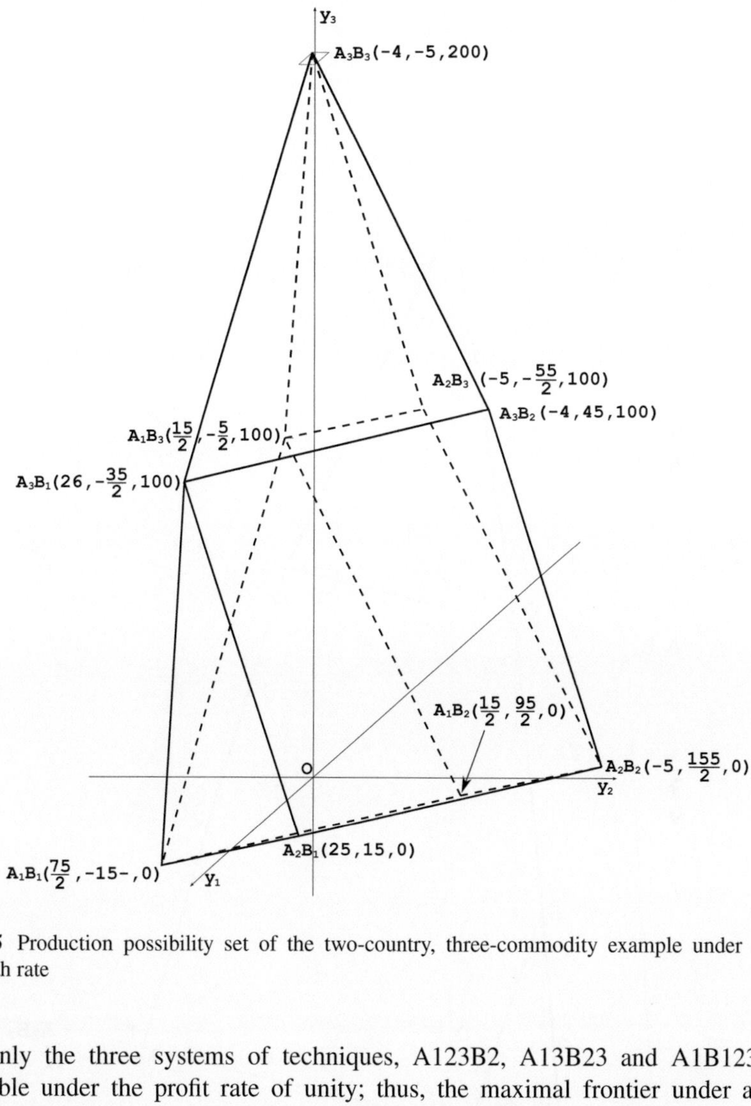

Fig. 5 Production possibility set of the two-country, three-commodity example under a zero growth rate

Only the three systems of techniques, A123B2, A13B23 and A1B123, are feasible under the profit rate of unity; thus, the maximal frontier under a zero growth rate consists of triangles $(A_1B_2)(A_2B_2)(A_3B_2)$ and $(A_1B_1)(A_1B_2)(A_1B_3)$ and parallelogram $(A_1B_2)(A_1B_3)(A_3B_3)(A_3B_2)$ in Fig. 5, the non-negative section of which is shown as UVW, UYZ and UWXY in Fig. 7. This frontier is below MNP, QRS and MPQST, and not concave to the origin, as in the two-commodity case in Fig. 4.

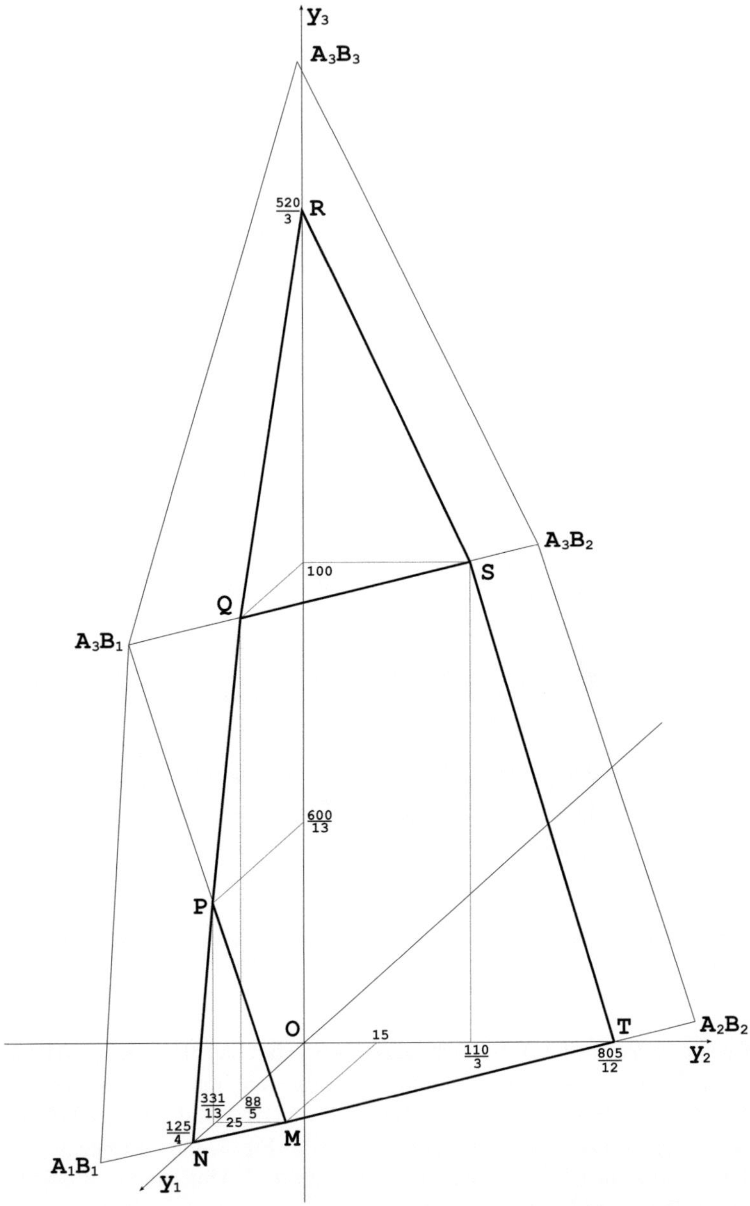

Fig. 6 Production possibility frontier of the two-country, three-commodity example under a zero growth rate: the non-negative section

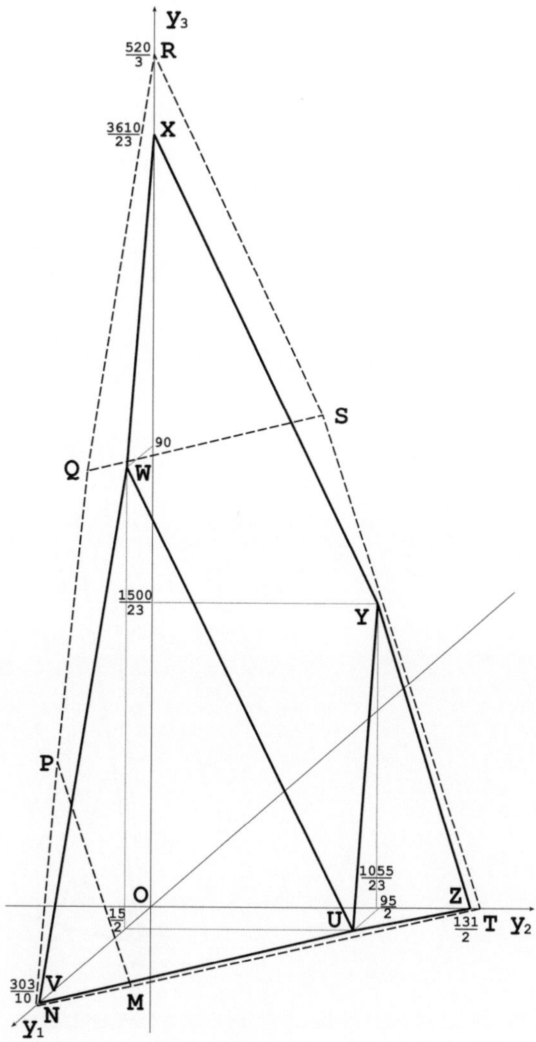

Fig. 7 Actual production possibility frontier of the trade economy with zero growth rate and unity profit rate

6 Relation Between Final Demand and Value

The non-negative sections of the three production possibility frontiers are assembled in Fig. 8. GHIJKL is the production possibility frontier under the hypothetical growth rate equal to the profit rate (here unity). This is the same one referred to as an 'R-efficient locus' (Takamasu 1991, p. 65; Mirlees 1969). We denote this REL. MNPQRST is the physical maximal frontier (PMF) under the actual growth rate of zero. PMF is above REL. Both are concave to the origin, but they will never appear; REL is a hypothetical one, and PMF can be realized only by the systems of

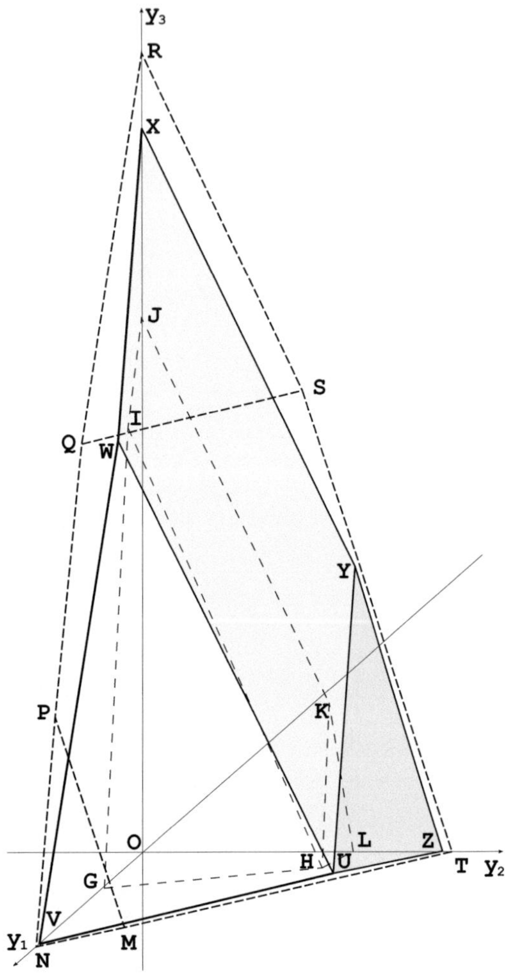

Fig. 8 Three production possibility frontiers of trade economy

techniques that never become feasible under the profit rate of unity or only under international values that are not admissible or that are not non-negative under a profit rate of unity.

UVWXYZ, which is not concave to the origin, is the production possibility frontier realized under a profit rate of unity. Let us call it the 'capitalistically feasible frontier (CFF)'. If labour is fully employed, actual products must be on the CFF. That the CFF is below the PMF indicates inefficiency in capitalistic production.[7]

When demand requires net products on, for example, the tetragon UWXY, production A13B23 has to be carried out. In order for that to happen, the international

[7]In Fig. 4, PQR is the RFF, TWV is the PMF, and TUV is the CFF.

value must meet $(p_1, p_2, p_3, w_A, w_B) = (9/17, 13/85, 13/170, 77/17, 1)$. The price vector (9/17, 13/85, 13/170) is normal to the plane HIJK, not to UWXY, on which the net products meeting the demand will exist. When world demand moves to a point on, say, triangle UVW, the system of techniques must become A1B123, and the international value must be $(p_1, p_2, p_3, w_A, w_B) = (3, 2/5, 1/5, 28, 1)$, which is normal to GHI.

Final demand is linked to international value in this manner. This relationship of demand to value is parallel with that in Ricardo and Sraffa's theory of rent (Ricardo 1951, p. 70; Sraffa 1960, pp. 75–76). In that theory, demands impose constraints on production techniques—growth in demand for foods requires techniques that produce more corn per acre with higher cost per unit of product—and prices are adjusted to make such techniques competitive. At the same time, rents emerge. In the new theory of international values also, demands impose constraints on production techniques—including the determination of which countries produce which commodities— and prices are adjusted to realize such patterns on division of production, accompanied by adjustments in wages.

Let us express the relation between demand and value in a general formula. We have mentioned above that net output coefficients are related to gross output coefficients and input coefficients, as

$$a_j^\tau = b_j^\tau - c_j^\tau(1 + r^\tau).$$

Since simple production is assumed, any technique produces only one commodity. Let us denote the commodity produced by technique τ as $\gamma(\tau)$; thus, $b^\tau_{\gamma(\tau)} > 0$, and $b_i^\tau = 0$ for $i \neq \gamma(\tau)$. Let us define $h_j^\tau (j = 1, 2, \cdots, N), l_k^\tau (k = 1, 2, \cdots, M)$, and $\delta_j^\tau (j = 1, 2, \cdots, N)$ as

$$h_j^\tau = \frac{c_j^\tau}{b^\tau_{\gamma(\tau)}}, \quad l_k^\tau = \frac{u_k^\tau}{b^\tau_{\gamma(\tau)}}, \quad \begin{cases} \delta^\tau_{\gamma(\tau)} = 1 \\ \delta_j^\tau = 0 \quad (j \neq \gamma(\tau)), \end{cases}$$

and H, L, and I as

$$H = \begin{bmatrix} h_1^1 \; h_2^1 \cdots h_N^1 \\ h_1^2 \; h_2^2 \cdots h_N^2 \\ \vdots \quad \vdots \qquad \vdots \\ h_1^T \; h_2^T \cdots h_N^T \end{bmatrix}, \quad L = \begin{bmatrix} l_1^1 \; l_2^1 \cdots l_M^1 \\ l_1^2 \; l_2^2 \cdots l_M^2 \\ \vdots \quad \vdots \qquad \vdots \\ l_1^T \; l_2^T \cdots l_M^T \end{bmatrix}, \quad I = \begin{bmatrix} \delta_1^1 \; \delta_2^1 \cdots \delta_N^1 \\ \delta_1^2 \; \delta_2^2 \cdots \delta_N^2 \\ \vdots \quad \vdots \qquad \vdots \\ \delta_1^T \; \delta_2^T \cdots \delta_N^T \end{bmatrix}.$$

Note that any low of L has only one component that has a positive value; if technique τ belongs to country $\mu(\tau)$, then $l^\tau_{\mu(\tau)} > 0$, $l_k^\tau = 0\,(k \neq \mu(\tau))$. Assuming the rate of profit for technique τ is r^τ and defining matrix $\tilde{H}$ as

$$\tilde{H} = \begin{bmatrix} h_1^1(1+r^1) & h_2^1(1+r^1) & \cdots & h_N^1(1+r^1) \\ h_1^2(1+r^2) & h_2^2(1+r^2) & \cdots & h_N^2(1+r^2) \\ \vdots & \vdots & & \vdots \\ h_1^T(1+r^T) & h_2^T(1+r^T) & \cdots & h_N^T(1+r^T) \end{bmatrix},$$

an admissible international value $(\boldsymbol{p}, \boldsymbol{w})'(> 0)$ must satisfy

$$(I - \tilde{H})\boldsymbol{p} \leqq L\boldsymbol{w}. \tag{4}$$

When $(\boldsymbol{p}, \boldsymbol{w})'$ is admissible, there exists $\tilde{\boldsymbol{y}}$ satisfying

$$\langle \tilde{\boldsymbol{y}}, \boldsymbol{p} \rangle = \boldsymbol{s}(I - \tilde{H})\boldsymbol{p} = \boldsymbol{s}L\boldsymbol{w} = \langle \boldsymbol{q}, \boldsymbol{w} \rangle, \tag{5}$$

which implies that, if $s^\tau > 0$, the τ-th component of (4) is satisfied with equality:

$$p_{\gamma(\tau)} - \langle \tilde{\boldsymbol{h}}(\tau), \boldsymbol{p} \rangle = l^\tau_{\mu(\tau)} w_{\mu(\tau)},$$

where $\tilde{\boldsymbol{h}}(\tau) = (h_1^\tau(1 + r^\tau), h_2^\tau(1 + r^\tau), \cdots, h_N^\tau(1 + r^\tau))$.

Suppose that the production using technique τ grows at rate g^τ and that matrix $\hat{H}$ is defined as

$$\hat{H} = \begin{bmatrix} h_1^1(1+g^1) & h_2^1(1+g^1) & \cdots & h_N^1(1+g^1) \\ h_1^2(1+g^2) & h_2^2(1+g^2) & \cdots & h_N^2(1+g^2) \\ \vdots & \vdots & & \vdots \\ h_1^T(1+g^T) & h_2^T(1+g^T) & \cdots & h_N^T(1+g^T) \end{bmatrix}.$$

Then, $\boldsymbol{y}$ that satisfies $\boldsymbol{y} = \boldsymbol{s}(I - \hat{H})$ is on a facet that is part of the CFF.

Conversely, if a final demand vector $\boldsymbol{y}$ is given on the CFF, $\boldsymbol{s}$ is determined to satisfy

$$\begin{cases} \boldsymbol{y} = \boldsymbol{s}(I - \hat{H}) \\ \boldsymbol{s}L = \boldsymbol{q}, \end{cases}$$

and such international value $(\boldsymbol{p}, \boldsymbol{w})'$ is selected as satisfies (4) and (5) for this $\boldsymbol{s}$. This is how demand affects value in the new theory.

Multiplying $\boldsymbol{y} = \boldsymbol{s}(I - \hat{H})$ by $\boldsymbol{p}$ to the right and comparing it with (5), we have

$$\langle \boldsymbol{y}, \boldsymbol{p} \rangle = \boldsymbol{s}L\boldsymbol{w} + \boldsymbol{s}(\tilde{H} - \hat{H})\boldsymbol{p}.$$

Here, the left-hand side represents the value of net products, or of final demand, and the first and second terms of the right-hand side represent labourers' consumption and capitalists' consumption, respectively.

7 Conclusion

I have illustrated the meanings of the new theory of international values using a two-country, two-commodity model and a two-country, three-commodity model in an RS trade economy and have examined the relation between values and demand in the new theory. It is found that three production possibility frontiers should be distinguished: an R-efficient locus, a physical maximal frontier and a capitalistically feasible frontier. When profit rates are given, an R-efficient locus is determined, each facet of which has a combination of prices the vector of which is normal to the facet. The combination of prices has a combination of wages that makes the techniques that can produce the points on the facet under the hypothetical growth rate equal to the profit rate competitive. When the REL is given, a capitalistically feasible frontier is determined, a facet of which will be chosen according to the final demand. The chosen facet on the CFF determines an international value, the price vector of which is normal to the corresponding facet of the REL and not normal to the facet on the CFF. Any point on the CFF is feasible by the full employment of labour but is below the physical maximal frontier under the actual growth rate.

References

Graham, F. D. (1923). The theory of international values: Re-examined. *Quarterly Journal of Economics, 28*, 54–86.

Graham, F. D. (1948). *The theory of international values*. Princeton: Princeton University Press.

Mirlees, J. A. (1969). The dynamic nonsubstitution theorem. *Review of Economic Studies, 39*, 67–76.

Ricardo, D. (1951). On the principles of political economy and taxation. In P. Sraffa (Ed.), *The works and correspondence of David Ricardo* (Vol. I). Cambridge: Cambridge University Press.

Shiozawa, Y. (2007). A new construction of Ricardian trade theory—a many-country, many-commodity case with intermediate goods and choice of production techniques. *Evolutionary and Institutional Economics Review, 3*(2), 141–187.

Shiozawa, Y. (2014, in Japanese). *A final solution of Ricardo problems on international values*. Tokyo: Iwanami-Shoten.

Sraffa, P. (1960). *Production of commodities by means of commodities: Prelude to a critique of economic theory*. Cambridge: Cambridge University Press.

Takamasu, A. (1986). Production possibility frontier in a Sraffa-Leontief economy. *The Manchester School, 59*, 202–207.

Takamasu, A. (1991, in Japanese). *Neo-Ricardian trade theory*. Tokyo: Sobunsha.

Analysis of Production-Efficient Patterns of Specialization Allowing Intermediate Inputs: The Meaning of Shiozawa's Model from the Viewpoint of Modern Economics

Takeshi Ogawa

Abstract Two rare analyses on the theory of international economics with linear economics exist that have different lines of thought but similar model specifications. One is the analysis of (production-)efficient patterns of specialization that allows intermediate goods with the Ricardo–Leontief model and that belongs to the field of modern economics. The other is the Sraffa model extended to international economy, which does not belong to the field of modern economics. In the model setting, the difference between the two analyses is whether the rate of profit exists or not, although the meanings whether the rate of profit exists or not are very different. However, at least in the era of Deardorff (2005a), only the definition of comparative advantage, including intermediate inputs, is not determined and has been the focus since McKenzie (1954a, b, 1955) and Jones (1961) analyzed the pattern of specialization in the multi-country, multi-good Ricardo–Graham model. Shiozawa (2007) made progress on this subject by extending the Sraffa model internationally on the evolutionary economics front but not in modern economics. In this subject, the solution to the problem which these analyses focus on is the production-efficient pattern of specialization; however, there are two problems with this approach. First, in the case where the number of goods is larger than that of countries, the efficient pattern of specialization essentially does not exist. Focusing on this case, Shiozawa (2007) showed the extended concept of pattern of specialization, i.e., "shared pattern of specialization," and pointed out the importance of the case in the real-world economy. Second, as in Higashida (2005a, Japanese), which uses illustrations of price and specialization traditionally presented in Amano (1966) and Ikema (1993, Japanese), the (production-)efficient pattern of solution is not unique in the case of Jones' (1961) setting allowing intermediate goods. Jones (1961) focused on the "production assignment problem" between technological parameters to determine (production-)efficient pattern of specialization. To solve the problem, Jones (1961)

T. Ogawa (✉)
School of Economics, Senshu University, Room No.9710, 7th floor, 9th Building, Ikuta-Campus, 2-1-1, Higashi-Mita, Tama-Ward, Kawasaki-City, Kanagawa-Prefecture, 214-8580, Japan
e-mail: takeshi.ogawa.123@gmail.com

Y. Shiozawa et al. (eds.), *A New Construction of Ricardian Theory of International Values*, Evolutionary Economics and Social Complexity Science 7,
DOI 10.1007/978-981-10-0191-8_4

uses the method of the Hawkins–Simon theorem, where the concept of Z-matrix is the easier treatable concept of the linear complementarity problem. Higashida's (2005b) result means that the (production-)efficient pattern of specialization cannot be determined easily with only a simple extension to Jones' (1961) way. Considering the solution in the case allowing intermediate inputs, the more difficult concept—the S-matrix—which does not have the equivalent concept, must be used. Thus, the final solution, i.e., the necessary and sufficient technological parameters' condition that determines the (production-)efficient pattern of specialization, may not exist. Shiozawa (2007) showed the general existence of a solution, considering the case allowing the number of goods is larger than that of countries (which may become the last meaningful progress of the model analysis), if the simple and meaningful economic condition like Jones' inequality does not appear. Shiozawa (2014, Japanese) saw through this and positioned the result as a "final solution," giving historical meaning to evolutionary economics. However, this progress has implications for not only evolutionary economics but also modern economics. This chapter discusses the significance of Shiozawa's progress in terms of modern economics and in the context of historical illustrations.

Keywords Intermediate inputs' trade • Production assignment problem • Amano-Ikema's illustration • Ricardo-Mill-Jones point • Existence of solutions

1 Introduction: Why Should the Chapter Focus on Shiozawa's (2007) Finding?

After Shiozawa's (2007) result and Shiozawa and Aruga's (2014) intuitions based on evolutionary economics, Shiozawa (2014) announced a "final solution" of Ricardian trade model, with deep interpretations based on evolutionary economics. The arguments posited are profound, but this work's influence has reached only the area of evolutionary economics, even though the model formulation is quite similar to the Ricardo–Leontief model (or Ricardo model including intermediate goods) in modern economics (i.e., non-Marxist economics) and the Ricardo–Sraffa (or neo-Ricardian) model in evolutionary economics (influenced by Marxist economics). The essential difference of model formulation between the Ricardo–Leontief and Ricardo–Sraffa models is the existence of a profit rate. The former model usually does not include a profit rate, whereas the latter usually includes a profit rate as important exogenous parameter. However, the backgrounds and intuitions of these two models are very different as are the explanations of model formulation.

The present author thinks there are several reasons why Shiozawa's result has not had an impact on modern economics. The first is that Shiozawa and Aruga (2014), Shiozawa (2014), and the researchers that followed—e.g., Oka (2015)—focused on the importance on meanings, intuitions, and interpretations within evolutionary economics; however, these topics have been of little interest to people in modern economics. The second is that as a result, Shiozawa (2007) comprehensively has an economic meaning in not evolutionary economics but modern economics, whereas Shiozawa (2007) is focused only on evolutionary economics. Shiozawa (2014) focused more on engaging in evolutionary economics than Shiozawa (2007), because of using Takamasu's (1991) interpretations. Therefore, many researchers in evolutionary economics place greater importance on Shiozawa (2014), with Shiozawa (2007) considered to only provide a platform for the later work, albeit an important one. The third is that Shiozawa (2007) and Shiozawa (2014) present discussions under the framework of evolutionary economics, which is quite different to that of modern economics given its distinctive background and model-building process from the viewpoints of evolutionary economics. Therefore, most researchers based in evolutionary economics have scarcely explained Shiozawa's (2007) result and meaning on the basis of modern economics.[1]

The present author has a modern economics base and is majoring in the determination of patterns of specialization in the Ricardo model including intermediate goods and joint productions, which is based on modern economics. With Leontief's (1936) model called input–output analysis, the Ricardo model including intermediate goods (Ricardo–Leontief model) has the same formulation as the Ricardo model with joint production using Koopmans' (1951) activity analysis, not Uekawa's (1984) joint output model formulations and so on. These are many-country, many-good models with linear production functions that use labor as only one kind of essential input. The difference between the Ricardo–Leontief model and the Ricardo model with joint production is the essential coefficient technology parameters. On the one hand, with Koopmans' (1951) activity analysis, the Ricardo model with joint production has nonnegative technology parameters expressing by-products.[2] On the other hand, the Ricardo–Leontief model has nonpositive technology parameters expressing intermediate inputs. With Ogawa (2011a, 2013b, 2014), the essential model formulation of these is the same without the point above. However, when the present author reviewed Ogawa (2011a) as part of his doctoral thesis, the author has not understood the importance of Shiozawa (2007). However, with the passage of time, the author feels inclined to express the importance of Shiozawa's (2007) result with the formulation of modern economics as the author's work. On the basis of the author's modern economics base, the current chapter explains Shiozawa's (2007) result, i.e., the "existence of solutions" with modern economics' formulations, and discusses some relevant results in the Ricard–Leontief model.

Shiozawa's (2007) importance in modern economics is in proving the proof of existence of solutions to the production assignment problem in the n-country, m-good Ricardo–Leontief model when $n \leq m$. Ogawa (2013b) essentially expanded Shiozawa's (2007) explicit result of the proof of solutions when $n > m$. However, even if the author mentioned only the words "existence of solutions," the impor-

[1]Shiozawa (2014) presents in Chaps. 3 and 5 another mathematical formulation of the results of Shiozawa (2007), but this is also written according to evolutionary economics formulations.

[2]For the importance of joint production, see OECD (2001) and Kainou (2012).

tance of Shiozawa's (2007) result is not understood. Therefore, before the model formulations, the author should stress the history of previous studies in this area.

Before beginning the next section, some important points must be mentioned. First, this chapter discusses as "translations" in one meaning, i.e., from evolutionary economics to modern economics; however, the chapter does not deem one to be superior to the other. Such an evaluation would be deviating from the central purpose of the chapter; please refer to other papers or chapters in the book. Second, the chapter only briefly mentions Shiozawa's (2007) results; it does not delve into the proof of the core result.

The next section mentions the history of previous research in the area, and the third section comprises illustrations based on the areas, that is, the many-country, many-good Ricardo model with the intermediate inputs or joint production. The fourth section presents the model and basic results, analyzing the model and an interpretation of "final solution" that is different from Shiozawa's (2014) original meaning. The fifth section concludes.

2 Previous Research

Graham's (1923, 1932, 1948) studies, which formulate the many-country, many-good Ricardo model, constitute the pioneering research in the area, that is, the many-country, many-good Ricardo model with the intermediate inputs or joint production. Graham (1923, 1932, 1948) essentially explained multilateral production cost theory with numerical examples through the Ricardo–Graham model. Before Graham's (1948) important book, Haberler (1936) treated the two-country, many-good case, and Viner (1937) and Becker (1952) analyzed the many-country, two-good case; however, these two cases (two-country, many-good and many-country, two-good) are directly expanded from Ricardo's two-country, two-good model, i.e., bilateral comparative production cost theory. Graham's (1948) book impacted the area because it is essentially a different multilateral production cost theory from the orthodox Ricardian bilateral production cost theory, with many numerical examples.[3] After Graham (1948), many researchers appeared in the area, such as Metzler (1950), Whitin (1953), and Schumann and Todt (1957). In particular, the general formulation of the Ricardo–Graham model was made by McKenzie (1954a, b, 1955). McKenzie (1954a) also expressed the efficient facet of the production possibility frontier. From McKenzie (1954a, b, 1955), the production efficiency, i.e., the method in which the production point is composed on the frontier, is focused on the main viewpoint in the area. Before McKenzie (1954a, b, 1955), the method of activity analysis by Koopmans (1951), which treats joint production, and the method of Leontief's (1936) model allowing intermediate

[3] For the difference between bilateral Ricardian comparative production cost theory and multilateral Ricardian comparative production cost theory, see Minabe (1971).

goods with linear economics (for input–output analysis) were formulated. Thus, until McKenzie (1954a, b, 1955), the analysis of not only the multi-country, multi-good Ricardo model (Ricardo–Graham model) but also the Ricardo–Leontief model allowing intermediate goods model was an open question.

Following McKenzie's (1954a, b, 1955) general formulations of Ricardo–Graham model, Jones (1961) proposed the production assignment problem, i.e., the way to determine which country should work which good (or which production process). In particular, Jones (1961) devised the production assignment that satisfies the condition that each country produces different goods, so *i-i* assignment, i.e., the world production assignment wherein any $i = 1, 2, \ldots, n$, the *i*-th country produces the *i*-th good, is considered in Jones (1961). With permutation theory, the generalization of *i-i* assignment can be directly generated for each production assignment, where each country produces different goods. On the basis of the production assignment problem, Jones (1961) essentially proposed the inequality (called Jones' inequality) to answer the problem in the *n*-country, *n*-good Ricardo–Graham model with Hawkins and Simon's (1949) theorem excluding "ties." Jones' (1961) approach is to compare *i-i* assignment with any assignment where each country produces different goods except *i-i* assignment, i.e., where other countries' world production assignments are not production efficient. In this way, Jones (1961) showed the uniqueness of the solution of the production assignment model in the *n*-country, *n*-good Ricardo–Graham model.[4] Jones' (1961) contribution was elaborated by Kuhn (1968) in two parts. The first part is the universal solution, where Jones' inequality is extended to allow "ties" as inequality with an equal sign. The other part is the strict (or unique) solution with the concept of the extreme point on the cone, where Jones' (1961) strict inequality usually treats excluding "ties."

After Jones (1961), which presented linear economics including many-country, many-good, neither the Ricardo–Graham model excluding intermediate goods nor the Ricardo–Leontief model including intermediate goods has been little focused on mainstream international trade theory (at least in modern economics), as useful and solvable results have already appeared even if open questions exist.[5] With continuous goods, the Ricardo model's main focus formulation has been moved to two-country (or multi-country), continuous-good, Ricardo model, Rudiger et al. (1977) model, Jonathan and Samuel (2002) model, and so on. The Ricardo–Graham model has received little focus on international trade textbooks after Takayama (1972), and it has also been scarcely mentioned in international trade handbooks after Jones and Neary (1984). Therefore, at least until Deardorff (2005a), the definition of Ricardian comparative advantage with intermediate inputs was not fixed, and even Deardorff (2005a) provided multiple patterns of definitions. In contrast with modern economics, the Ricardo model including intermediate goods

[4]For a general explanation including before Jones' (1961) research, see, e.g., Chipman (1965), Minabe (1995), Higashida (2005a), Deardorff (2005b), and Shiozawa (2007).

[5]For linear economic models, see, e.g., Gale (1960).

has evolved and integrated to Sraffa's (1960) model in evolutionary economics.[6] The main difference is the existence of profit rates. The Sraffa model has profit rate as exogenous parameter, whereas in the Ricardo–Sraffa model, research existed even prior to Shiozawa (2007), e.g., Takamasu (1991).

Recently, Jones' (1961) result has been elaborated through several steps. First, Jones (1961) did not show the existence of solutions of production assignment problem. At this point, the two types of completion appeared. The first type is from Shiozawa (2007), i.e., only the existence of solutions. From Shiozawa (2007), one way to prove the existence of a universal solution is applied from Su's (1999) rental harmony theorem, which is near the area of Sperner's lemma. In addition, Shiozawa (2007) proved the existence of the unique solution with the model extended n-country, m-good model satisfying $n \leq m$. To extend the model with $n \leq m$, Shiozawa (2007) proposed the concept of "shared patterns of specialization." The shared patterns of specialization is the extension of patterns of specialization, i.e., a world production assignment where each country makes different goods. Moreover, Shiozawa's (2007) two results of existence of solutions are originally the Ricardo–Sraffa model without profit rates (which is equivalent to the Ricardo–Leontief model). Therefore, these results can be applied not only to the Ricardo–Graham model but also the Ricardo–Leontief model allowing intermediate inputs. The second step is from Shiozawa (2015), which uses Helly's theorem in convex set theory—a more essential approach in linear economics.[7] Shiozawa (2015) also elaborated the theory of the Ricardo–Graham model mathematically with (arranged) tropical algebra.[8]

Third, Jones (1961) focused only on the n-country, n-good model. The extension of the result in the n-country, n-good model to n-country, m-good model was essentially conducted by Ogawa (2013a, b). Likewise, the extension of the result in the n-country, n-good Ricardo–Graham model to n-country, m-good Ricardo–Graham model was essentially completed by Ogawa (2013a). In evolutionary economics theory, the condition $n \leq m$ is real, so the case $n > m$ is not argued explicitly.[9] However, the model that gives the most important implication in Graham (1948) is the four-country, three-good model, which satisfies $n > m \geq 3$. Therefore, at least in theory, the extension has meaning. Ogawa (2013a) considers the extension from two viewpoints. On the one hand, in the case of $n < m$, to divide one country with several areas, the chapter can arrange an m-area, m-good Ricardo–Graham model. On the other hand, in the case of $n > m$, to treat the same goods as different commodities when the country making the same goods is different, the chapter can arrange an n-country, n-commodity Ricardo–Graham model. Moreover, the extension is essentially expanded by Ogawa (2013b) from the Ricardo–Graham model to the Ricardo–Leontief model allowing intermediate inputs. Ogawa (2013b)

[6]For more about the Ricardo–Sraffa formulation, see, e.g., Takamasu (1991).

[7]For more about Helly's theorem in convex set theory, see, e.g., Ludwig et al. (1963).

[8]For more about tropical algebra, see, e.g., Ilia et al. (2009).

[9]Comparing to Shiozawa (2007) and Ogawa (2013a), the notations of n and m are converse.

also expanded the model with different numbers of production processes in each country. Therefore, Shiozawa's (2007) result of existence of the solution of production assignment problem in the Ricardo–Leontief model is expanded by Ogawa (2013b) in the case of $n > m$.

In studies following Jones (1961), the development of illustrations in the three-good model cannot be missed. First, Amano (1966) provided the original illustrations between the world real commodity price vector and specialization in the two-country, three-good Ricardo–Leontief model allowing intermediate inputs. Using Amano's (1966) illustrations, Ikema (1993) essentially expressed the benchmark theory in the three-country, three-good Ricardo–Graham model excluding intermediate inputs. Amano's (1966) and Ikema's (1993) illustration are two sides of the same coin as McKenzie's (1954a) efficient facet.[10] Thus, Ogawa (2012a) connected two figures. McKenzie's (1954a) efficient facet allows easy comprehension of the whole production possibility frontier even if it is difficult to draw from only the information of technology parameters (and labor endowments). Conversely, Amano's (1966) and Ikema's (1993) illustrations are easy to draw when numerical examples of the information of technology parameters are given, even if it is difficult to procure the information of whole production possibility frontier. Two figures are essentially in three-good model. On the basis of Shiozawa (2007), Ogawa (2012a) showed the way of drawing McKenzie's (1954a, b) efficient facet from Ikema's (1993) illustrations between the world real commodity price vector and specialization in the three-country, three-good Ricardo–Graham model. Additionally, in the three-country, three-good Ricardo–Graham model, Ogawa (2012a) showed the frontier's information that can be revealed by McKenzie's (1954a) efficient facet (from Ikema's (1993) illustrations) but not the production assignment problem.

On the basis of Amano's (1966) illustrations, Higashida (2005b) showed the numerical example where two solutions of production assignment problem exist and the illustration where three solutions of the production assignment problem exist in the three-country, three-good Ricardo–Leontief model including intermediate inputs. In the three-country, three-good model, the number of world production assignments where each country produces different goods is 6 (= 3!). However, the Hawkins–Simon theorem can be used only for the matrix satisfying nonpositive off-diagonals, which can be applied to the Ricardo–Graham model excluding intermediate inputs and joint production. In the general case allowing intermediate goods or by-products, the matrix of positive off-diagonal elements appears when the

[10] On the basis of Ikema (1993), Minabe (2001) illustrated the wage and good produced in the three-country model. However, Minabe's (2001) illustration is difficult to extend to the Ricardo–Leontief model allowing intermediate goods, at least in general. Minabe's (2001) illustration is applied in Shiozawa (2015).

i-i assignment is compared with another world production assignment.[11] Moreover, there are two patterns to change the permutation changing any country's role. Therefore, the method of showing the uniqueness of the solution to the production assignment problem in Jones (1961) cannot be used when three solutions of the production assignment problem exist. Ogawa (2014) completed Higashida's (2005b) illustrations to give the numerical examples of the three solutions. Moreover, different to Ikema's (1993) illustration excluding intermediate inputs and joint production, Amano's (1966) illustration had no basifications where the illustration can be drawn in the general cases. Ogawa (2014) provided sufficient conditions where Amano's (1966) illustration can be illustrated in the three-country, three-good Ricardo–Leontief model allowing intermediate inputs. In the conditions, Ogawa (2014) also focused on the conditions where an autarky economy exists in any country. In that context, autarky means that all goods can be produced in the country itself for each country. The conditions existing in the autarky economy are included in the shadow condition in Shiozawa (2007).

In joint production case illustrations, Ogawa (2015) also shows similar conditions existing in the autarky economy in the case including intermediate inputs to show the existence of a universal solution to the production assignment problem. The existence of a strict solution can be shown by referring to Shiozawa (2007) and Ogawa (2013b, 2015). In addition, Ogawa (2015) shows the same sufficient conditions where Amano's (1966) illustration can be drawn in the joint production case, and it is shown that the sufficient condition is included in the similar conditions existing in the autarky economy. Moreover, in the joint production, Ogawa (2011b) shows that including by-products, the world efficient production pattern can be changed. Ogawa (2011b) also shows the multiple solutions to the production assignment problem allowing the joint production. Additionally, Ogawa (2012b) shows the possibility of different patterns of main production processes from the exported good's pattern. Moreover, Ogawa (2015) showed 48 types of Amano's (1966) illustrations in the two-country, three-good model with joint productions.

In the Ricardo–Sraffa–Shiozawa model in evolutionary economics, the essential model formulation is two-country, three-good model because the determination of a good's price vector within production part is important. Therefore, Ogawa's (2015) results require mention. Conversely, in the Ricardo–Leontief model in modern economics, the freedom of a good's price is important because each country's utility function and welfare formulations have freedom.

Considering previous research in the area, Shiozawa's (2007) results of existence of (universal and strict) solutions in the production assignment problem have importance in not only evolutionary economics but also modern economics.

[11]To treat similar conditions of Hawkins–Simon theorem with matrices allowing positive off-diagonal elements, the theory of linear complementarity problem is required. See the linear complementarity problem's orthodox textbook, e.g., Cottle et al. (1992), but the same (i.e., necessary and sufficient) conditions as in the Hawkins–Simon theorem with matrices allowing positive off-diagonal elements do not exist in Cottle et al. (1992).

3 Illustrations of Benchmarks: Three-Good Case

In the session, through the benchmark case, the basic formulation and illustrations are shown.[12]

Following Ogawa (2011b, 2015), the chapter considers a Ricardo–Graham model with a linear production function involving three countries, three goods, three production processes, and one element (labor), permitting joint production. Let $L^i(>0)$ be the labor endowment for country i and $L^i_j\ (\geq 0)$ the labor input for country i and production process j, where $i\,,j=1\,,2\,,3$. The constraint on labor availability can be expressed as

$$L^i_1 + L^i_2 + L^i_3 = L^i.$$

In a world of joint production, it has been a common practice since Koopmans' (1951) analysis activity to represent the amount that one unit of labor can produce as a production coefficient. Firstly, the chapter defines net production coefficients. Let a^i_{jk} be the amount of good k that can be produced with one unit of labor using production process j in country i, where $i\,,j\,,k=1\,,2\,,3$.[13] When the model allows joint production without intermediate inputs, the conditions $a^i_{jk}\ \geq\ 0$ and $a^i_{jj}\ >\ 0$ hold for any $i\,,j\,,k=1\,,2\,,3$, where the condition $a^i_{jj}\ >\ 0$ follows that for any country i, the main output of production process j is good j. Similarly, in a world including intermediate inputs without joint production, for any country i, using production process j, good j is produced with the amount of $a^i_{jj}\,(>0)$ from one unit of labor and $a^i_{jk}\,(\leq 0)$ unit of intermediate inputs $k(\neq j)$.[14] In the usual formulation of intermediate inputs, as presented in Higashida (2005b), the formulation of one unit of main output is usually used in an activity. However, in this chapter, the formulation of one unit of labor is used in an activity, as seen in Shiozawa (2007), assuming that labor is a necessary input.[15] Two reasons exist. First, to write this formulation, the model with intermediate inputs and that with joint production can be treated with the same formulation. Second, comparing each production process for higher marginal (or average) net revenue, the difference of

---

[12]In the model formulation, this section is essentially based on Ogawa (2014, 2015)—i.e., before the three-good, two-country, two-good model is needed—however, this chapter omits two-country, two-good model.

[13]In reality, the difference of each coefficient may be explained as other factors that cannot move, such as the endowment of immobile capital or land.

[14]More generally, the formulation of joint production allowing intermediate inputs requires multiple kinds of outputs, but neither the nonnegative condition of each coefficient nor the naming of production process in Ogawa (2011b) is required. However, this is a benchmark case, so in the context, the formulation is used. Moreover, this chapter uses this formulation which is simple but a little strong assumption because Ogawa's (2013b) meaning is that the readers can generalize the general formulation only arguing the simple formulation essentially.

[15]For more about relaxing the assumption that labor is necessary for production, see Hosoda (2007).

technology coefficients with one unit of labor is a key concept. When the chapter uses the formulation of one unit of main product in an activity, the condition seems more complex to divide each coefficient of main product than that when the formulation of one unit of labor is in an activity. To express that the main output of production process k is good k for any country i, $a^i_{kk} > \max\left\{0, a^i_{jk}\right\}$ is assumed for any $j(\neq k)$. The activity for country i's production process j can be written as $\left(a^i_{j1}, a^i_{j2}, a^i_{j3}\right)^T$.[16] Total production of good k by country i can be written as X^i_k, where $X^i_k = a^i_{1k}L^i_1 + a^i_{2k}L^i_2 + a^i_{3k}L^i_3$. The set W of world production possibilities can be written as[17]

$$W := \left\{ X = (X_1, X_2, X_3)^T \geq 0 \left| \begin{array}{l} X_k \leq X^1_k + X^2_k + X^3_k,\ X^i_k = a^i_{1k}L^i_1 + a^i_{2k}L^i_2 \\ \qquad +a^i_{3k}L^i_3, \\ \\ L^i_1 + L^i_2 + L^i_3 = L^i \end{array} \right. \right\},^{18}$$

The elements of W are called *production points*, and their inclusion in W is called a *production possibility*. This chapter now uses the definition of efficiency in production.

Definition Production point $X^* \in W$ is efficient when for any production point $X \in W$, satisfying $X \geq X^*$ is in fact equal to X^*. The set of all efficient production points is called frontier of the production possibility set.

In considering the definition of primary and secondary products, the chapter addresses the problem of maximizing world output (net revenue). Given a goods pricing vector $P = (p_1, p_2, p_3)$ $(>>0)$, this problem can be expressed as

$$\max_{L^i_j \geq 0} \sum_k p_k \sum_{i,j} a^i_{jk} L^i_j \quad \text{s.t.} \quad \sum_j L^i_j \leq L^i. \tag{P1}$$

The multiplier in this problem reduces wages to $w = (w^1, w^2, w^3)$ $(>>0)$. This chapter omits the proof of the following lemma because it is straightforwardly derived and the essence is in Takamasu (1991).

Lemma 1 The following two statements are equivalent for $X \in W$, where $X >> 0$.

1. X is efficient and on the frontier.
2. There exists a positive price system $P(>>0)$ for which X maximizes world output.

This section is essentially based on Ogawa (2014), which discusses the intermediate input model, and Ogawa (2015), which includes the joint production model. For each country, it is important to find the region of the price system where specializing in a particular production process maximizes that country's production value. The

---

[16] T indicates transposition.

[17] In this section, the free disposal of goods with positive value is permitted but not for those with negative value.

[18] The chapter represents vector inequality with $\geq, >, >>$.

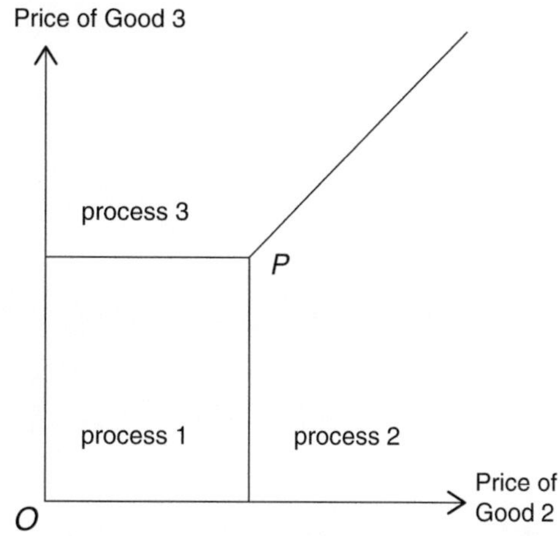

Fig. 1 World price and specialization

convex region of the price system where country *i* specializing in production process *j* maximizes that country's production value is represented as follows:

$$\left\{(p_1, p_2, p_3) > 0 | a_{j1}^i p_1 + a_{j2}^i p_2 + a_{j3}^i p_3 > a_{k1}^i p_1 + a_{k2}^i p_2 + a_{k3}^i p_3, \text{ for all } k \neq j\right\}.$$

If production process 1 maximizes the country's output in the given world price, then this region of price area is labeled production process 1 in Fig. 1, where good 1 is the numeraire ($p_1 = 1$). Similarly, the regions of the price system where country 1 specializing in production process 2 and production process 3 maximizes country 1's production value are labeled production processes 2 and 3, respectively, in Fig. 1 (Amano's (1966) illustration showing the connection between world price vector and specialization). In Fig. 1, *P* shows price vector where any marginal (or average) net revenue is the same in every production process.

The boundary shared by the region of country *i* specializing in processes 1 and 3 is as follows:

$$a_{11}^i + a_{12}^i p_2 + a_{13}^i p_3 = a_{31}^i + a_{32}^i p_2 + a_{33}^i p_3 \qquad \therefore p_3 = \frac{a_{12}^i - a_{32}^i}{a_{33}^i - a_{13}^i} p_2 + \frac{a_{11}^i - a_{31}^i}{a_{33}^i - a_{13}^i}.$$

Because $a_{11}^i > a_{31}^i$ and $a_{33}^i > a_{13}^i$, on this boundary when good 2 price $p_2 = 0$, then good 3 price $p_3 > 0$, giving it a positive intercept on the good 3 price axis. A similar analysis demonstrates that on the boundary between the price regions for specializing in process 1 and process 2, if $p_3 = 0$, then $p_2 > 0$, giving a positive intercept on the good 2 prices axis. By the same token, the boundary between the price regions for specializing in process 2 and process 3 is

$$a_{21}^i + a_{22}^i p_2 + a_{23}^i p_3 = a_{31}^i + a_{32}^i p_2 + a_{33}^i p_3 \qquad \therefore p_3 = \frac{a_{22}^i - a_{32}^i}{a_{33}^i - a_{23}^i} p_2 + \frac{a_{21}^i - a_{31}^i}{a_{33}^i - a_{23}^i},$$

giving this boundary a positive slope.

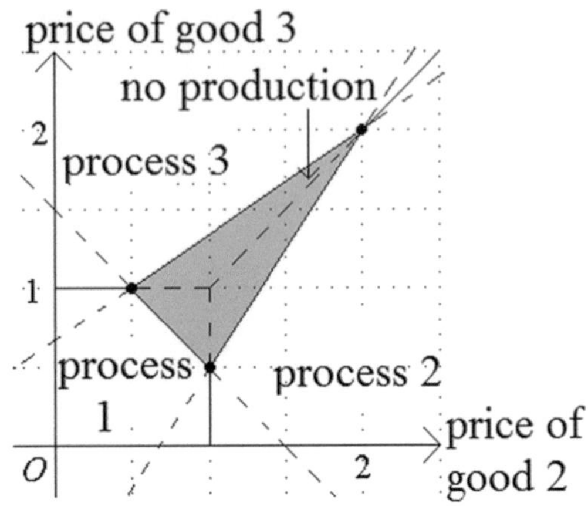

Fig. 2 Possibility of no production area

In addition, given some assumptions $a^i_{kk} > a^i_{jk} \quad (j \neq k)$, a price for good 2 much higher than the others yields a specialization in process 2, and a price for good 3 much higher than the others yields a specialization in process 3. In the case that prices for both goods 2 and 3 are very low, the country specializes in process 1.

However, in the case of the intermediate input model, if self-provisioning and self-sufficiency are impossible in a country, there exists a price vector area where no production is chosen as in Fig. 2. The following example shows Fig. 2:

Example: $$\begin{bmatrix} a^i_{11} & a^i_{21} & a^i_{31} \\ a^i_{12} & a^i_{22} & a^i_{32} \\ a^i_{13} & a^i_{23} & a^i_{33} \end{bmatrix} = \begin{bmatrix} 3 & -2 & -2 \\ -2 & 3 & -2 \\ -2 & 3 & 3 \end{bmatrix}.$$

Therefore, the assumption of existing autarky equilibrium, i.e., the assumption that self-provisioning and self-sufficiency are possible, is required in the chapter.

Assumption 1 (Existence of Autarky Economy for Intermediate Input Model) To guarantee that self-provisioning and self-sufficiency are possible for any country, for any i, the principal minors of the following matrix are all positive (self-provisioning and self-sufficiency are possible):

$$\begin{bmatrix} a^i_{11} & a^i_{21} & a^i_{31} \\ a^i_{12} & a^i_{22} & a^i_{32} \\ a^i_{13} & a^i_{23} & a^i_{33} \end{bmatrix}.$$

This chapter uses Assumption 1 and Hawkins and Simon's (1949) theorem.

Hawkins–Simon Theorem (Hawkins and Simon 1949) The following three propositions are equivalent for an off-diagonal, nonpositive matrix M:

1. For any $x >> 0$, there exists a $v >> 0$ such that $Mv = x$. (M is an S-matrix.)
2. The leading principle minors of M are all positive.
3. The principle minors of M are all positive.

With Assumption 1 and the Hawkins–Simon theorem, the following lemma holds straightforwardly.

Lemma 2 Under Assumption 1, self-provisioning and self-sufficiency are possible for any country; i.e., adjusting for labor input, any good produces a positive amount in the meaning of net.

Now, the intersection of the three boundaries of three goods with marginal productivity as measured by value satisfies the following:

$$\begin{cases} a_{11}^i + a_{12}^i p_2 + a_{13}^i p_3 = a_{21}^i + a_{22}^i p_2 + a_{23}^i p_3 \\ a_{11}^i + a_{12}^i p_2 + a_{13}^i p_3 = a_{31}^i + a_{32}^i p_2 + a_{33}^i p_3 \end{cases} \Longleftrightarrow \begin{bmatrix} a_{22}^i - a_{12}^i & a_{23}^i - a_{13}^i \\ a_{32}^i - a_{12}^i & a_{33}^i - a_{13}^i \end{bmatrix} \begin{bmatrix} p_2 \\ p_3 \end{bmatrix} = \begin{bmatrix} a_{11}^i - a_{21}^i \\ a_{11}^i - a_{31}^i \end{bmatrix}$$

. If the chapter writes this as

$$H := \left(a_{22}^i - a_{12}^i\right)\left(a_{33}^i - a_{13}^i\right) - \left(a_{23}^i - a_{13}^i\right)\left(a_{32}^i - a_{12}^i\right),$$

then for consistency with the case of no joint production or no intermediate inputs, $H > 0$ must be satisfied. Using Cramer's rule, this solution may be written as

$$p_2 = \frac{1}{H}\left\{\left(a_{11}^i - a_{21}^i\right)\left(a_{33}^i - a_{13}^i\right) - \left(a_{23}^i - a_{13}^i\right)\left(a_{11}^i - a_{31}^i\right)\right\},$$
$$p_3 = \frac{1}{H}\left\{\left(a_{22}^i - a_{12}^i\right)\left(a_{11}^i - a_{31}^i\right) - \left(a_{32}^i - a_{12}^i\right)\left(a_{11}^i - a_{21}^i\right)\right\}.$$

Let us consider what determines the sign of this model. This chapter can analyze this condition consistently if this price (point P in Fig. 1) is positive, as done by Ikema (1993). The foregoing logic can be summarized in Proposition 1. For this purpose, the chapter posits Assumption 2.

Assumption 2 The following inequalities hold: $\begin{cases} \left(a_{22}^i - a_{12}^i\right)\left(a_{33}^i - a_{13}^i\right) > \\ \quad \left(a_{23}^i - a_{13}^i\right)\left(a_{32}^i - a_{12}^i\right), \\ \left(a_{11}^i - a_{21}^i\right)\left(a_{33}^i - a_{13}^i\right) > \\ \quad \left(a_{23}^i - a_{13}^i\right)\left(a_{11}^i - a_{31}^i\right), \\ \left(a_{22}^i - a_{12}^i\right)\left(a_{11}^i - a_{31}^i\right) > \\ \quad \left(a_{32}^i - a_{12}^i\right)\left(a_{11}^i - a_{21}^i\right). \end{cases}$.

This assumption is satisfied in the case of no joint productions or no intermediate inputs.

Proposition 1 Given Assumption 2, there exists a pricing system in the first quadrant (point P in Fig. 1), where all marginal productivities as measured by value are equal.

Fig. 3 Multiple-assignment price

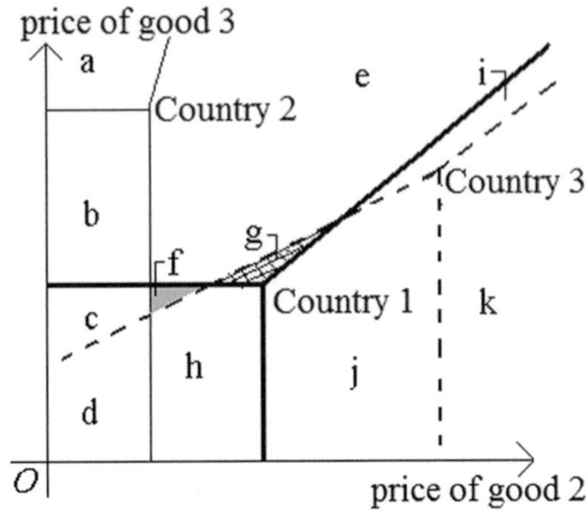

Proposition 1 shows that Amano's (1966) illustration can be used in an orthodox manner under some conditions.

Next, following Ogawa (2011a, 2012a, 2014), we illustrate McKenzie's (1954a) efficient facet from Amano's (1966) illustrations about the following example (from Ogawa 2014) satisfying the assumptions above.

Example:
$$\begin{bmatrix} a_{11}^1 \\ a_{12}^1 \\ a_{13}^1 \end{bmatrix} = \begin{bmatrix} 3 \\ 0 \\ 0 \end{bmatrix}, \begin{bmatrix} a_{21}^1 \\ a_{22}^1 \\ a_{23}^1 \end{bmatrix} = \begin{bmatrix} 0 \\ 3 \\ 0 \end{bmatrix}, \begin{bmatrix} a_{31}^1 \\ a_{32}^1 \\ a_{33}^1 \end{bmatrix} = \begin{bmatrix} 0 \\ 0 \\ 5 \end{bmatrix},$$
$$\begin{bmatrix} a_{11}^2 \\ a_{12}^2 \\ a_{13}^2 \end{bmatrix} = \begin{bmatrix} 3 \\ 0 \\ 0 \end{bmatrix}, \begin{bmatrix} a_{21}^2 \\ a_{22}^2 \\ a_{23}^2 \end{bmatrix} = \begin{bmatrix} 0 \\ 6 \\ 0 \end{bmatrix}, \begin{bmatrix} a_{31}^2 \\ a_{32}^2 \\ a_{33}^2 \end{bmatrix} = \begin{bmatrix} 0 \\ 0 \\ 2 \end{bmatrix},$$
$$\begin{bmatrix} a_{11}^3 \\ a_{12}^3 \\ a_{13}^3 \end{bmatrix} = \begin{bmatrix} 3 \\ 0 \\ 0 \end{bmatrix}, \begin{bmatrix} a_{21}^3 \\ a_{22}^3 \\ a_{23}^3 \end{bmatrix} = \begin{bmatrix} 0 \\ 2 \\ 0 \end{bmatrix}, \begin{bmatrix} a_{31}^3 \\ a_{32}^3 \\ a_{33}^3 \end{bmatrix} = \begin{bmatrix} 0 \\ -1 \\ 6 \end{bmatrix}.$$

In the example, Amano's (1966) illustration can be drawn as Fig. 3 above. Considering Ogawa's (2012a) method, McKenzie's (1954a) efficient facet can be drawn as Fig. 4 from Fig. 3's information. From Fig. 4, the production possibility frontier can be drawn as Fig. 5. The points f and g are shown as the extreme points on the production possibility frontier, which show the strict solutions of production assignment problem. Like Higashida's (Higashida 2005b) explanation, the uniqueness of solution of production assignment problem is broken in the three-country, three-good model including intermediate inputs, where the similar result has appeared in the model with joint production.

Moreover, the following example from Ogawa (2014) shows that three strict solutions of the production assignment problem exist as extreme points. The example satisfies the assumptions above: Fig. 6 is Amano's (1966) illustration,

Fig. 4 Efficient facet

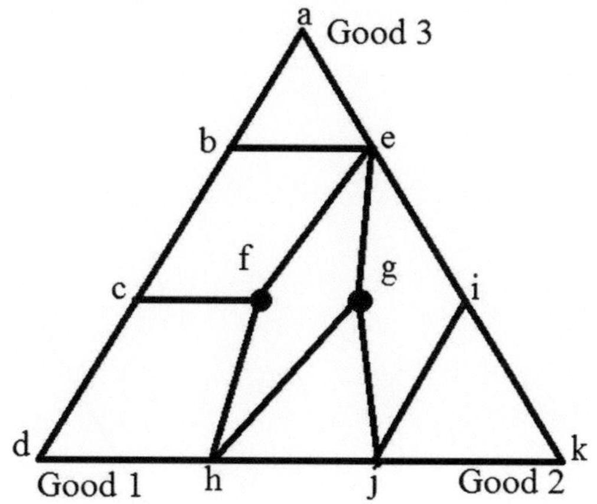

Fig. 5 World frontier

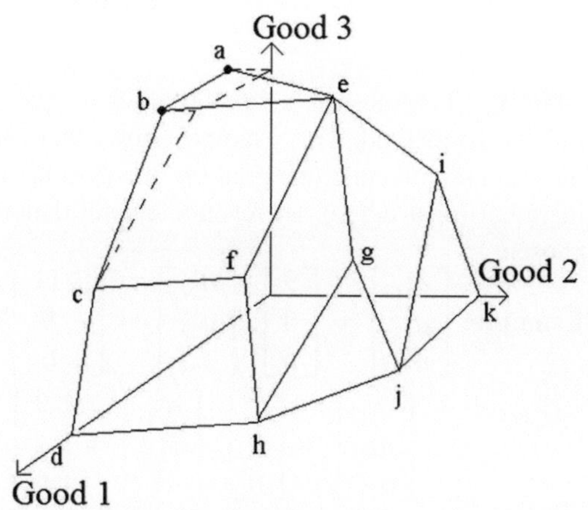

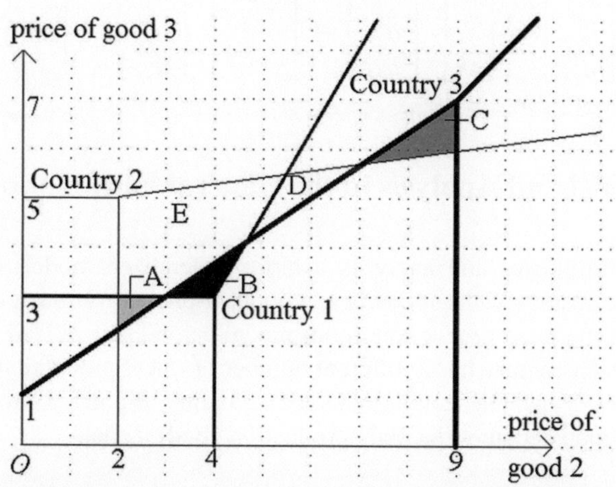

Fig. 6 Triple-assignment world price

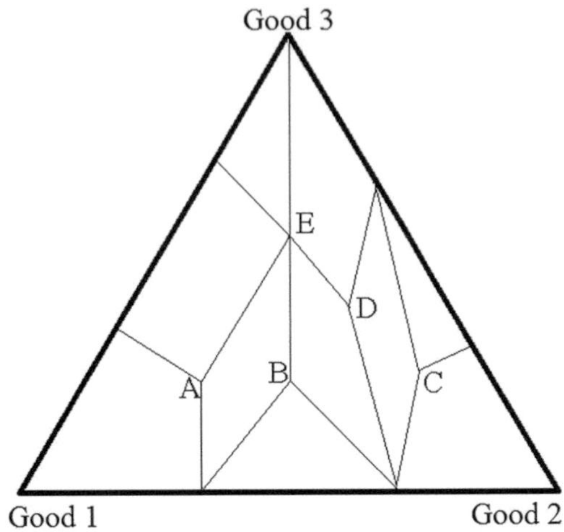

Fig. 7 Triple-assignment efficient facet

and Fig. 7 is McKenzie's (1954a) efficient facet derived from Fig. 6 with Ogawa's (2012a, b) method. This example shows that Jones' (1961) method of comparing only an assignment's marginal (or average) net revenue with another assignment's marginal (or average) net revenue cannot demonstrate that one assignment is not chosen.

Example:
$$\begin{bmatrix} a_{11}^1 \\ a_{12}^1 \\ a_{13}^1 \end{bmatrix} = \begin{bmatrix} 9 \\ 0 \\ 0 \end{bmatrix}, \begin{bmatrix} a_{21}^1 \\ a_{22}^1 \\ a_{23}^1 \end{bmatrix} = \begin{bmatrix} -11 \\ 5 \\ 0 \end{bmatrix}, \begin{bmatrix} a_{31}^1 \\ a_{32}^1 \\ a_{33}^1 \end{bmatrix} = \begin{bmatrix} 0 \\ 0 \\ 3 \end{bmatrix},$$
$$\begin{bmatrix} a_{11}^2 \\ a_{12}^2 \\ a_{13}^2 \end{bmatrix} = \begin{bmatrix} 2 \\ 0 \\ 0 \end{bmatrix}, \begin{bmatrix} a_{21}^2 \\ a_{22}^2 \\ a_{23}^2 \end{bmatrix} = \begin{bmatrix} 0 \\ 1 \\ 0 \end{bmatrix}, \begin{bmatrix} a_{31}^2 \\ a_{32}^2 \\ a_{33}^2 \end{bmatrix} = \begin{bmatrix} -33 \\ 0 \\ 7 \end{bmatrix},$$
$$\begin{bmatrix} a_{11}^3 \\ a_{12}^3 \\ a_{13}^3 \end{bmatrix} = \begin{bmatrix} 9 \\ 0 \\ 0 \end{bmatrix}, \begin{bmatrix} a_{21}^3 \\ a_{22}^3 \\ a_{23}^3 \end{bmatrix} = \begin{bmatrix} -18 \\ 3 \\ 0 \end{bmatrix}, \begin{bmatrix} a_{31}^3 \\ a_{32}^3 \\ a_{33}^3 \end{bmatrix} = \begin{bmatrix} 0 \\ -6 \\ 9 \end{bmatrix}.$$

4 The Model and Analysis in the General Formulations

This section formulates and analyzes a more generalized model. From Ogawa (2013a, b), the results of the n-country, n-good, n-process (in each country, each good becomes the main production good) model can be generalized as n-country, m-good, and each country has a different number of linear production processes in almost all cases. Thus, the section treats the n-country, n-good, n-process (in each country, each good becomes the main production good) model.

In country i, a^i_{jk} is good k's net output coefficient, with production process j's technology parameter for any $i, j, k = 1, 2, \ldots, n$. In the joint production model, a^i_{jk} (≥ 0) is good k's output amounts in process j with one unit of labor in country i, and a^i_{jj} (> 0) is the main product, i.e., good j's amount in process j with one unit of labor in country i. In the intermediate good model, a^i_{jj} (> 0) is good j's positive output amounts with one unit of labor and a^i_{jk} (≤ 0) unit of good k's intermediate input, where $k \neq j$ in country i. In country i, the activity of process j is $\left(a^i_{j1}, a^i_{j2}, \ldots, a^i_{jn}\right)^T$. Labor is only one essential input in each process in each country, and $L^i(>0)$ is the positive labor endowment in country i. L^i_j (≥ 0) is labor input of process j in country i, satisfying the labor endowment constraint as $\sum_j L^i_j \leq L^i$ for any i.

Now, let us define "pattern of specialization."

Definitions (Pattern of Specialization)

1. For country i to specialize in production process, j means that $L^i_j = L^i$; i.e., the entire labor force is devoted to production process j. This chapter abbreviates this as specialization.
2. For the world economy to be completely specialized means that each country specializes in one production process, and the chapter calls this phenomenon the pattern of specialization.
3. For the world economy to have one-to-one assignment means a pattern of specialization in which each country specializes in a different production process.
4. For the world economy to have $i - i$ assignment means one-to-one assignment of country i in production process i.

This section supposes several assumptions:

Assumption 3 (For Intermediate Input Model)
For any $\sigma \in S_n$, the principal minors of the non-diagonal nonpositive matrix created on the basis of production coefficients for the one-to-one assignment involving country i specializing in production process $\sigma(i)$ are uniformly positive:

$$\begin{bmatrix} a^1_{\sigma(1)\sigma(1)} & a^2_{\sigma(2)\sigma(1)} & \cdots & a^n_{\sigma(n)\sigma(1)} \\ a^1_{\sigma(1)\sigma(2)} & a^2_{\sigma(2)\sigma(2)} & \cdots & a^n_{\sigma(n)\sigma(2)} \\ \vdots & \vdots & \ddots & \vdots \\ a^1_{\sigma(1)\sigma(n)} & a^2_{\sigma(2)\sigma(n)} & \cdots & a^n_{\sigma(n)\sigma(n)} \end{bmatrix}.$$

Here, S_n is an n-th degree permutation set, i.e.,
$S_n := \{\sigma : \{1, 2, \cdots, n\} \rightarrow \{1, 2, \cdots, n\} | \sigma \text{ is a bijective function}\}$.

Assumption 4 (Existence of Autarky Economy)
For any $i=1,2,\cdots,n$, the principal minors of the following technology matrix A_i are all positive (when rewritten for intermediate goods, self-provisioning and self-sufficiency are possible):

$$A_i := \begin{bmatrix} a^i_{11} & a^i_{12} & \cdots & a^i_{1n} \\ a^i_{21} & a^i_{22} & \cdots & a^i_{2n} \\ \vdots & \vdots & \ddots & \vdots \\ a^i_{n1} & a^i_{n2} & \cdots & a^i_{nn} \end{bmatrix}.$$

$P=(p_1,p_2,\cdots,p_n)>>0$ is a positive world price vector. The world net output of good k is $X_k = \sum_{i,j} a^i_{jk} L^i_j$. The closed and convex world production set W is as follows:

$$W = \left\{ X = (X_1, X_2, \cdots, X_n) \geq 0 \,\middle|\, X_k \leq \sum_{i,j} a^i_{jk} L^i_j, \quad \sum_j L^i_j \leq L^i, \quad L^i_j \geq 0 \right\}.$$

$0 \leq X_k \leq \sum_{i,j} a^i_{jk} L^i_j$ implies that the free disposal of goods with positive value is permitted, but disposal of goods with negative value is prohibited. The world net revenue, i.e., revenue removed intermediate input cost, maximizes (P1) as follows:

$$\max_{L^i_j \geq 0} \sum_k p_k \sum_{i,j} a^i_{jk} L^i_j \quad \text{s.t.} \quad \sum_j L^i_j \leq L^i. \tag{P1}$$

In orthodox evolutionary economics, the formulation of optimization is not used, whereas in orthodox modern economics, the formulation of optimization is usually used. $w^i(\geq 0)$ is a wage multiplier of the labor constraint $\sum_j L^i_j \leq L^i$ for country i. Lagrange function $\Phi\left(L^i_j, w^i\right)$ becomes

$$\Phi\left(L^i_j, w^i\right) := \sum_k p_k \sum_{i,j} a^i_{jk} L^i_j + \sum_i w^i \left(L^i - \sum_j L^i_j \right).$$

With Assumption 4, the first-order necessary conditions can be written as follows:

$$\begin{cases} \sum_k p_k a^i_{jk} \leq w^i, \quad L^i_j \geq 0, \quad \left(\sum_k p_k a^i_{jk} - w^i\right) L^i_j = 0, \\ w^i > 0, \quad \sum_j L^i_j = L^i. \end{cases}$$

If $L^i_j > 0$ at the solution of (P1), $\sum_k p_k a^i_{lk} \leq \sum_k p_k a^i_{jk} = w^i$ holds, which can be rewritten as

$$\sum_k \left(a^i_{jk} - a^i_{lk}\right) p_k \geq 0. \tag{1}$$

From Assumption 4, the following lemma derives straightforwardly from the Hawkins–Simon theorem.

Lemma 3 (Existence of Autarky Economy)

1. There is a labor input distribution in any country i $\left(L_1^i, L_2^i, \cdots, L_n^i\right) (\geq 0)$ satisfying that all goods are produced with positive amount, i.e.,

$$\begin{bmatrix} a_{11}^i & a_{21}^i & \cdots & a_{n1}^i \\ a_{12}^i & a_{22}^i & \cdots & a_{n2}^i \\ \vdots & \vdots & \ddots & \vdots \\ a_{1n}^i & a_{2n}^i & \cdots & a_{nn}^i \end{bmatrix} \begin{bmatrix} L_1^i \\ L_2^i \\ \vdots \\ L_n^i \end{bmatrix} = \begin{bmatrix} \sum_j a_{j1}^i L_j^i \\ \sum_j a_{j2}^i L_j^i \\ \vdots \\ \sum_j a_{jn}^i L_j^i \end{bmatrix} >> 0.$$

2. If all world prices are positive, i.e., $P = (p_1, p_2, \cdots, p_n) >> 0$, at the solution of (P1), $w := (w^1, w^2, \cdots, w^n) >> 0$ (positive wage) and $\sum_j L_j^i = L^i$ (complete employment) hold. Any labor input chooses some production process at the solution.

Here, recall that Hawkins and Simon's (1949) theorem is for only an off-diagonal, nonpositive matrix.

In the case of joint production, Lemma 3.2 is obvious, but for the existence of the positive world price vector, similar conditions like the existence of autarky economy (Assumption 4) are useful. Following Ogawa (2015), the maximization problem (P1) can be rewritten as (P2), with both having the same solution:

$$\max_{L_j^i \geq 0} \sum_k p_k \sum_{i,j} \left(a_{jk}^i - \max_{l \neq k} a_{lk}^i \right) L_j^i \quad \text{s.t.} \quad \sum_j L_j^i \leq L^i. \tag{P2}$$

With (P2), Assumptions 3 and 4 can be changed as follows in the case of joint production.

Assumption 3′ (For Joint Production Model)

For any $\sigma \in S_n$, the principal minors of the non-diagonal nonpositive matrix created on the basis of production coefficients for the one-to-one assignment involving country i specializing in production process $\sigma(i)$ are uniformly positive:

$$\begin{bmatrix} a^1_{\sigma(1)\sigma(1)} - \max\limits_{j \neq \sigma(1)} a^1_{j\sigma(1)} & a^2_{\sigma(2)\sigma(1)} - \max\limits_{j \neq \sigma(1)} a^2_{j\sigma(1)} & \cdots & a^n_{\sigma(n)\sigma(1)} - \max\limits_{j \neq \sigma(1)} a^n_{j\sigma(1)} \\ a^1_{\sigma(1)\sigma(2)} - \max\limits_{j \neq \sigma(2)} a^1_{j\sigma(2)} & a^2_{\sigma(2)\sigma(2)} - \max\limits_{j \neq \sigma(2)} a^2_{j\sigma(2)} & \cdots & a^n_{\sigma(n)\sigma(2)} - \max\limits_{j \neq \sigma(2)} a^n_{j\sigma(2)} \\ \vdots & \vdots & \ddots & \vdots \\ a^1_{\sigma(1)\sigma(n)} - \max\limits_{j \neq \sigma(n)} a^1_{j\sigma(n)} & a^2_{\sigma(2)\sigma(n)} - \max\limits_{j \neq \sigma(n)} a^2_{j\sigma(n)} & \cdots & a^n_{\sigma(n)\sigma(n)} - \max\limits_{j \neq \sigma(n)} a^n_{j\sigma(n)} \end{bmatrix}.$$

Assumption 4′ (Similar Case of the Existence of Autarky Economy)

For any $i = 1, 2, \cdots, n$, the principal minors of the following adjusted technology matrix are all positive (when rewritten for intermediate goods, self-provisioning and self-sufficiency are possible):

$$\begin{bmatrix} a^i_{11} - \max\limits_{l\neq 1} a^i_{l1} & a^i_{21} - \max\limits_{l\neq 1} a^i_{l1} & \cdots & a^i_{n1} - \max\limits_{l\neq 1} a^i_{l1} \\ a^i_{12} - \max\limits_{l\neq 2} a^i_{l2} & a^i_{22} - \max\limits_{l\neq 2} a^i_{l2} & \cdots & a^i_{n2} - \max\limits_{l\neq 2} a^i_{l2} \\ \vdots & \vdots & \ddots & \vdots \\ a^i_{1n} - \max\limits_{l\neq n} a^i_{ln} & a^i_{2n} - \max\limits_{l\neq n} a^i_{ln} & \cdots & a^i_{nn} - \max\limits_{l\neq n} a^i_{ln} \end{bmatrix}.$$

Lemma 3 and so on can be written similarly. Note that each $\sigma \in S_n$ corresponds one-to-one with world production assignment that for any $i = 1, 2, \cdots, n$, country i specializes process $\sigma(i)$. If we define $id \in S_n$, then id means the $i - i$ assignment; i.e., country i specializes in process i for any $i = 1, 2, \cdots, n$. Any one-to-one assignment can be rewritten as the $i - i$ assignment, so we focus only on the case that the $i - i$ assignment is the solution of production assignment problem, i.e., the problem determining which country should work with which process (or good). To compare the $i - i$ assignment with another (not limited to only one-to-one) assignment $\tau \in F_n := \{\{1, 2, \cdots, n\} \to \{1, 2, \cdots, n\}\}$ where country i specializes in process $\tau(i)$, (1) can be written as follows:

$$B_\tau P \geq 0, \text{ where } B_\tau := \begin{bmatrix} a^1_{11} - a^1_{\tau(1)1} & a^1_{12} - a^1_{\tau(1)2} & \cdots & a^1_{1n} - a^1_{\tau(1)n} \\ a^2_{21} - a^2_{\tau(2)1} & a^2_{22} - a^2_{\tau(2)2} & \cdots & a^2_{2n} - a^2_{\tau(2)n} \\ \vdots & \vdots & \ddots & \vdots \\ a^n_{n1} - a^n_{\tau(n)1} & a^n_{n2} - a^n_{\tau(n)2} & \cdots & a^n_{n2} - a^n_{\tau(n)2} \end{bmatrix} \text{ and } P = \begin{bmatrix} p_1 \\ p_2 \\ \vdots \\ p_n \end{bmatrix}. \tag{2}$$

Next, the section defines the frontier and the extreme point.

Definition Production point $X^* \in W$ is efficient only if for production points $X \in W$ where $X \geq X^*$, we have $X = X^*$, and efficient production points are called points on the *frontier.*

The basic important propositions are as follow. The former proposition is derived straightforwardly and Takamasu (1991) also treated it, so the proof is omitted here.

Proposition 2 (Equivalent of Universal Solutions) The following two statements are equivalent about labor distributions $L^i_j (\geq 0)$ and world production $X_k = \sum_{i,j} a^i_{jk} L^i_j$.

1. The world production point $X = (X_1, X_2, \cdots, X_n) \in W$ is on the frontier.
2. There is a positive price vector $P = (p_1, p_2, \cdots, p_n) >> 0$ under which labor distribution $L^i_j (\geq 0)$ is a solution of (P1), i.e., L^i_j maximizes the world. Moreover, the production point X composed L^i_j maximizes (P1) under the existing condition of $P >> 0$.

Next, to guarantee the existence of universal (or weak) solutions of production assignment problem, the chapter uses Su's (1999) rental harmony theorem, following Shiozawa (2007).[19]

Rental Harmony Theorem (Su 1999)
Suppose n housemates in an n bedroom house seek to decide who gets which room and for what portion of the total rent. Suppose also that the following conditions hold:

1. *Good house*: In any partition of the rent, each person finds some room acceptable.
2. *Miserly tenants*: Each person always prefers a free room (one that costs no rent) to a non-free room.
3. *Closed preference sets*: A person who prefers a room for a convergent sequence of prices prefers that room at the limiting price.

Then, there exists a partition of the rent such that each person prefers a different room. The chapter can adapt this model, assuming people as countries, rooms as production processes, and rent as wages. The chapter assumes that wages are uniquely determined by the prices of goods for a one-to-one assignment, which is an efficient pattern of specialization.

Shiozawa (2007) did not argue about the method of using Su's (1999) rental harmony theorem. However, to apply Su's theorem and to apply condition 2 (miserly tenants), an additional assumption is required.

Assumption 5 Every good requires at least the final good or intermediate input.

With these assumptions, the following proposition for the existence of a universal solution to the production assignment problem holds.

Proposition 3 (Existence of Universal Solution of Production Assignment Problem)
1. Suppose that under world price, required intermediate inputs can be obtained. There is a positive wage vector $w=(w^1, w^2, \cdots, w^n)>>0$ and the one-to-one assignment chosen.
2. Suppose that under the $i-i$ assignment, all goods can produce positive amounts of net output. Moreover, suppose that the one-to-one assignment chosen is the $i-i$ assignment. There is a positive world price vector $P=(p_1, p_2, \cdots, p_n)>>0$ under which the world production point $X*$ based on the $i-i$ assignment maximizes the world net revenue problem (P1); i.e., the $i-i$ assignment is a solution of (P1) under the price vector P.

The sketch of proof of (1) is as follows.

Sketch of Proof of (1) Consider a choice of $w=(w^1, w^2, \cdots, w^n)>>0$ under $\sum_i w^i = 1$. Condition 3 (closed preference sets) is straightforwardly satisfied.

[19]In Shiozawa (2007), patterns of specialization are extended as a "shared pattern of specialization" when the number of the countries is smaller than the number of goods.

Condition 1 (good house) is satisfied under Assumption 4 (or 4′). Condition 2 (miserly tenants) is satisfied because the number of countries and the number of processes are the same (or the number of country is smaller) under Assumption 5. Therefore, the three conditions are satisfied; thus, using Su's (1999) rental harmony theorem, condition 1 is proved.

After proof of (1) from Assumption 3 (or 3′), (2) is proved from the Hawkins–Simon theorem.

Shiozawa (2007) proved the existence of a universal solution to the production assignment problem in several ways. One of the important points of Shiozawa (2007) is this since Kuhn (1968).

Next, regarding the strict solution of production assignment problem, this section defines the extreme point with reference to Kuhn (1968).

Definition For any positive amount of world production point $X^*(>>0) \in W$ on the frontier, X is the extreme point; if $X^* = tA + (1-t)B$, then $A = B = X^*$ for any $A \in W$, $B \in W$, and $0 < t < 1$.

With the extreme point, Proposition 2 can be rewritten as follows.

Proposition 4 (Equivalent of Strict Solutions): The following two statements are equivalent about the $i-i$ assignment, and $X = (X_1, X_2, \cdots, X_n)(>>0) \in W$ composed the $i-i$ assignment.

1. The world production point $X = (X_1, X_2, \cdots, X_n)(>>0) \in W$ is an extreme point.
2. There is a positive price vector $P = (p_1, p_2, \cdots, p_n) >> 0$ under which the $i-i$ assignment is a unique solution of (P1); i.e., the $i-i$ assignment uniquely maximizes the world. Moreover, the method of making X is unique, and the production point X that composes the $i-i$ assignment uniquely maximizes (P1) under the existing condition of$P >> 0$.

For the proof, the following theorem in convex cone theory is used. The argument is from Nikaido (1968).

Theorem (Closed and Convex Cone's Extreme Point)
Suppose that $X(>>0) \in W$ is an extreme point and origin and the pointed end of a closed and convex cone C. (The closed and convex cone C with origin as the point end is pointed if $X(\neq 0) \in C$ then $-X \notin C$.) Then, there exists $q \neq 0$ satisfying $q \cdot Z > 0$ for any$Z(\neq 0) \in C$.

Using the theorem, Proposition 4 can be proven as follows.

Sketch of Proof from (2) to (1) Using the contradiction method, suppose that $X(>>0) \in W$ is not an extreme point. Thus, there are $(A, B, t) \in W \times W \times (0, 1)$ satisfying $X = tA + (1-t)B$ and $A \neq X$. Hence, the following inequalities show that $P(>>0)$ does not hold, under which X maximizes (P1): $P \cdot A \lesseqgtr P \cdot \{tA + (1-t)B\} \iff P \cdot A \lesseqgtr P \cdot B \iff P \cdot \{tA + (1-t)B\} \lesseqgtr P \cdot B$.

Sketch of Proof from (1) to (2) First, if $X(>>0)$ is an extreme point, then each country specializes. Second, if $X(>>0)$ composing the $i-i$ assignment is an extreme point, then X cannot be composed by another world complete specialization (one-to-one assignment). After the steps, we show that we can use the conditions of Theorem of Closed and Convex Cone's Extreme Point above.

Shiozawa's (2007) main result is the existence of the strict solution of production assignment problem. The original formulation is as follows[20]:

> *Let the number of commodities n be at least equal to the number of countries m. Techniques are assumed to satisfy conditions (1) linear production techniques, (2) simple production-type techniques, and (3) existence of a productive system. Suppose, in addition, that for any non-negative wage rate vector there is at least one industry for which one country is strongly competitive. Then, there exists an m-dimensional convex cone, in the interior of which the wage rate vector induces a strongly shared pattern of specialization.*

To rewrite Shiozawa's (2007) main result in the chapter, the following argument holds.

Claim (Existence of Strict Solution to Production Assignment Problem)

1. Suppose that under world price required, intermediate inputs can be gotten. There is a positive wage vector $w = (w^1, w^2, \ldots, w^n) >> 0$ and the one-to-one assignment is chosen.
2. Suppose that under the $i - i$ assignment, all goods can produce positive amounts of net output. Moreover, suppose that the one-to-one assignment chosen is the $i - i$ assignment. *Under some conditions excluding "ties,"* there is a positive world price vector $P = (p_1, p_2, \ldots, p_n) >> 0$ under which the world production point X^* based on the $i - i$ assignment is uniquely composed and X^* uniquely maximizes the world net revenue problem (P1); i.e., the $i - i$ assignment is a solution of (P1) under the price vector P.

These claims mean the existence of a strict solution to the production assignment problem. The existence of the problem is a very important step even in modern economics, so this chapter adequately addresses it.

5 Conclusion and Future Difficulty

Shiozawa's (2007) results present very useful implications for this chapter because, at least in modern economics, the result may become a final meaningful result. In this area, the uniqueness of the production assignment problem has already been revealed by Higashida (2005b). The Hawkins–Simon theorem, as used by Jones (1961), cannot be applied here; thus, the extended Hawkins–Simon theorem is required (in the linear complementarity problem). The necessary and sufficient conditions do not exist in the linear complementarity problem, at least in orthodox textbooks such as Cottle et al. (1992). Moreover, from Ogawa (2014), only the

[20] Some words are complemented.

way to compare the $i-i$ assignment to another assignment cannot show the non-efficiency, at least in the viewpoint of marginal (or average) net revenue. Moreover, to formulate the theorem, Helly's theorem in convex set theory (see, e.g., Ludwig et al. 1963) is needed as in Shiozawa (2015), but because by-product or intermediate input is included, the conditional forms may seem unnatural compared with Jones' (1961) inequality. The formulated theorem can be written in mathematical formulation but is unnatural.

Therefore, Shiozawa's (2007) results may have become essentially the "final" result in the area because after Shiozawa (2007), the meaningful large results may not appear in the future, at least in modern economics. Shiozawa (2014) has offered very useful implications in evolutionary economics; however, its results are less meaningful in modern economics and do not overcome Shiozawa's (2007) result. The main reason for this is that while the Shiozawa's stress is on evolutionary economics, the author continues to research "evolution" in the area in modern economics.

Acknowledgment First, the author is indebted to Makoto Tawada. The author also thanks Makoto Hanazono, Ryuhei Okumura, Makoto Okamura, Yukio Karasawa-Ohtashiro, Keisaku Higashida, Hiroshi Ohta, Toshihiro Ichida, and Ken Urai for their useful comments. The author acknowledges the organizers of the presentations for JEEA 2009 Spring (Nagoya University), JSRSAI 2009 (Hiroshima University), JSRSAI 2010 (GRIPS), NIESG-IDE/JETRO Joint Workshop 2013 Bangkok, and JSME Kinki 2014 Summer (Osaka University). In particular, the author acknowledges the support of JSRSAI. Moreover, the author appreciates Yoshinori Shiozawa, Toshihiro Oka, Taichi Tabuchi, and so on for inviting the author, who has a background in modern economics, at the Workshop on the Theory of International Values, which focuses on evolutionary economics. The authors would like to thank Enago (www.enago.jp) for the English language review. This work was financially supported by JSPS KAKENHI Grant-in-Aid for Young Scientists (B) Number JP24730206 and JP15K21547. The author is solely responsible for any errors.

References

Amano, A. (1966). Intermediate goods and the theory of comparative advantage: A two-country, three-commodity case. *Weltwirtschaftliches Archiv, 96*, 340–345.

Becker, G. S. (1952). A note on multi-country trade. *American Economic Review, 42*(4), 558–568.

Chipman, J. S. (1965). A survey of the theory of international trade: Part 1, The classical theory. *Econometrica, 33*(3), 477–519.

Cottle, R. W., Jong-Shi, P., & Stone, R. E. (1992). *The linear complementarity problem*. Philadelphia: SIAM. p. 148.

Deardorff, A. V. (2005a). Ricardian comparative advantage with intermediate inputs. *North American Journal of Economics and Finance, 16*(1), 16–34.

Deardorff, A. V. (2005b). How robust is comparative advantage? *Review of International Economics, 13*(5), 1004–1016.

Gale, D. (1960). *The theory of linear economic models*. Chicago: The University of Chicago Press.

Graham, F. D. (1923). The theory of international values re-examined. *Quarterly Journal of Economics, 38*(1), 54–86.

Graham, F. D. (1932). The theory of international values. *Quarterly Journal of Economics, 46*(4), 581–616.

Graham, F. D. (1948). *The theory of international values*. Princeton: Princeton University Press.
Haberler, G. (1936) *Theory of international trade*. German edition: Haberler, Gottfried (1933) (trans: A. Stonier and F. Benham). London: William Hodge and Co.
Hawkins, D., & Simon, H. A. (1949). Some conditions of macroeconomic stability. *Econometrica, 17*(3), 245–248.
Higashida, K. (2005a). Survey of chapter 5: Geometric analysis of many-country, many- good ricardo model (in Japanese). In I. Jota & F. Taiji (Eds.), *Development of international trade theory (Kokusai Boeki Riron no Tenkai)* (pp. 243–247).
Higashida, K. (2005b). Intermediate goods and international production specialization patterns: A multi-country multi-good model (in Japanese). In I. Jota & F. Taiji (Eds.), *Development of International Trade Theory (Kokusai Boeki Riron no Tenkai)* (pp. 289–302). Chapter 17
Hosoda, E. (2007). *Environmental restriction and reproducibility of an economy* (in Japanese). Tokyo: Keio University Press.
Ikema, M. (1993). Determination of the patterns of specialization in international production: A multi-country, multi-commodity case (in Japanese). *Hitotsubashi Ronso, 110*, 873–894.
Ilia, I., Grigory, M., & Eugenii, S. (2009). *Tropical algebraic geometry* (Second ed.). Germany: Birkhauser.
Jonathan, E., & Samuel, K. (2002). Technology, geography, and trade. *Econometrica, 70*(5), 1741–1779.
Jones, R. W. (1961). Comparative advantage and the theory of tariffs: A multi-country, multi-commodity model. *The Review of Economic Studies, 28*(3), 161–175.
Jones, R. W., & Neary, J. P. (1984). The positive theory of international trade. In R. W. Jones & P. B. Kenen (Eds.), *Handbook of international economics volume 1* (pp. 1–62). Amsterdam: North-Holland. Chapter 1
Kainou, K. (2012). *Description of the general energy statistics* (in Japanese). RIETI, available at http://www.rieti.go.jp/users/kainou-kazunari/download/index.html. Accessed on 21 April 2013.
Koopmans, T. C. (1951). Analysis of production as an efficient combination of activities. In T. C. Koopmans (Ed.), *Activity analysis of production and allocation* (pp. 33–97). New York: Wiley.
Kuhn, H. W. (1968). Lectures on mathematical economics. In *Mathematics of the decision science part 2*. Providence: American Mathematical Society.
Leontief, W. W. (1936). Quantitative input and output relations in the economic systems of the united states. *The Review of Economic Statistics, 18*(3), 105–125.
Ludwig, D., Branko, G., & Victor, K. (1963). Helly's theorem and it's relatives. In K. Victor (Ed.), *Convexity: Proceedings of symposia in pure mathematics* (Vol. 7, pp. 101–177).
McKenzie, L. W. (1954a). Specialization and efficiency in world production. *The Review of Economic Studies, 21*(3), 165–180.
McKenzie, L. W. (1954b). On equilibrium in Graham's model of world trade and other competitive systems. *Econometrica, 22*(2), 147–161.
McKenzie, L. W. (1955). Specialization in production and the production possibility locus. *The Review of Economic Studies, 23*(1), 56–64.
Metzler, L. A. (1950). Graham's theory of international values. *American Economic Review, 40*(3), 301–322.
Minabe, N. (1971). *Pure theory of international trade*. Tokyo: Kazama Shobo (in Japanese).
Minabe, N. (1995). *Production and international trade*. Osaka: Otemon Gakuin University Press.
Minabe, N. (2001). Many-countries, many-good trade model (in Japanese). In M. Oyama (Ed.), *Horizon of international economics theory* (pp. 221–241). Tokyo: Toyo-Keizai Shinpo-Sha. Chapter 15
Nikaido, H. (1968). *Convex structures and economic theory (Mathematics in Science & Engineering)* (p. 43). New York: Academic Press.
OECD. (2001) *Multifunctionality towards an analytical framework*, from http://www.oecd.org/dataoecd/43/31/1894469.pdf. Accessed on 16 Oct 2010.

Ogawa, T. (2011a). A theoretical analysis of patterns of specialization in the ricardian model with joint prodution and intermediate goods. *Doctoral Thesis* 2010, School of Economics, Nagoya University.
Ogawa, T. (2011b). Analysis of patterns of specialization in the ricardo model with joint production (in Japanese). *Studies in Regional Sciences, 41*(2), 331–344.
Ogawa, T. (2012a). Classification of the frontier in the three-country, three-good ricardian model. *Economics Bulletin, 32*(1), 639–647.
Ogawa, T. (2012b). A consideration of exporting by-products and in a pure exchange economy (in Japanese). *Keizai Kagaku Kenkyu (Hiroshima Shudo University), 16*(1), 27–42.
Ogawa, T. (2013a). Application of Jones' inequality to the *n*-country, *m*-good Ricardo–Graham Model. *Economics Bulletin, 33*(1), 379–387.
Ogawa, T. (2013b). From *n*-country, *n*-good model to *n*-country, *m*-good model (in Japanese). *Keizai Kagaku Kenkyu (Hiroshima Shudo University), 17*(1), 79–89.
Ogawa, T. (2014). Possibility using illustration in Ricardian=Leontief model (in Japanese). *Keizai Kagaku Kenkyu (Hiroshima Shudo University), 18*(1), 57–73.
Ogawa, T. (2015). Three-good Ricardian model with joint production: A schematic reconsideration. In M. Kitahara, & K. Okamura (Eds.), *New solutions in legal informatics, economic sciences and mathematics* (Series of monographs of contemporary social systems solutions; Vol. 6) (pp. 17–46). Chapter 2, printed in Japan, available at http://dx.doi.org/10.2139/ssrn.2242057. Accessed on 9 July 2016.
Oka, T. (2015). The meaning and task of Ricardo=Sraffa=Shiozawa trade model (in Japanese). *Research of Economics and Management in Fukui Prefectural University, 32*, 15–32.
Rudiger, D., Stanley, F., & Samuelson, P. A. (1977). Comparative advantage, trade, and payments in a ricardian model with a continuum of goods. *American Economic Review, 67*(5), 823–839.
Schumann, J., & Todt, H. (1957). Classical theory, Graham's theory, and linear programming in international trade: Comment. *Quarterly Journal of Economics, 71*(3), 464–469.
Shiozawa, Y. (2007). A new construction of Ricardian trade theory—A many-country, many-commodity case with intermediate goods and choice of production techniques—. *Evolutionary and Institutional Economics Review, 3*, 141–187.
Shiozawa, Y. (2014) *A final solution of Ricardo problem of international values* (in Japanese). Tokyo: Iwanami Shoten.
Shiozawa, Y. (2015). International trade theory and exotic algebra. *Evolutionray and Instutional Economics Review, 12*, 177–212.
Shiozawa, Y., & Aruga, Y. (2014). *Rebuilding economics –Evolutional economics and classical value theory- (Keizaigaku wo Saikensuru –Shinka Keizaigaku to Koten-Ha Kachi-ron-)* (in Japanese). Hachiōji: Chuo University Press.
Sraffa, P. (1960). *Production of commodities by means of commodities: Prelude to a critique of economic theory*. Cambridge: Cambridge University Press.
Su, F. E. (1999). Rental harmony: Sperner's Lemma in fair division. *The American Mathematical Monthly, 106*(10), 930–942.
Takamasu, A. (1991). *Neo-Ricardian trade theory –Overcoming non-equivalent trade theory- (Neo-Ricardian no Boueki Riron –Futouka-Koukan-Ron wo Koete-)* (in Japanese). Japan: Soubun-Sha.
Takayama, A. (1972). *International trade*. New York: Holt, Rinehart and Whinston, Inc..
Uekawa, Y. (1984). Some theorem of trade with joint production. *Journal of International Economics, 16*, 319–333.
Viner, J. (1937). *Studies in the theory of international trade*. London: George Allen & Unwin.
Whitin, T. M. (1953). Classical theory, Graham's theory, and linear programming in international trade. *The Quarterly Journal of Economics, 67*(4), 520–544.

The Neo-Ricardian Trade Theory and the New Theory of International Values

Akira Takamasu

Abstract Ricardo's (On the principles of political economy, and taxation, 1817) theory of comparative advantage is the first rigorous theory that demonstrates that free trade benefits every country. He explained his theory using a numerical example of two countries and two commodities. However, the fact that the theory cannot be true when we expand his model to the multicountry and multicommodity case, or to the model that assumes intermediate goods, became clear. Following the study by Graham (Q J Econ 46:581–616, 1932) and McKenzie (Rev Econ Studies 21: 165–180, 1954), the neo-Ricardian theories of international trade as developed by Steedman (Fundamental issues in trade theory, Macmillan, London, 1979) reconsidered gains from trade and showed the possibility of losses from free trade. Recently, Shiozawa (Evol Inst Econ Rev 3: 141–187, 2007) indicated the differences in the number of countries and goods and analyzed cases in which prices did not depend on demand but were determined by production cost. This chapter surveys the development of trade theories and analyzes the gains from trade using the most generalized model. Furthermore, it also considers how the new theory of international values proposed by Shiozawa (Evol Inst Econ Rev 3: 141–187, 2007) provides a new horizon to the previous results.

Keywords Neo-Ricardian • Trade theory • Gains from trade • Sraffa • New theory of international values

1 Introduction

Ricardo's (1817) comparative advantage theory is considered to be one of the few theories that is accepted as a "correct theory" by almost all schools in economics. The simple and clear conclusion of this theory is as follows. First, in free trade, every country has at least one commodity that can be produced at a lower price than that of its trade partners. Second, every country achieves gains from trade by

A. Takamasu (✉)
Faculty of Sociology, Kansai University, 3-3-35 Yamate-cho, Suita-shi, Osaka, 564-8680, Japan
e-mail: akira@takamasu.net

Y. Shiozawa et al. (eds.), *A New Construction of Ricardian Theory of International Values*, Evolutionary Economics and Social Complexity Science 7,
DOI 10.1007/978-981-10-0191-8_5

specializing in producing and exporting these lower-price commodities. Until now, this theory has been a basic doctrine to support the free trade policy.

However, comparative advantage theory has some difficulties in both its theoretical ground and its applicability to the real world. As a pure theory, it has two problems: (1) international prices cannot be determined inside its system and (2) the theory crucially depends on the assumptions of two countries, two commodities, and no intermediate products. The theory also faces problems in the real world: as world trade expands, the income gap between developed and developing countries widens, and developing countries seem to suffer losses from trade, which is inconsistent with the conclusion of the theory.

Numerous studies have been conducted on such problems.[1] Some problems were solved and others were verified as unsolvable. At present, problems still exist that remain unsolved. Thus, this chapter systematically explains the development of Ricardo's (1817) theory using the most generalized model. Furthermore, it summarizes what is proven and what is not.

New claims on Ricardo's (1817) theory have recently emerged. Among these, Shiozawa (2007) indicated that the number of commodities is much larger than the number of countries and that the same commodity can be produced in many countries. In this case, the international price is not influenced by the world demand for that commodity. This assertion becomes known as "the theory of new international values." In this book, the meanings and the development of this theory are discussed in various ways. This chapter examines the theoretical meaning of the theory.

The composition of this chapter is as follows. Section 2 presents the formulation of Ricardo's (1817) comparative advantage theory through a two-country and two-commodity model and verifies Ricardo's assertions.

Section 3 introduces a utility function into Ricardo's (1817) basic model and demonstrates how Mill (1852) solved the problem of determining international prices. This section also considers the plausibility that Mill's (1852) "reciprocal demand theory" became the fundamental principle of neoclassical economics, namely, that "price is determined by supply and demand."

Section 4 expands the model to a multicommodity and multicountry case and examines the difficulties that arise in that case. Ricardo's (1817) criterion of comparative advantage was proven to remain true in the two-commodity multicountry case and two-country multicommodity case. However, if we assume that both the number of commodities and number of countries are more than three, this advantage cannot hold true. Further, we confirm a counter example presented by Graham (1932) and discuss its implication.

Section 5 introduces the intermediate goods and considers their effects on Ricardo's (1817) comparative advantage theory. McKenzie (1954) showed that

[1] Chipman (1965) is a survey article on the development of pure theory after Ricardo (1817). Emmanuel (1973) is the most famous book that criticizes the applicability of the comparative advantage theory to the real world.

Ricardo's (1817) criterion does not hold true even in the case of two countries and three commodities. The plausibility of his example is confirmed using an example in which I modify the model presented by Amano (1966).

The neo-Ricardian economic theory assumes that production cannot be completed instantaneously or that the rate of profit is positive. The economic model which has this property was called "the time-phased Ricardian economy" by Samuelson (1975). In that situation, the existence of intermediate goods causes far more difficult problems. Section 6 examines the plausibility of the neo-Ricardian trade theory offered by Steedman (1979), who addressed this situation. They showed the possibility that the comparative advantage in terms of production prices may differ from it in terms of labor values. Furthermore, they showed that in such a situation, some countries may suffer losses from trade.[2] On this point, Smith (1979) presented a counterargument that the equilibrium should satisfy a condition of intertemporal efficiency. We rigorously formulate the generalized model and evaluate the meaning and the limit of the neo-Ricardian trade theory.

Section 7 presumes a model in which the number of commodities is much larger than the number of countries, namely, "the new theory of international values," and considers new findings that can be added by this theory to the traditional trade theories.

The final section summarizes the contents of this paper and provides prospects for the future development of such studies.

2 Ricardo's Comparative Advantage Theory

In this section, instead of presenting a numerical example as Ricardo (1817) did in his book, I formulate a general mathematical model and confirm the correctness of Ricardo's (1817) argument.

We assume that two countries, A and B, produce the same two commodities. The prices of the two commodities in the two countries are given by the following equation:

$$\begin{aligned} 1 &= \left(1 + r^A\right) w^A l_1^A & \qquad 1 &= \left(1 + r^B\right) w^B l_1^B \\ p^A &= \left(1 + r^A\right) w^A l_2^A & \qquad p^B &= \left(1 + r^B\right) w^B l_2^B. \end{aligned} \tag{1}$$

Here, p^h indicates the price of commodity 2 in terms of commodity 1, r^h indicates the profit rate, w^h indicates the wage rate, and l_j^h indicates the labor input coefficient of the jth commodity in country h $(h = A, B)$.

[2] Important articles are collected in the study by Steedman (1979).

Let us assume that commodity 1 is relatively cheaper in country A than in country B. Thus, we have

$$p^A > p^B. \tag{2}$$

In this case, Ricardo's (1817) principle teaches us that country A specializes in producing commodity 1 and that country B specializes in producing commodity 2. If two countries specialize in such a manner, the international prices of the two commodities are given by Eq. (3).

$$\begin{aligned} 1 &= \left(1 + r^{AT}\right) w^{AT} l_1^A \\ p^T &= \left(1 + r^{BT}\right) w^{BT} l_1^B \end{aligned} \tag{3}$$

In this equation, superscript T indicates that the variable is in a free trade situation. The production in each country that is conducted in this specialization pattern indicates that the unused production processes are not profitable. Moreover, in these production processes, the production cost evaluated by the rate of profit, wage rate, and international prices under free trade exceeds its international price. Thus, we have

$$\begin{aligned} p^T &< \left(1 + r^{AT}\right) w^{AT} l_2^A \\ 1 &< \left(1 + r^{BT}\right) w^{BT} l_1^B. \end{aligned} \tag{4}$$

From Eqs. (1), (2), (3), and (4), we have the following relations:

$$p^A = \frac{l_2^A}{l_1^A} > p^T > \frac{l_2^B}{l_1^B} = p^B \tag{5}$$

Thus, the price of commodity 2 in terms of commodity 1 under free trade p^T must be determined between the prices in the two countries in an autarky. In other word, the international price should be determined in Ricardo's limbo.

In the next step, using this relation, we show that trade certainly brings gains to both countries and that free trade can create an efficient production pattern in the world. Let L^A and L^B denote labor endowment in countries A and B, respectively. Then, in an autarky, the quantities of produced commodities in each country should satisfy the following labor constraints:

$$\begin{aligned} l_1^A X_1^A + l_2^A X_2^A &\leqq L^A \\ l_1^B X_1^B + l_2^B X_2^B &\leqq L^B \end{aligned} \tag{6}$$

Here, X_j^h indicates the production of the jth commodity in country h ($h = A, B$). These labor constraints are shown in Figs. 1a and 1b. In Figs. 1a and 1b, the solid lines denote the inequalities (6), and the southwest area of the solid line represents the production possibility set in each country. We easily understand

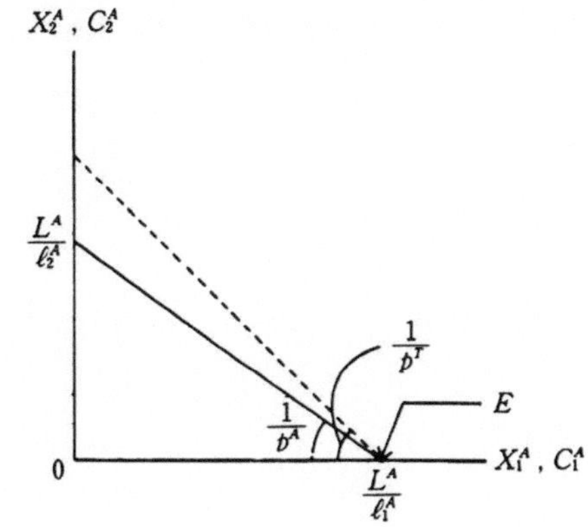

Fig. 1a Production possibility set and consumption possibility set in country *A*

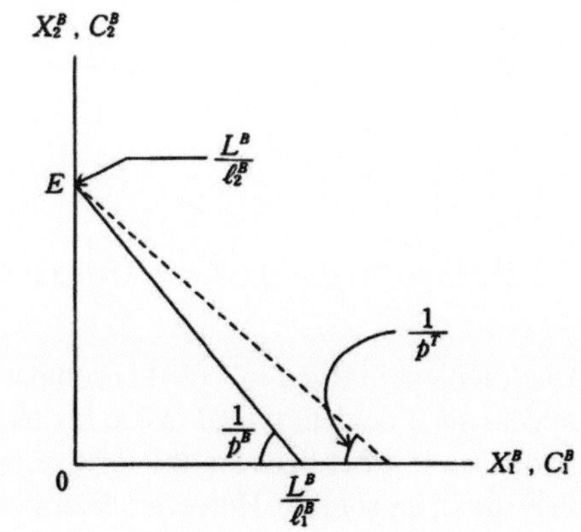

Fig. 1b Production possibility set and consumption possibility set in country *B*

that the inclination of the line equals the relative price of the two commodities in each country. From these two figures, the world production possibility set can be depicted, as shown in Fig. 2.

When both countries open trade, country *A* specializes in producing commodity 1 and country *B* specializes in producing commodity 2. The production of country *A* is depicted at point *E* in Fig. 1a, and the production of country *B* is at point *E* in Fig. 1b. The combination of the production in countries *A* and *B* is depicted at point *E* in Fig. 2. We see that point *E* is situated northeast of point *F*, where country *A* specializes in commodity 2 and country *B* in commodity 1. Thus, we confirm that production is efficiently conducted under free trade. We also see that international price p^T exists between p^A and p^B, and the consumption possibility set—the southeast area of the dotted line—is larger than the set before trade in both countries and represents the gains from trade. Thus, in Ricardo's (1817) model, when capitalists specialize in production to maximize their profits, an efficient production is realized in the world, and both countries gain from trade in the sense that they consume more commodities.

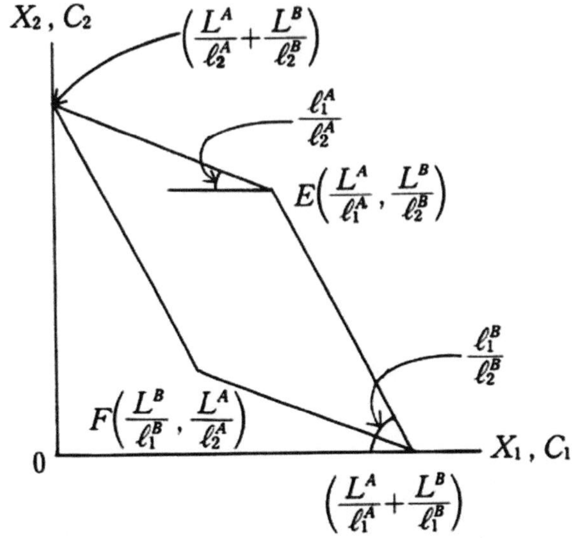

Fig. 2 Production possibility set in the world

3 Determination of the International Price by Mill

One problem in Ricardo's (1817) comparative advantage theory is that the terms of trade cannot be determined inside his model. In a free trade situation, the terms of trade should exist between the relative price in country *A* and in country *B*, when they are in an autarky. However, determining the definite level of the terms of trade is impossible.

To determine the terms of trade, we should specify the demands of both countries on the two commodities. Thus, we should introduce the demand functions to Ricardo's (1817) analysis. The economist who first made such an introduction was Mill (1852). In Section 6–8, which was added in the third edition of chapter XVIII of *Principle*, Mill (1852) assumed a demand function in a specific form and showed how international prices and consumption of both countries are determined. Using the terminology of modern economics, Mill (1852) can be said to have formulated Ricardo's (1817) model as a general equilibrium model and derived the solutions. In this section, we reformulate Mill's (1852) analysis by expanding his model to a more generalized one and consider the plausibility of his analysis.

First, let us briefly explain Mill's (1852) demand functions. Mill (1852) provided the following explanation in his book:

> As the simplest and most convenient, let us suppose that in both countries any given increase of cheapness produces an exactly proportional increase of consumption or, in other words, that the value expended in the commodity, the cost incurred for the sake of obtaining it, is always the same, whether that cost affords a greater or a smaller quantity of the commodity. Mill (1965, p. 609)

In other words, Mill assumes that the elasticity of demand with price is equal to one and the cross-elasticity of demand is equal to zero. The necessary and sufficient condition for the demand function having this property is that the utility function is a type of Eq. (7).

$$U = \varnothing \left(C_1^{\alpha} C_2^{1-\alpha}\right) \tag{7}$$

Next, let us see how the terms of trade is determined in the model that assumes this type of demand function. We assume that the supply side of the model is the same as in the previous model. Therefore, Eq. (1) is on hold in an autarky. In this case, country A specializes in producing commodity 1, and country B specializes in producing commodity 2.

Let us assume that the demand function in country A is

$$U^A = \left(C_1^A\right)^{\alpha}\left(C_1^A\right)^{1-\alpha} \tag{8}$$

Here, $1 > \alpha > 0$ is a parameter. Then, country A faces the following maximization problem.

$$\begin{aligned} \text{Max.} \quad & U^A = \left(C_1^A\right)^{\alpha}\left(C_1^A\right)^{1-\alpha} \\ \text{s.t.} \quad & C_1^A + p^T C_2^A \leqq \tfrac{L^A}{l_1^A} = X_1^A \end{aligned} \tag{9}$$

From the necessary condition of the maximization problem, we have

$$\alpha p^T C_2^A = (1-\alpha)\, C_1^A \tag{10}$$

If we assume that the utility function of country B is given by Eq. (11),

$$U^B = \left(C_1^B\right)^{\beta}\left(C_1^B\right)^{1-\beta} \tag{11}$$

Here, $1 > \beta > 0$ is a parameter. Then, we have the following equation.

$$\beta p^T C_2^B = (1-\beta)\, C_1^B \tag{12}$$

From Eqs. (10) and (12) and the budget constraints of country A and B, we have the following demand and supply equalities.

$$\begin{aligned} C_1^A + p^T C_2^A &= \tfrac{L^A}{l_1^A} = X_1^A \\ C_2^B + p^T C_2^B &= \tfrac{p^T L^B}{l_2^B} = p^T X_2^B \end{aligned} \tag{13}$$

Because the quantities of commodity supply are determined by the labor endowments of both countries and labor coefficients, we have

$$\begin{aligned} C_1^A + C_1^B &= \frac{L^A}{l_1^A} \\ C_2^A + C_2^B &= \frac{L^B}{l_2^B}. \end{aligned} \tag{14}$$

Thus, the international price is

$$p^T = \frac{(1-\alpha)\, l_2^B L^A}{\beta l_1^A L^B} \tag{15}$$

and the production and consumption of two commodities are given by

$$\begin{array}{ll} C_1^A = \alpha L^A / l_1^A & C_1^B = (1-\alpha)\, L^A / l_1^A \\ C_2^A = \beta L^B / l_2^B & C_2^B = (1-\beta)\, L^B / l_2^B \\ X_1^A = L^A / l_1^A & X_1^B = 0 \\ X_2^A = 0 & X_2^B = L^B / l_2^B \end{array} \tag{16}$$

Thus, we determine the terms of trade and the consumption of the two commodities in two countries in the model to which we additionally introduce the utility functions.

However, depending on the demand volumes, the terms of trade might not be left in limbo. Before analyzing this case, we first return to Mill's (1852) *Principle* and see how Mill (1852) treated this case. Mill stated the following:

> Let it be supposed that in England 100 yards of cloth, previously to the trade, exchanged for 100 of linen, but that in Germany 100 of cloth exchanged for 200 of linen. When the trade was opened, England would supply cloth to Germany, Germany linen to England. Mill (1965, p. 609)

Thus, our model completely accords with the example in Mill's (1852) *Principle* if we change names as follows: from country A to England, country B to Germany, commodity 1 to cloth, and commodity 2 to linen. We also assume the price in an autarky $p^A = 1$ and $p^B = 1/2$.

In this case, England has a comparative advantage in cloth and specializes in producing it. Germany specializes in linen. As was previously shown, if the demand function has the property that Mill (1852) assumes, the proportion of total income spend on the consumption of each commodity in each country is constant after opening trade. Thus, the quantity of consumption of the commodity for which England and Germany specialize in production after establishing trade is the same as the consumption before trade. If demand equals supply for two commodities, the quantity of the commodity produced in England and that is not consumed in the domestic market is exchanged for the quantity of the commodity produced in Germany and that is not consumed in the country as an equivalent value. Mill stated the following:

> Let the quantity of cloth which England can make with the labor and capital withdrawn from the production of linen, be $= n$. Let the cloth previously required by Germany (at the German cost of production) be $= m$.

> Then n of cloth will always exchange for exactly $2m$ of linen.
> Mill (1965, p. 611)

If England specializes in producing cloth, the production of cloth is $2n$ and the quantity of cloth not consumed in England is n. Similarly, because the price of linen in terms of cloth in Germany is 1/2, the quantity of linen that can be produced with the labor withdrawn from the production of linen is 2 m. Thus, n of cloth always exchanges for 2 m of linen. In this case, can England and Germany gain from the benefits of trade? Mill stated the following:

> If $n = 2m$, the whole advantage will be on the side of Germany.
> If n be greater than m, but less than $2m$, the two countries will share the advantage; England getting $2m$ of linen where she before got only n; Germany getting n of cloth where she before got only m.
> Mill (1965, p. 611)

We can easily confirm the resembling result in our generalized Mill model. If we assume

$$p^B > \frac{(1-\alpha)\, l_2^B L^A}{\beta l_1^A L^B}, \tag{17}$$

then $p^T = p^B$, and country A specializes in producing commodity 1, and country B produces both commodities. The consumption and production of two commodities in two countries are given by

$$\begin{array}{ll} C_1^A = \alpha L^A / l_1^A & C_1^B = \beta L^B / l_1^B \\ C_2^A = (1-\alpha)\, l_1^B L^A / l_1^A l_2^B & C_2^B = (1-\beta)\, L^B / l_2^B \\ X_1^A = L^A / l_1^A & X_1^B = \beta L^B / l_1^B - (1-\alpha)\, L^A / l_1^A \\ X_2^A = 0 & X_2^B = (1-\alpha)\, l_1^B L^A / l_2^B l_1^A + (1-\beta)\, L^B / l_2^B \end{array} \tag{18}$$

Here, $X_1^B = C_1^A + C_1^B - X_1^A$, which is positive from (16). Thus, if the sum of the demand for commodity 1 in two countries is larger than the production of commodity 1 in country A—in other words, if country A is a relatively small country—this situation might happen, and all of the benefits of trade will be on the side of country A.

As we have seen so far, Mill's (1852) analysis is almost perfect as a general equilibrium analysis although its defect is using a specific type of demand function. We cannot criticize his analysis even from the viewpoint of contemporary economics. Thus, we can state that his analysis was ahead of his time or was too advanced. For precisely that reason, Chipman (1965) stated in his article that Mill's contribution was not correctly understood for a long time.

Mill seemed to lead economics from the classical price theory, which states that prices are determined by the production cost of a commodity to the neoclassical theory, which states that prices are determined by an equilibrium between a commodity's demand and supply. If two countries specialize in producing a commodity for which a country has a comparative advantage under free trade, then prices are certainly not proportional to labor input and depend on demand. In this case, the neoclassical approach seems more appropriate than the classical approach for explaining price determination. In that sense, we state that Mill killed classical economics, and Chipman (1965) and Negishi (1981, 1983) evaluated this point.

However, we should be careful about this subject. First, in the case of incomplete specialization, prices are in accord with production costs in a large country, and classical economics is restored. Second, is classical economics defined as the economic doctrine that assumes that prices are independent of demand or that price is proportional to its labor value correct? The dependency of prices on demand also occurs when we consider rent.[3] However, should we really believe that dependency means the end of classical economics and the rise of neoclassical economics? When we regard classical economics in a broader context and define it as economics that stresses the importance of analyzing the economy from the viewpoint of reproducibility, the most important point in Ricardo's analyses should be considered to be the existence of intermediate goods rather than the dependence of prices on demand. However, this point had not been considered after Ricardo until McKenzie (1954) and Jones (1961) analyzed it using the modern linear programming method in the 1950s.

4 Many Commodities and Many Countries

Another problem of comparative advantage theory is that this theory crucially depends on three assumptions, namely, two countries, two commodities, and the nonexistence of intermediate goods. What types of difficulties arise if we relax these assumptions and generalize the theory to the situation of a multicommodity, a multicountry, and the existence of intermediate goods?

It is confirmed that the theory is robust if we increase only the number of countries from two to many, assuming that the number of commodities is two. The theory is also robust if we increase only the number of commodities.

However, if we increase both of the number of countries and commodities, a difficulty arises. Let us consider an example. We assume that three countries, *A*, *B*, and *C*, produce three types of commodities before trade. The necessary labor input to produce a unit of each commodity and the labor endowment in the three countries are shown in Table 1. In Table 1, l_j indicates a labor input to produce a unit of the *j*th commodity and *L* indicates a labor endowment.

In this example, which production specialization pattern satisfies the principle of Ricardo's (1817) comparative advantage theory and which pattern is efficient? First, let us consider the pattern ○, which indicates that country *A* specializes in producing commodity 2, country *B* specializes in commodity 1, and country *C* specializes in commodity 3. As is easily confirmed, the pattern ○ satisfies the standard of

[3] See, for example, Montani (1975), Kurz (1978), and Takamasu (1983).

Table 1 Counterexample to Ricardo, three countries, three commodities case

	Country A	Country B	Country C
l_1	•100	∘100	100
l_2	∘ 50	70	• 30
l_3	40	• 30	∘ 20
L	4500	4500	3000

Ricardo's (1817) comparative advantage theory because every country specializes in producing the commodity with a comparative advantage for any pair of two countries and two commodities. For example, when we check for countries A and B and commodities 1 and 2, we have inequality (19) and country A has a comparative advantage in commodity 2.

$$\frac{50}{100} = \frac{l_2^A}{l_1^A} < \frac{l_2^B}{l_1^B} = \frac{70}{100} \tag{19}$$

This statement is also true for the country B and country C pair, and for the country A and country C pair. Thus, the pattern ∘ is consistent with Ricardo's (1817) standard in comparative advantage theory.

However, this production pattern cannot be compatible with a competitive equilibrium. Let us show the incompatibility. As was seen in the section that explains Ricardo's (1817) comparative advantage theory, for unused production processes, the cost of producing a unit of the commodity measured using current prices, the wage rate, and the profit rate exceeds the price. In contrast, the cost equals the price for the actually operating production process. Thus, we have the following inequalities and equalities for three commodities.

$$\begin{array}{lll} p_1^T < \left(1 + r^{AT}\right) 100 w^{AT} & p_1^T = \left(1 + r^{BT}\right) 100 w^{BT} & p_1^T < \left(1 + r^{CT}\right) 100 w^{CT} \\ p_2^T = \left(1 + r^{AT}\right) 50 w^{AT} & p_2^T < \left(1 + r^{BT}\right) 70 w^{BT} & p_1^T < \left(1 + r^{CT}\right) 30 w^{CT} \\ p_1^T < \left(1 + r^{AT}\right) 40 w^{AT} & p_1^T < \left(1 + r^{BT}\right) 30 w^{BT} & p_1^T = \left(1 + r^{CT}\right) 20 w^{CT} \end{array} \tag{20}$$

By eliminating the profit rate and the wage rate in Eq. (20), we have Eq. (21).

$$\begin{array}{lll} p_1^T < \frac{100}{50} p_2^T & p_2^T < \frac{70}{100} p_1^T & p_1^T < \frac{100}{20} p_3^T \\ p_3^T < \frac{40}{50} p_2^T & p_3^T < \frac{30}{100} p_1^T & p_2^T < \frac{30}{20} p_3^T \end{array} \tag{21}$$

When we start from the upper left of Eq. (21) and use the lower right and lower middle, we have

$$p_1^T < 2 p_2^T < 3 p_3^T < \frac{90}{100} p_1^T \tag{22}$$

Obviously, no positive price exists, and the profit rate and the wage rate satisfy (22).

Table 2 A three-country and three-commodity case in which Ricardo's comparative advantage theory does not hold

	Pattern ∘	Pattern •
Commodity 1	45	45
Commodity 2	90	100
Commodity 3	150	150

The production pattern ∘ can also be verified as not being efficient in the sense that no Pareto-dominant production pattern exists for that pattern. Table 2 shows the outputs of each commodity for the production patterns ∘ and •. The outputs of commodities 1 and 3 in the world are the same, and the output of commodity 2 is larger in the pattern •. Thus, we see that the pattern ∘ is not efficient.

Is the pattern • truly efficient and a competitive equilibrium? McKenzie (1954) and Jones (1961) clarified this point. McKenzie (1954) showed that a competitive equilibrium in free trade is an internationally efficient production pattern. We provide proof of this equivalency in a generalized model in Sect. 6. Before proceeding to the proof, we consider the meaning of intermediate goods in an open economy in Sect. 5. For efficient production patterns, Jones (1961) showed that an efficient production pattern is one that minimizes the product of labor inputs of the produced commodities, such as $l_1^A l_2^B l_3^C$ if the number of countries equals the number of commodities and each country specializes in producing only one commodity.

5 Intermediate Goods

The case in which Ricardo's (1817) comparative advantage theory does not hold also exists in the situation in which we assume intermediate goods. McKenzie (1954) showed this phenomenon in the case of three countries and three commodities. Amano (1966) also showed this phenomenon in the case of two countries and three commodities.

Following Amano (1966), we make an example of two countries and three commodities for which ordering the comparative advantage in an autarky and in free trade does not accord. The method for providing an explanation is slightly different from that of Amano (1966).

Let us assume that two countries, country A and B, exist, and both countries produce three commodities. The production technique of both countries is assumed to be as follows.

Country A

$$\begin{array}{llll} a_{11}^A = 0 & a_{21}^A = 0 & a_{31}^A = 0 & l_1^A = 100 \\ a_{12}^A = 0 & a_{22}^A = 0 & a_{32}^A = 0.8 & l_2^A = 50 \\ a_{13}^A = 0 & a_{23}^A = 0 & a_{33}^A = 0 & l_3^A = 200 \end{array}$$

Table 3 Labor directly or indirectly required to produce a commodity, namely, labor value in countries *A* and *B*

	Country *A*	Country *B*
v_1	100	100
v_2	210	140
v_3	200	100

Country B

$$\begin{array}{llll} a_{11}^B = 0 & a_{21}^B = 0 & a_{31}^B = 0 & l_1^B = 100 \\ a_{12}^B = 0 & a_{22}^B = 0 & a_{32}^B = 0.4 & l_2^B = 100 \\ a_{13}^B = 0 & a_{23}^B = 0 & a_{33}^B = 0 & l_3^B = 100 \end{array}$$

Here, $a_{ij}^h(h = A, B)$ is the quantity of the *i*th commodity required to produce one unit of the *j*th commodity, and l_j^h is the labor input required to produce one unit of the *j*th commodity in country *h*. When we assume that the profit rate in countries *A* and *B* equals zero, prices in countries *A* and *B* are calculated as in Eq. (23).

$$\begin{array}{ll} p_1^A = 100w^A & p_1^B = 100w^B \\ p_2^A = 0.8p_3^A + 50w^A & p_2^B = 0.4p_3^B + 100w^B \\ p_3^A = 200w^A & p_3^B = 100w^B \end{array} \tag{23}$$

The quantity of labor directly and indirectly required to produce a unit of a commodity can be calculated, as shown in Table 3. In Table 3, v_j indicates the labor input required to produce one unit of each commodity or labor value.

Because the commodity price is proportional to the labor value if the profit rate is zero, we have the following relationships.

$$\frac{p_1^B}{p_1^A} > \frac{p_2^B}{p_2^A} > \frac{p_3^B}{p_3^A}$$

Hence, country *A* should have a comparative advantage against country *B* in the order of commodity 1, commodity 2, and commodity 3. Thus, in free trade, country *A* must specialize in producing commodity 1.

However, we easily show that the production specialization pattern for which country *A* specializes in commodity 1 and country *B* in commodities 2 and 3 cannot be compatible with a competitive equilibrium. As was previously shown, in a competitive equilibrium, we have the following equalities and inequalities.

$$\begin{array}{ll} p_1^T = 100w^{AT} & p_1^T < 100w^{BT} \\ p_2^T < 0.8p_3^T + 50w^{AT} & p_2^T = 0.4p_3^T + 100w^{BT} \\ p_3^T < 200w^{AT} & p_3^T = 100w^{BT} \end{array} \tag{24}$$

Substituting the middle left of Eq. (24) for the upper left and lower right, we have

$$p_2^T < 80w^{BT} + 0.5p_1^T$$

Considering the upper right of Eq. (24), we have

$$p_2^T < 130w^{BT} \tag{25}$$

However, from the middle right and lower right of Eq. (24), we have

$$p_2^T = 140w^{BT}$$

which is inconsistent with Eq. (25). Thus, no nonnegative prices enable this production specialization pattern.

In addition, we show that the same phenomena occur even in the case of two countries and two commodities if the rate of profit is positive. Thus, if we assume the intermediate goods, the order of a comparative advantage in an autarky does not coincide in general with the order in free trade.

6 Intermediate Goods and a Positive Profit Rate: Critique by the Neo-Ricardian

In the analysis of Sect. 4, the assumption is that no intermediate goods are required to produce a commodity and the production period is the same for every commodity. Thus, the labor hours required directly or indirectly to produce one unit of a commodity, namely, labor value, equal the production price for every commodity. In Sect. 5, we introduce intermediate goods. However, because we assume that the rate of profit is zero, the labor value or labor required directly or indirectly to produce a unit of commodity still equals its price.

When we assume that the production periods differ from each other, or assume that intermediate goods are required to produce commodities and the rate of profit is positive, the labor value is not proportional to its price. In that case, ordering the comparative advantage in terms of the production price could not be in accord with ordering in terms of the labor value.

In that case, can every country still gain benefits from trade? This situation was analyzed by the neo-Ricardian trade theory. Instead of assuming intermediate goods, Steedman and Metcalfe (1973) presented an example in which the production periods differ from each other and the ordering of the comparative advantage in terms of price and labor values is different. In contrast, Takamasu (1991, pp. 44–49) assumes intermediate production goods and a positive profit rate and presents a similar example. Following Takamasu (1991), we provide an example that has the same property and consider the type of results that will ensue.

Let us assume that country A has the following input coefficients.

Country A

$$a_{11}^A = 0.4 \quad a_{21}^A = 0 \quad l_1^A = 60$$
$$a_{12}^A = 0.2 \quad a_{22}^A = 0 \quad l_2^A = 100$$

In this case, the direct or indirect labor to produce a unit of commodities 1 and 2 in country A can be calculated using the following equations.

$$\begin{aligned} 0.4v_1^A + 60 &= v_1^A \\ 0.2v_1^A + 100 &= v_2^A \end{aligned} \tag{26}$$

Solving this Eq. (26), we have $v_1^A = 100$ and $v_2^A = 120$. These values are the same as in Ricardo's example.

Next, let us calculate the prices of commodities 1 and 2 in country A. When we assume that wages are paid after production, commodity prices can be given by the following equations.

$$\begin{aligned} 0.4\left(1 + r^A\right) + 60w^A &= 1 \\ 0.2\left(1 + r^A\right) + 100w^A &= p^A \end{aligned} \tag{27}$$

As is evident by comparing Eq. (27) with Eq. (26), the price of a commodity is not proportional to the labor value if the rate of profit is positive. When we give $r^A = 1$, we have $p^A = 11/15$.

To make a comparison with the argument in Sect. 2, let us suppose that the labor endowment of country A is 4800 units. Then, we derive the consumption possibility set of country A. Provided that X_1^A and X_2^A denote the gross outputs of commodity 1 and 2 in country A, respectively, then we have the following labor constraint inequality.

$$60X_1^A + 100X_2^A \leqq 4800 \tag{28}$$

Because the net outputs of commodities 1 and 2, Y_1^A and Y_2^A, are the gross outputs minus the inputs for the production in the next period, we have

$$\begin{aligned} Y_1^A &= X_1^A - \left(0.4X_1^A + 0.2X_2^A\right) \\ Y_2^A &= X_2^A \end{aligned} \tag{29}$$

Solving (29) with respect to X_1^A and X_2^A, and by assigning (28), we have

$$100Y_1^A + 120Y_2^A \leqq 4800 \tag{30}$$

Thus, the consumption possibility set is the same as that of Ricardo's (1817) original example.

Then, let us derive the labor values, the commodity prices, and the consumption possibility frontier of country B. Let us assume the production technique of country B as follows.

Country B

$$a_{11}^B = 0.3 \quad a_{21}^B = 0 \quad l_1^B = 63$$
$$a_{12}^B = 0.3 \quad a_{22}^B = 0 \quad l_2^B = 53$$

Then, v_1^B and v_2^B are calculated from (31).

$$\begin{aligned} 0.3v_1^B + 63 &= v_1^B \\ 0.3v_1^B + 53 &= v_2^B \end{aligned} \tag{31}$$

Solving (31), we have $v_1^B = 90$ and $v_2^B = 80$. These values are also the same as in Ricardo's example. The production prices can be calculated from (32).

$$\begin{aligned} 0.3\left(1 + r^B\right) + 63w^B &= 1 \\ 0.3\left(1 + r^B\right) + 53w^B &= p^B \end{aligned} \tag{32}$$

When we give $r^B = 1$ in (32), we have $p^B = 59/63$ and $w^B = 2/315$. The consumption possibility set in country B is given by

$$90Y_1^B + 80Y_2^B \leqq 3600 \tag{33}$$

Thus, excluding the commodity prices, everything is the same as in Ricardo's (1817) numerical example.

Comparing the relative price of commodity 1 in terms of commodity 2 in country A with that in country B, the relative price is smaller in country A than in country B.

$$p^A = \frac{11}{15} < \frac{59}{63} = p^B \tag{34}$$

We note that the direction of the inequality is opposite to the direction of Ricardo's example, in which prices are assumed to be proportional to the labor values.

When the capitalists in country A maximize profits, they specialize in producing commodity 2 and importing commodity 1.

Suppose that the international price p^T is 5/6 ($p^A = \frac{11}{15} < \frac{5}{6} < \frac{59}{63} = p^B$). Then, the sets of consumable commodities when country A specializes in producing commodity 2 and country B specializes in producing commodity 1 can be calculated as follows. Let us calculate for country A first. When country A uses all of its 4800 units of labor to produce commodity 2, the country produces 48 units of commodity 2. To produce 48 units of commodity 2, 48/5 units of commodity 1 are required. Consequently,

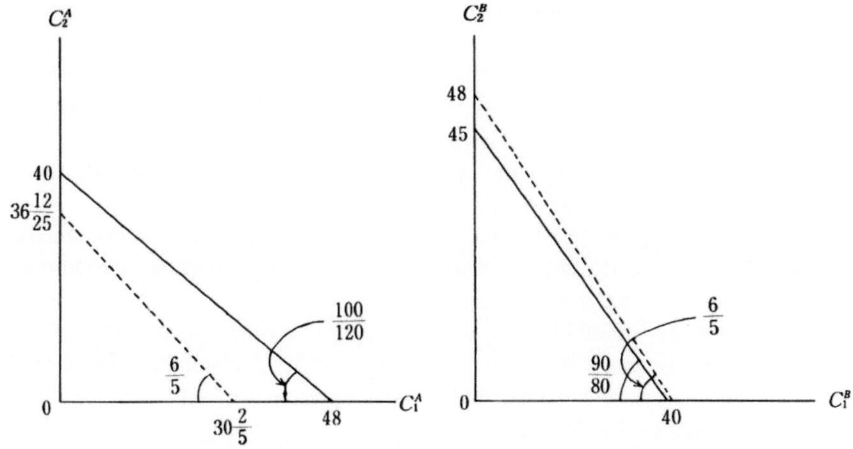

Fig. 3a Consumption possibility set when price is not proportional to labor value ($p^T = 5/6$)

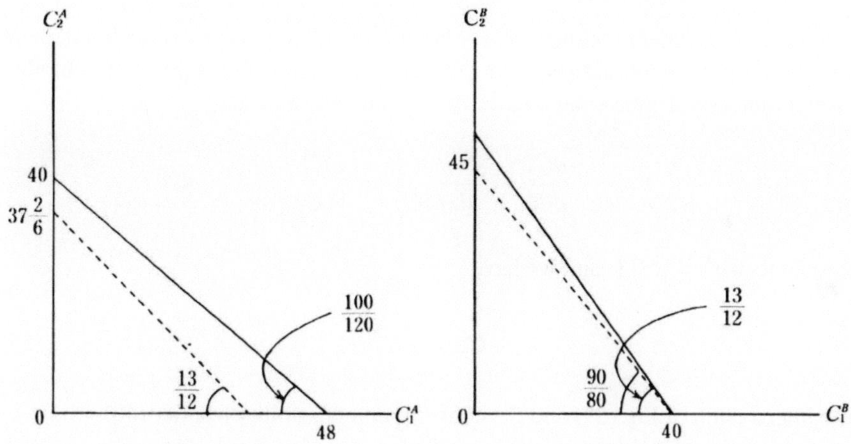

Fig. 3b Consumption possibility set when price is not proportional to labor value ($p^T = 13/122$)

$$C_1^A + \frac{5}{6}C_2^A \leqq 48 \times \frac{5}{6} - \frac{48}{5} = 30\frac{2}{5} \tag{35}$$

is the set of consumable commodities in country A. For country B, the set of consumable commodities in country B can be calculated using (36).

$$C_1^B + \frac{5}{6}C_2^B \leqq 40 \tag{36}$$

As is evident from Fig. 3a, the set of consumable commodities in country A is made smaller by opening trade. If we assume that the international price p^T is 12/13, as shown in Fig. 3b, both consumable sets in countries A and B shrink by opening trade.

Therefore, when price is not proportional to labor value, one or both of two countries is shown as possibly suffering from trade in the sense that some countries shrink their consumption possibility set. Thus, the neo-Ricardian trade theory created fundamental doubt over the benefits of free trade as proven by Ricardo's (1817) comparative advantage theory.

However, Smith (1979) indicated that the transition periods between autarky and trade are not considered in this argument and claimed that the intertemporal optimality can be proven, even in these examples.

We consider this point in our model. To determine the consumption of two commodities, we assume a utility function for a country, as we conducted in Sect. 3. Let us assume that every consumer has the same preference for two commodities, which is characterized by the utility function (37).

$$U = \left(C_1^A\right)^{6/17}\left(C_2^A\right)^{11/17} \tag{37}$$

When each consumer maximizes his or her utility under the budget constraint, he or she purchases commodities such that the marginal rate of substitution of commodity 1 for commodity 2 equals the relative price. Thus, in an autarky,

$$\frac{dC_2^A}{dC_1^A} = \frac{\partial U/\partial C_1^A}{\partial U/\partial C_2^A} = \frac{6C_2^A}{11C_1^A} \tag{38}$$

is equal to $1/p^A = 15/11$, and we have

$$C_2^A = \frac{5}{2}C_1^A \tag{39}$$

The intersection of Eq. (39) and the consumption possibility frontier (40)

$$100C_1^A + 120C_2^A = 4800 \tag{40}$$

is $\left(C_1^A,\ C_2^A\right) = (12, 30)$. Thus, in this combination, consumers maximize their utility and full employment is realized.

Next, let us consider the transition period of country A from autarky to free trade. Capitalists in country A specialize in producing commodity 2. To produce 48 units of commodity 2, 48/5 units of commodity 1 should be prepared. Because the gross outputs of commodities 1 and 2 in an autarky are (30, 30), the quantities of commodities that can be consumed are (102/5, 30). Thus, in a transition period, consumers maximize their utility under constraint (41).

$$C_1^A + p^T C_2^A \leqq \frac{102}{5} + 30p^T \tag{41}$$

If we assume that the international price of commodity 1 in terms of commodity 2, p^T is 5/6, (41) can be rewritten as

$$C_2^A \leqq \frac{6}{5} C_1^A + 54\frac{12}{25} \tag{42}$$

In contrast, because of the marginal rate of substitution of commodity 2 for commodity 1, $-dC_2^A/dC_1^A$ is given by

$$-\frac{dC_2^A}{dC_1^A} = \frac{6C_2^A}{11C_1^A} \tag{43}$$

which equals the international price $1/p^T = 6/5$, and we have

$$C_2^A = \frac{11}{5} C_1^A \tag{44}$$

Solving the equalized form of inequality (42) and equality (44), we have $\left(C_1^A, C_2^A\right) = (16.02, 35.25)$, which is the consumption vector in a transition period.

For the periods after two countries completely transfer to free trade, we have the consumption possibility set

$$C_2^A \leqq -\frac{6}{5} C_1^A + 36\frac{12}{25} \tag{45}$$

From (45) and (44), the consumption vector in free trade is $\left(C_1^A, C_2^A\right) = (10.72, 23.60)$.

The streams of the consumption of two commodities in an autarky and in free trade are shown in Table 4.

Although the consumption of the two commodities in an open economy is smaller than that in an autarky after opening trade, the consumption of both commodities in the transition period is certainly larger than in an autarky. To compare these two consumption streams, let us evaluate the values in terms of the international price $p^T = 5/6$ and use the rate of profit as the discount rate ($r = 1$). Then, the present discounted value of the consumption stream

Table 4 Consumption stream in time-phased Ricardian economy

	0	1	2	3	……
	Before transition	Transition	After transition		
Autarky					
C_1^A	12	12	12	12	……
C_2^A	30	30	30	30	……
Open economy					
C_1^A	12	16.02	10.72	10.72	……
C_2^A	30	35.25	23.60	23.60	……

$$C = \sum_{t=1}^{\infty} \left(C_{1t}^{A} + p^{T} C_{2t}^{A}\right) / (1 + r)^{t} \tag{46}$$

is 37.9 in an open economy and is larger than 37 in an autarky. This intertemporal efficiency of free trade is claimed by neoclassical economists.

7 The New Theory of International Values

In this section, we introduce a basic model of the new theory of international values developed by Shiozawa (2007) and others and show theorems derived from this model. I change some economic notations from Shiozawa's (2007) original ones to the notations used in ordinary Sraffian economics or neo-Ricardian trade theory.

First, production prices in an autarky are given by the next equation, which is a standard Sraffian model.

$$p^{h} = \left(1 + r^{h}\right) p^{h} A^{h} + w^{h} l^{h} \quad h = A, B, C, \ldots, N \tag{47}$$

Here, p^h denotes a price vector in country h, r^h denotes the rate of profit in country h, A^h denotes a commodity input coefficient matrix in country h, w^h denotes the wage rate in country h, and l^h denotes a labor input coefficient vector.

Each country must satisfy the following labor constraint in an autarky and in free trade.

$$l^{h} x^{h} \leqq L^{h} \quad h = A, B, C, \ldots, N \tag{48}$$

Here, x^h denotes the column vector of output in country h, and L^h denotes the labor endowment of country h. When countries open their trade, international prices and the wage rate of each country must satisfy (49).

$$q^{T} = \left(p^{T}, w^{T}\right) = \left(p_1^{T}, p_2^{T}, \cdots, p_n^{T}, w^{TA}, w^{TB}, \cdots, w^{TN}\right)$$
$$q^{T} \begin{bmatrix} I - \left(1 + r^{T}\right) A^{A} & I - \left(1 + r^{T}\right) A^{B} & \cdots & I - \left(1 + r^{T}\right) A^{N} \\ -l^{A} & 0 & \cdots & 0 \\ 0 & -l^{B} & \ddots & \vdots \\ \vdots & \ddots & \ddots & 0 \\ 0 & \cdots & 0 & -l^{N} \end{bmatrix} \leqq 0 \tag{49}$$

Here, p^T denotes the row vector of the international price, r^T denotes the rate of profit, and w^{Th} denotes the wage rate of country h after trade.

The gross output vector $x = (x^A, x^B, \ldots x^N)' = (x_1^A, \ldots, x_n^A, \ldots, x_1^N, \ldots, x_n^N)'$ should satisfy Eq. (50), which means that the price is equal to the cost for the production process actually used, and the cost is higher than the price for the unused production process.

$$\begin{aligned} &p_j^T = (1 + r^T) \textstyle\sum_i a_{ij}^h p_i^T + w_j^{Th} l_j^h \rightarrow x_j^h \geqq 0 \\ &p_j^T < (1 + r^T) \textstyle\sum_i a_{ij}^h p_i^T + w_j^{Th} l_j^h \rightarrow x_j^h = 0 \\ &l^h x^h \leqq L^h \quad h = A, \ldots, N \end{aligned} \tag{50}$$

When q^T and x satisfy (49) and (50), the situation is called a competitive equilibrium in an open economy.

We can prove some theorems for this equilibrium. First, let us check the efficiency of the equilibrium. In an economy in which the rate of profit is positive, we should consider the R-efficient locus as proposed by Mirrlees (1969) instead of the production possibilities frontier. A production vector $\widehat{x}$ is called R-efficient when no x exists that satisfies the labor constraint and (51).

$$\begin{aligned} &\left[I - (1 + r^T) A^A \;\; I - (1 + r^T) A^B \;\; \ldots \;\; I - (1 + r^T) A^N\right] x \\ &\quad \geq \left[\mathrm{I} - (1 + r^T) A^A \; \mathrm{I} - (1 + r^T) A^B \ldots \mathrm{I} - (1 + r^T) A^N\right] \widehat{x} \end{aligned} \tag{51}$$

Here, for the convenience of a subsequent argument, let us define the column vector y by the following equation, which is the time-phased economy version of the net output vector in the world.

$$y = \left[I - (1 + r^T) A^A \; I - (1 + r^T) A^B \cdots I - (1 + r^T) A^N\right] x$$

When we introduce the notion of R-efficient production, we prove the following Theorem 1.

Theorem 1 An equilibrium is an R-efficient production.

Proof Let us assume that $\widehat{x}$ is not an R-efficient production. Then, a gross output vector x exists that satisfies (51). Multiplying p^T to (51) from the left-hand side and deducing $w^T L$, we have

$$\begin{aligned} &p^T \left[I - (1 + r^T) A^A \; I - (1 + r^T) A^B \ldots I - (1 + r^T) A^N\right] x - w^T L \\ &> p^T \left[I - (1 + r^T) A^A \; I - (1 + r^T) A^B \ldots I - (1 + r^T) A^N\right] \widehat{x} - w^T L = 0 \end{aligned} \tag{52}$$

Here, $L = (L^A, \cdots, L^N)$, which is a contradiction of $(p^T, \; w^T)$, and $\widehat{x}$ is an equilibrium.

Next, let us show that a price vector exists that is part of an equilibrium for an R-efficient production.

Theorem 2 Prices and a wage vector q^T exist that are part of an equilibrium and are compatible with the R-efficient production $\widehat{y}$.

For this proof, we need Lemma 1.

Lemma 1[4] For a matrix C, if

$$Cz \geq 0 \qquad z \geqq 0 \tag{53}$$

have no solution, then the following inequalities

$$qC \leqq 0 \qquad q > 0 \tag{54}$$

have a solution.

Proof We first show that the matrix on the left-hand side of (55) has the property of the matrix C in Lemma 1.

$$\begin{bmatrix} -\widehat{y} & I-\left(1+r^T\right)A^A & I-\left(1+r^T\right)A^B & \cdots & I-\left(1+r^T\right)A^N \\ L^A & -l^A & 0 & \cdots & 0 \\ L^B & 0 & -l^B & \ddots & \vdots \\ \vdots & \vdots & \ddots & \ddots & 0 \\ L^N & 0 & \cdots & 0 & -l^N \end{bmatrix} \begin{bmatrix} \lambda \\ x^A \\ x^B \\ \vdots \\ x^N \end{bmatrix} = \begin{bmatrix} v^1 \\ v^2 \end{bmatrix} \geq 0 \tag{55}$$

Let us assume that this matrix has a nonnegative solution $(\lambda, x^A, \cdots, x^N)' \geq 0$. We can rewrite this equation as the equivalent form.

$$\begin{aligned} &-\lambda\widehat{y}+\left[I-\left(1+r^T\right)A^A\right]x^A+\left[I-\left(1+r^T\right)A^B\right]x^B+\cdots+\left[I-\left(1+r^T\right)A^N\right]x^N = v^1 \\ &\lambda L^h - l^h x^h = v^{2h} \qquad h = A, B, \ldots, N \end{aligned} \tag{56}$$

Let $\left(\widehat{x}^A, \cdots, \widehat{x}^N\right)'$ be the gross output vector corresponding to $\widehat{y}$. Then, consider $\overline{x} = (x+\widehat{x})\,/\,(1+\lambda)$. As shown in (57), $\overline{x}$ satisfies the labor constraint.

$$L^h - l^h \frac{x^h + \widehat{x}^h}{1+\lambda} = L^h - \frac{l^h x^h}{1+\lambda} - \frac{L^h}{1+\lambda} = \frac{\lambda L^h - l^h x^h}{1+\lambda} = \frac{v^{2h}}{1+\lambda} \geqq 0 \tag{57}$$

[4] Nikaido (1961, pp. 157–158).

The net output vector $\bar{y}$, which corresponds to $\bar{x}$, is larger than $\hat{y}$ if $v^1 \geq 0$, as shown in (58).

$$\frac{y}{1+\lambda} + \frac{\hat{y}}{1+\lambda} = \frac{v^1}{1+\lambda} + \hat{y} \geq \hat{y} \tag{58}$$

Because this equation is contrary to the definition of efficiency, the nonexistence of a nonnegative solution is verified. If $v^1 = 0$, we increase an element of vector x^h, of which the element of $v^2 > 0$.

Thus, from Lemma 1, it is shown that (59)

$$\left(p^T, w^{TA}, w^{TB}, \ldots, w^{TN}\right) \begin{bmatrix} -\hat{y} & I-\left(1+r^T\right)A^A & I-\left(1+r^T\right)A^B & \cdots & I-\left(1+r^T\right)A^N \\ L^A & -l^A & 0 & \cdots & 0 \\ L^B & 0 & -l^B & \ddots & \vdots \\ \vdots & \vdots & \ddots & \ddots & 0 \\ L^N & 0 & \cdots & 0 & -l^N \end{bmatrix} \leqq 0 \tag{59}$$

has a positive price and wage rate vector.

Thus, if there exists an R-efficient output vector, then the existence of the competitive equilibrium in the model of the new theory of international values is certified.

Here, we note that we do not consider the efficiency of the net output but, instead, the R-efficiency proposed by Mirrlees (1969). Although the R-efficient frontier is concave to the origin, the production possibility frontier calculated from each point on the R-efficient locus is, in general, not concave to the origin. This point is argued in detail in Takamasu (1986).

We can prove that the equilibrium, which is a point on the R-efficient locus, is intertemporal efficient.

Theorem 3 A competitive equilibrium is an intertemporal efficient production.

Proof For the convenience of proving Theorem 3, let us define the commodity input coefficient matrix in the world and the labor coefficient vector in the world as follows.

$$A \equiv \left[A^A\, A^B \ldots A^N\right]$$
$$l \equiv \left(l^A, l^B, \ldots, l^N\right)$$

Using these notations, prices and commodity production in an autarky are given by

$$p_t = (1 + r_t)\, p_{t-1}A + w_t l$$
$$x_t = Ax_{t+1} + y_t$$
$$lx_t \leqq L_t$$

International prices and commodity production in free trade must satisfy (60).

$$\begin{aligned} p_t^T x_t^T &= \left(1 + r_t^T\right) p_{t-1}^T \mathrm{A} x_t^T + w_t^T l x_t^T \\ p_t^T x_t^T &= p_t^T \mathrm{A} x_{t+1}^T + p_t^T y_t^T \\ l x_t^T &\leqq L_t \end{aligned} \tag{60}$$

When we assume that the world economy is in an autarky at time 0 and transits to free trade at time 1, i.e., $x_0 = x_0^T$, the value of net outputs in an autarky evaluated by international prices is given by

$$\begin{aligned} Y_t = p_t^T y_t &= p_t^T x_t - p_t^T A x_{t+1} \\ &\leqq \left(1 + r_t^T\right) p_{t-1}^T A x_t + w_t^T l x_t - p_t^T A x_{t+1} \end{aligned} \tag{61}$$

In contrast, the value of net outputs in an open economy is given by

$$\begin{aligned} Y_t^T = p_t^T y_t^T &= p_t^T x_t^T - p_t^T A x_{t+1}^T \\ &= \left(1 + r_t^T\right) p_{t-1}^T A x_t^T + w_t^T l x_t^T - p_t^T A x_{t+1}^T \end{aligned} \tag{62}$$

Inequality (63) holds for the difference between the values of consumption in an autarky and in free trade.

$$Y_t^T - Y_t \geqq \left(1 + r_t^T\right) p_{t-1}^T A \left(x_t^T - x_t\right) - p_t^T A \left(x_{t+1}^T - x_{t+1}\right) \tag{63}$$

The differences from period 0 to period n are shown as

$$\begin{aligned} & Y_0^T - Y_0 \geqq \left(1 + r_0^T\right) p_{-1}^T A \left(x_0^T - x_0\right) - p_0^T A \left(x_1^T - x_1\right) \\ & \mathrm{Y}_1^\mathrm{T} - \mathrm{Y}_1 \geqq \left(1 + \mathrm{r}_1^\mathrm{T}\right) \mathrm{p}_0^\mathrm{T} \mathrm{A} \left(\mathrm{x}_1^\mathrm{T} - \mathrm{x}_1\right) - \mathrm{p}_\mathrm{t}^\mathrm{T} \mathrm{A} \left(\mathrm{x}_2^\mathrm{T} - \mathrm{x}_2\right) \\ & \vdots \\ & \mathrm{Y}_\mathrm{n}^\mathrm{T} - \mathrm{Y}_\mathrm{n} \geqq \left(1 + \mathrm{r}_\mathrm{n}^\mathrm{T}\right) \mathrm{p}_{\mathrm{n}-1}^\mathrm{T} \mathrm{A} \left(\mathrm{x}_\mathrm{n}^\mathrm{T} - \mathrm{x}_\mathrm{n}\right) - \mathrm{p}_\mathrm{n}^\mathrm{T} \mathrm{A} \left(\mathrm{x}_{\mathrm{n}+1}^\mathrm{T} - \mathrm{x}_{\mathrm{n}+1}\right) \end{aligned} \tag{64}$$

When we divide each inequality of (64) by $\left(1 + r_0^T\right)$, $\left(1 + r_0^T\right)\left(1 + r_1^T\right)$, … , and $\left(1 + r_0^T\right) \cdots \left(1 + r_n^T\right)$ and summate them, we have

$$\begin{aligned} & \left(\mathrm{Y}_0^\mathrm{T} - \mathrm{Y}_0\right) / \left(1 + \mathrm{r}_0^\mathrm{T}\right) + \left(\mathrm{Y}_1^\mathrm{T} - \mathrm{Y}_1\right) / \left(1 + \mathrm{r}_0^\mathrm{T}\right)\left(1 + \mathrm{r}_1^\mathrm{T}\right) \ldots \\ & \geq p_t^T A \left(x_{n+1}^T - x_{n+1}\right) / \left(1 + r_0^T\right) \ldots \left(1 + r_n^T\right) \end{aligned} \tag{65}$$

If we increase n to infinity, the left-hand side of (65) converges to 0, and the value of net outputs in an open economy evaluated by the international prices and the rate of profit in the open economy is evaluated as being larger than that in an autarky.

What new findings or new theorems can we derive from this theory when we consider that the number of commodities is much larger than the number of countries?

The simultaneous equations contain $n + N–1$ unknowns that determine the international equilibrium. For n prices, N wage rates, and one rate of profit, if we take one commodity as a numeraire and assume the rate of profit given, the number of unknowns is $n + N–1$. In contrast, because the number of price equations is n, prices can be determined without depending on demand if more than $N–1$ commodities are produced in the same countries.

Let us confirm this concept using an example of two countries and three commodities. We assume two countries A and B, with commodities 1 and 2 produced in country A and commodities 2 and 3 produced in country B. In this situation, the following equations hold true.

$$\begin{aligned}
p_1^T &= \left(1 + r^T\right)\left(p_1^T a_{11}^A + p_2^T a_{21}^A + p_3^T a_{31}^A\right) + w^{TA} l_1^A \\
p_2^T &= \left(1 + r^T\right)\left(p_1^T a_{12}^A + p_2^T a_{22}^A + p_3^T a_{32}^A\right) + w^{TA} l_2^A \\
p_2^T &= \left(1 + r^T\right)\left(p_1^T a_{12}^B + p_2^T a_{22}^B + p_3^T a_{32}^B\right) + w^{TB} l_2^B \\
p_3^T &= \left(1 + r^T\right)\left(p_1^T a_{13}^B + p_2^T a_{23}^B + p_3^T a_{33}^B\right) + w^{TB} l_3^B
\end{aligned} \tag{66}$$

From (66), if we assume the price of commodity 1 as a numeraire and that the rate of profit is given, all prices and the wage rate of the two countries can be determined. Note that if production does not change, a tradeoff exists among the profit rate, the wage rate of country A, and the wage rate of country B.

The condition under which at least one commodity exists that is produced in more than one country is, approximately, that the world demand of that commodity is larger than the quantity of production that can be produced in one country. However, we have not analyzed the details of this condition. This task is left for future research.

8 Concluding Remarks

In this paper, we examined the development of Ricardo's (1817) comparative advantage theory subsequent to his work using the most generalized model. Through our analyses, we clarified that Ricardo's theory crucially depends on the assumptions of two countries, two commodities, and the nonexistence of intermediate goods. We also showed that Mill's argument on the determination of international prices depends on the assumption of perfect specialization.

Thus, we should extend Ricardo's analysis to a model that assumes multicountries, multicommodities, and the existence of intermediate goods. The positive rate of profit or the existence of a production period is also important for analyzing international trade. In this situation, prices do not depend on demand and may be determined by the production cost. When we consider intertemporal efficiency, we cannot say that trade may damage some countries. However, we should be more careful about the benefits of trade.

These assumptions are more similar to reality; because they may change the results of traditional trade theories, we should accept them and attempt to develop the analyses using them. Such analyses will be conducted by numerous researchers in the future. I am pleased if this chapter provides some assistance to these researchers.

References

Amano, A. (1966). Intermediate goods and the theory of comparative advantage: A two-country, three-commodity case. *Weltwirtshaftliches Archiev, 96*, 342–345.

Chipman, J. S. (1965). A survey of the theory of international trade: Part 1, the classical theory. *Econometrica, 33*, 132–147.

Emmanuel, A. (1973). *Unequal Exchange*. New Left Books.

Graham, F. D. (1932). The theory of international value. *Quarterly Journal of Economics, 46*, 581–616.

Jones, R. W. (1961). Comparative advantage and the theory of tariffs: A multi country model. *Review of Economic Studies, 28*, 161–175.

Kurz, H. D. (1978). Rent theory in a multisectoral model. *Oxford Economic Papers, 30*, 16–37.

McKenzie, R. W. (1954). Specialization and efficiency in world production. *Review of Economic Studies, 21*, 165–180.

Mill, J. S. (1852), *Principles of Political Economy with Some Applications to Social Philosophy*, Third Edition, London. The Collected Works of John Stuart Mill, Volume II – The Principles of Political Economy with Some of Their Applications to Social Philosophy (Books I-II), ed. John M. Robson, introduction by V. W. Bladen (Toronto: University of Toronto Press, London: Routledge and Kegan Paul, 1965).

Mirrlees, J. A. (1969). The dynamic non-substitution theorem. *Review of Economic Studies, 36*, 67–76.

Montani, G. (1975). Scarce natural resources and income distribution. *Metroeconomica, 27*, 68–101.

Negishi, T. (1981). *Classical economics and modern economics*. Iwanami Publishing. (in Japanese).

Negishi, T. (1983). *History of economics*. Toyokeizaishinposha. (in Japanese)

Nikaido, H. (1961). *Linear mathematics for economics*. Baifukan. (in Japanese)

Ricardo, D. (1817), On the Principles of Political Economy, and Taxation.

Samuelson, P. A. (1975). Trade pattern reversals in time-phased Ricardian systems and international efficiency. *Journal of International Economics, 5*, 309–363.

Shiozawa, Y. (2007). A new construction of Ricardian trade theory a many-country, many-commodity case with intermediate goods and choice of production techniques. *Evolutionary and Institutional Economics Review, 3*, 141–187.

Smith, M. A. M. (1979). Intertemporal gains from trade. *Journal of International Economics, 9*, 239–248.

Steedman, I. (1979). *Fundamental issues in trade theory*. London: Macmillan.

Steedman, I. & J.S. Metcalfe (1973), On foreign trade, *Economia Internazionale*, pp. 516–528, reprinted in Steedman (1979).

Takamasu, A. (1983). Neo-Ricardian rent theory. *Keizai Ronso, 132*, 119–140. (in Japanese)

Takamasu, A. (1986). Production possibility frontier in a Sraffa-Leontief economy. *The Manchester School, 59*, 202–207.

Takamasu, A. (1991). *Neo-Ricardian trade theory*. Sobunsha. (in Japanese)

Application of Normal Prices to Trade Analysis: National Self-Sufficiency and Factors of Competition

Yoshitaka Hirano

Abstract The Ricardo–Sraffa trade economy model is notable for the vital role it plays in determining international normal prices as well as link commodities. It has also laid a new foundation for the study of employment conditions in trade analyses. After comparing the views of Keynes ("National Self Sufficiency" in (1982) The Collected Writings of John Maynard Keynes, vol 21, Cambridge University Press, Macmillan, 1933), Parrinello (The notion of national competitiveness in a global economy In: Vint J, Metcalfe S, Kurz H, Samuelson P, Salvadori N (eds) Economic theory and economic thought : essays in honour of Ian Steedman. Routledge, London, 2009), and Shiozawa (Evol Inst Econ Rev 3(2):141–187, 2007, A final solution of Ricardo problems on international values. Iwanami-shoten, Tokyo (in Japanese), 2014) on the market mechanisms not eliminating unemployment, this study reviewed the elements regarding the long-term competitiveness of corporations and semiautonomous bodies.

Parrinello clearly shows that a bottom line of national competitiveness is established, that is, the condition that the international profit rate must be higher than the self-sufficiency profit rate during complete specialization. However, the RSte model shows that the complete specialization point does not occur in a more general international economic environment. Moreover, when full employment is not guaranteed by trade in the country, Keynes proposed spending time and carefully ascertaining the section that should be brought up in the country.

Finally, this study viewed the RSte model as a possible theoretical basis for the national self-sufficiency concept of Keynes. In this concept, Keynes indicated that some domestic industries should be preserved from a long-term viewpoint and not merely be regarded as a short-term cost consideration.

Keywords Link commodities • International normal prices • The long-term competitiveness of corporations and semi-autonomous bodies

Y. Hirano (✉)
Department of Liberal Arts and Sciences, Faculty of Engineering, Toyama Prefectural University, Imizu, Japan
e-mail: Hirano@pu-toyama.ac.jp

Y. Shiozawa et al. (eds.), *A New Construction of Ricardian Theory of International Values*, Evolutionary Economics and Social Complexity Science 7,
DOI 10.1007/978-981-10-0191-8_6

1 Introduction

To our knowledge, few examples of theoretical analysis persuasively illustrate the relation between trade and unemployment in the context of the global economy. A notable exception is the trade analysis model of Shiozawa (2007, 2014), which applies a Graham–Sraffa analysis (hereinafter referred to as the Ricardo–Sraffa trade economy or RSte model). This study explores whether trade and employment issues can be analyzed by extrapolating RSte to the national self-sufficiency vision put forward by Keynes (1933). This study is organized as follows. First, we explore how Ricardo and Keynes viewed trade and then describe the characteristics of Parrinello's (2009) model and the RSte model in this context. Next, we compare and contrast how Keynes (1933), Parrinello (2009), and Shiozawa (2007, 2014) viewed the issue of market mechanisms not eliminating unemployment. Third, the market does not automatically eliminate unemployment when trade commences; thus, a nation's economic competitiveness will be a key factor that influences how trade commences. In this regard, we look at examples in Japan of the long-term competitiveness of corporations and semiautonomous bodies with respect to the economic competitiveness of the nation. Here, we focus on key factors in the industrial sector's long-term development decisions that are not supported by current cost calculations. Finally, we will draw some conclusions with regard to the relevance of the RSte model in terms of trade and employment issues.

2 Ricardo and Keynes

Ricardo believed that the volume of labor inputs determines domestic product prices. In trade between Portugal and England, however, the benefits of trade from exchange prices were gained from the difference between the home country's production costs (coefficients of labor inputs) for imported goods and the home country's production costs (coefficients of labor inputs) for its exported goods. Ricardo's hypothetical example never considered the possibility that the commencement of trade would cause unemployment. This factor influenced Ricardo's view of global economic conditions. The historical context in which Ricardo was operating was that higher profitability could probably be obtained by importing cheap Portuguese agricultural goods into England because this would lower both the prices of daily necessities and the wages, thereby keeping England's economy from stagnating and its profitability from deteriorating. In addition, Ricardo himself noted that restrictions on capital movements were essential for the law of comparative advantage to function (Kurz 2015). However, Ricardo's hypothetical example later developed into a principle of neoclassical trade theory as it stated that free trade, in which labor is only a factor of production, would result in profitability for both trading partners.

Meanwhile, Keynes, in his later years, doubted that free trade led to profits for both trading partners and struggled to obtain a new vision. Keynes contended that the production of goods leading to financial and domestic stability was distinct from international relations and that domestic employment could be maintained by setting up a system of limited domestic self-sufficiency (Keynes 1933). Even assuming that a decline in prices of daily necessities, as a result of division of labor, would lead to a decline in real wages, there is no appropriate proof that opening up trade would alleviate unemployment without any domestic unrest/resistance or that short-term wages decline when a country was already experiencing considerable unemployment.

3 National Competitiveness

Mainwaring (1974), Schefold (1989), and Parrinello (2009), among others, have applied Sraffa's price theory (known as normal price analysis) to the trade issues discussed above. In examining how the law of comparative advantage applies to choice-of-techniques analysis, Mainwaring (1974) showed that Ricardo's example of comparative advantage is not valid when the labor theory of value is inappropriate. While highlighting the analogy between the land factor and the labor factor in international trade, Schefold (1989) used agricultural product specialization as an example of extensive rent. He examined a technologically advanced country with a small labor force (where rents are manifested as surplus wages) and a country with a large labor force but outdated production technology (a country without rents and where wages are at the subsistence level). The subsistence-level wage country produces all goods, and the surplus-wage country produces only one good. If real wages rise in the subsistence-level wage country, that country's international profitability will decline, which will affect wages in the surplus-wage country. If countries are ranked by comparative advantage in descending order from the highest level of surplus wages, changes in profitability will alter the ranking of comparative advantage.

Although these studies were meant to highlight the limited validity of the principle of comparative advantage, they were not intended to raise the issue regarding the competitiveness of a nation in international economic circumstances. In fact, it was Paul Krugman who questioned this concept. According to Krugman (1994), countries "may be happy or unhappy with their economic performance, but they have no well-defined bottom line," and "countries do not compete with each other the way corporations do."

Unlike Krugman, Parrinello (2009) shows that a nation has a bottom line under the condition of capital free movement. Trading with countries in which the domestic self-sufficiency rate of profit is high in comparison with the trade equilibrium profit rate will result in a complete shutdown of the domestic capitalistic production process. Therefore, Parrinello contends that neoclassical trade analysis is premised on full employment, flexible wages, and a technology characterized by substitutability between capital and labor. This is the main reason why Krugman did not place importance on the concept of national competitiveness.

4 Parrinello Model

Parrinello (2009) uses choice-of-techniques analysis to show that, under trade conditions premised on fixing real wages with free movement of capital, absolute advantage could come into play and make it impossible for capitalistic economies to sustain themselves.

In his model, it is assumed that there are two countries (i.e., England and Portugal) and two goods (i.e., corn and silk), and these goods are produced by using labor and corn. In each country, the distinct real wages are fixed in real term and capital movement is free. Moreover, no restrictions are established regarding labor supply. In these circumstances, the self-sufficiency economic conditions (under which both goods are produced in the same country) and the complete specialized international economic conditions (under which different countries produce two goods) are compared. The choice-of-techniques analysis by Sraffa is applied in these cases, and the technique that achieves the highest profit rate rules the economy at the given wage. Several arrangements of a wage–profit curve (a wage frontier) are possible, but the case when a national bottom line by Parrinello can exist is considered.

Each of the following frontiers of wages can exist. For a given value of the wage rate in England, it is the Portuguese wage rate, the self-sufficiency profit rate curve, the two curves of Portuguese wage rate, the international specialized profit rate, and the British self-sufficiency profit rate line corresponding to the fixed wage rate. When the self-sufficiency profit rate in England is too high under this setting, capital movement from Portugal to England occurs, and the production process that uses the capital can no longer be performed in Portugal. As a result, Portugal goes out of business as a capitalistic economy. In addition, since no restrictions are established about labor supply, the wage rate does not increase in England as a result of the capital movement. Thus, no equilibrating force can help lead to a uniform rate of profit, with both countries engaged in production and trade.

More specifically, applying Sraffa's choice-of-techniques analysis and assuming that capital flows freely among nations and real wages become inflexible in real terms, it is more likely that the principle of absolute advantage can prevail over the effects of comparative advantage. In this case, instead of both trading partners benefiting from the situation, the result may be that one of these capitalistic trading partners may be unable to survive.

> Hence a meaningful bottom line for the national economy exists in a global economy. We would say that a whole capitalistic economy is not competitive if all its capital-using techniques are unprofitable at the international equilibrium prices. This result overrules the claim that "a country must always possess a comparative advantage in something." (Parrinello 2009, p. 52)

Thus, as Parrinello notes, for the rules of comparative advantage to function, the conditions must be present in which each country's reproducible rate of profit, as a protectionist economy, must be sufficiently low, compared with the international rate of profit after trade has commenced.

Parrinello's model assumes freedom of capital movement and fixed real wages in real terms. However, it is also premised on complete specialization in a two-country, two-good analysis, and it does not consider the international division of labor by E.D. Graham's "link commodities" to be the normal state of the global economy.

5 RSte Model

In this section, the RSte model is examined to shed light on the limited characteristics of Parrinello's analytical framework. In the RSte model, there are some link commodities in the general international circumstances in which the number of goods is larger than the stated number. Hence, a complete specialization point does not exist regarding the facet of production possibility set.

The RSte model is notable for how it extrapolates the qualities of the Graham–Sraffa analysis into trade analysis. This shows that to a considerable degree, trading nations independently establish international normal prices of global final demand. Each country has a different wage rate. If full employment were to be reached on a global scale, international normal prices would be largely independent of global final demand. Ricardo's law of comparative advantage very explicitly discusses the expansion of productivity through the division of labor, but it is a fact that the market mechanism will not automatically assure employment, except under extraordinary conditions. The RSte model does not merely expand on the production possibility set that assumes full employment using Sraffa's normal price analysis, which is not based on full employment. In its explicit consideration of choice of techniques, the model also verifies the existence of international normal prices, which is considerably separated from global final demand. The RSte model also implies that it is not valid to differentiate between comparative advantage and absolute advantage.

Neoclassical trade analysis relies on the market mechanism and is premised on full employment. In contrast, the RSte model presents a framework of trade analysis using normal prices and is premised on technology and income distribution. Neoclassical analysis contends that the benefits of free trade can be realized by simultaneously adjusting market prices and volumes. In the RSte model, however, sellers usually determine the prices, whereas buyers determine the volumes. Assuming that this is how companies behave, there will be the external imposition of different international and nationwide uniform mark-up rates and a production technology coefficient resulting from the physical flows in the division of labor, given a global final demand. This will limit the domain (facet of the production possibility set) of international normal prices (the normal price of each good and the wage rate specific to each country). This domain for international normal prices can vary with changes in the composition of global final demand. However, international normal prices will not change as long as the changes in demand remain within the scope of the given facet of the production possibility set. In this sense, international normal prices are independent of global final demand. This maintains the attributes of Sraffa's normal prices in a closed economy or in an open economy after the global economy has been fully homogenized.

We next discuss the correlation with Graham's trade analysis. In the modern global economy, it is normal to have many countries producing and competing in products that have the same functions. The law of comparative advantage is a law that indicates only the benefits of having division of labor under conditions of full employment. Insofar as economics can be summed up by the premise that real wages are adequately equalized in all countries, free trade will engender goodwill among trading partners. But such clear-cut specialized illustrations are divorced from the reality of the international economic situation. Instead, the appropriate theoretical assumption is that most goods are what E.D. Graham calls "link commodities" that many countries compete to produce. In the RSte model, all countries possess at least one link commodity. It is through link commodities, in which a number of countries produce and share in global final demand, that both the disparities in the countries' real wages and the normal prices of other goods are determined. Also, the existence of link commodities ensures the uniqueness of international normal prices, except those that are constants.[1]

Furthermore, the RSte model has a special quality with respect to employment analysis. The model determines facet of the production possibility set under the condition of full employment, but this does not mean that it assumes full employment in a comprehensive trade analysis. First, let us assume the time when global final demand happens to lead to full employment. In such an event, it illustrates a gradual trend toward the convergence in market prices and normal prices under circumstances, in which changes in production technology and demand conditions, such as wage disparities among countries, occur to the extent that normal prices remain within the bounds of the relevant facet of the production possibility set over a long period of time. Here, a long period of time is defined as the period during which the relevant facet of the production possibility set does not move, regardless of changes in assumptions. Although not emphasized by the RSte model, this is a market mechanism found in the classical analysis. Post-Sraffian analyses add the condition that the market has no mechanism for adjusting employment to achieve full employment. An even more interesting and important aspect of RSte lies in the fact that unemployment increases to such an extent that markets will be unable to resolve it using price-adjustment mechanisms when global final demand does not reach the point on the relevant facet or when global effective demand falls short of reaching global full employment. The opinion that the market's coordinated adjustment of prices and volumes will rectify insufficient demand by restoring full employment indicates nothing but the possibility that goods and labor could be substitutes. Also, experience has shown that coordinating productivity with demand through technological advances and improving the efficiency of the division of labor are not very likely. Cutting wages further when unemployment is already high will only inflame domestic unrest. If unemployment cannot be eliminated using the market's price-adjustment mechanisms, then measures to deal with the unemployment need to be taken by corporations, regional administrations, semiautonomous bodies that are neither corporate nor governmental, nonprofit organizations (NPOs), nongovernmental organizations (NGOs), and even each country's central government. This does not imply an either/or view of the global economy, in which the only alternative is free trade or protectionism. It simply means that we need to take a new approach to examining global economic situations.

6 Keynes's Vision of National Self-Sufficiency

In his final years, Keynes believed that the benefits of free trade are limited to situations of full employment. In such cases, the commencement of trade could lead to falling wages and potentially benefit from the international divisions of labor. Unless a given economy has already achieved full employment, it will not be able to deal with further wage cuts when trade commences, and as a result, it will not receive any benefits from trade. Therefore, an economy should already have achieved full employment if it is to reap any benefits from free trade. As a consequence, the governmental bodies of high-wage nations may sometimes need to deal with the possibility of serious domestic unemployment when they are trading with nations that have low wages and are competitive. In addition, low-wage nations do not necessarily reap unconditional trade benefits from the fact that their wages are relatively low.

Keynes clarified the limitations of reaping benefits from free trade and began to seek a new vision for the global economy that could adopt and alter the free trade paradigm. This was his vision of national self-sufficiency (Keynes 1933). Keynes's essay, "National Self-Sufficiency," while discussing whether he was a continuous or discontinuous proponent of free trade or protectionism, is a search for a vision that is neither one of free trade nor of protectionism. According to his vision of national self-sufficiency, trade specialization occurs as a result of wide differences in factors such as climate, natural resources, and population density, and is a situation in which many countries can competitively produce the same product, as considered the norm for the global economy. He also held the view that economic activities such as agriculture need to remain domestic to ensure the stability of a nation's autonomy. For instance, he contended that countries should take whatever steps they deem necessary to preserve their autonomy, even if it seems that such areas are serving as cost centers in the national accounts at a point in time. In this case, he asserted that it is the duty of governments to behave not like corporations, but as entities that protect the domestic arts. Furthermore, Keynes believed that the British economy should maintain its domestic automobile industry and certain other domestic industries.

However, this is where objections arise. Looking at the spread of automobiles in today's global economy, automobiles are considered goods that should be produced competitively in a number of countries through organizations that set up production and assembly lines overseas and consider those countries' cultural expectations. In any event, the Keynesian vision of national self-sufficiency (while containing many details that warrant reconsideration) is an effective viewpoint on trade that comprises a value judgment because this vision reflects each country's circumstances and is not an either/or choice between free trade and protectionism. In addition, the RSte model can be interpreted as an extrapolation of this vision. In particular, the model does not place much importance on Ricardo's easily misunderstood trade specialization. Furthermore, its perception of the international economic situation of Keynes's time aligns with the theory (which is the normal state of affairs for today's global economy) that agreement between global final demand and productivity accidentally occurs as well as the point that Graham's link commodities can be incorporated into trade analyses.

7 National Competitiveness Reconsidered

According to Keynes (1933), one should not take hasty protectionism. Taking into account the domestic employment situation, it is important to carefully assess and select which goods are suitable for domestic production. Moreover, this point should be reinterpreted as Keynes's national competitiveness, since it leads to the development of such competitiveness.

Parrinello's concept of national competitiveness depends on the comparison of the self-sufficiency profit rate and the international equilibrium profit rate during complete specialization. The RSte model, however, shows that a perfect specialization point does not normally appear in the facet of the production possibility set. Moreover, it does not appear in the general international economic environment in which the number of the goods exceeds the national number, as in the two-country, three-good case. As a result, a comparison by the profit rate, which implies state operation abeyance, may be impossible.

Conversely, according to Keynes (1933), when there is a possibility of an increase in unemployment in the country through trade, it is necessary to restrict free movement of capital and carefully ascertain which goods should be domestically produced. In his opinion, the latter must become a priority among policy objectives. In other words, when domestic employment is not guaranteed by free trade, it is important to maintain national competitiveness, while carefully observing domestic employment circumstances.

If such measures are taken, then domestic employment is protected from sudden changes, and national competitiveness is maintained or even improved. Overall, the conditions for national competitiveness by Keynes are regarded as those that complement Parrinello's concept of national competitiveness. And the Japanese cases concerned with upbringing of the section which may not be supported by current cost calculations are taken up at below.

8 Fostering the Competitiveness of Corporations and Semiautonomous Bodies for the Short and Long Periods

Sraffa's theory states that the uniform rate of profits is inversely related to the wage rate. The uniform rate of profits implies that the right profit opportunity has been sought and enough time has elapsed for capital to move. This rate is usually regarded as an element in a long-period analysis. The RSte model, however, regards the rate of profit as a mark-up rate and uses it in short-term analysis.[2] Shiozawa has placed importance on short-term analysis. Yet, the RSte model can be extended to the long-period analysis. A long period is more than just an artificial concept. This example may be a bit odd; however, as discussed by Ricardo, Keynes, and Sraffa, a succession of day trades in the equity market, which are short-term transactions, does not turn into a long-period equity transaction. A long-period equity transaction in this case is premised on the expectation that the equity price will change in the following ways.

First, computing the theoretical value of a company through various means, such as using the company's financial statements, will indicate the path whereby the actual and theoretical equity prices will converge over a long period. Also, the company builds a track record of performance, and its theoretical value accordingly moves. In this situation, the psychology of participants in the equity market, which is akin to a beauty contest, will move around the path. Long-period equity transactions are transactions that are based on such company's track record of performance and expectations on it. In his youth, Keynes was an investor who preferred short-term trading in the futures and currency markets, but over time, he evolved to become a long-period equity investor (Wasik 2014). Of course, one can profit by concentrating on short-term equity trades. Still, the routine beauty contest-like market psychology basically has nothing to do with the outcome of long-period equity transactions. In short, a long-period cause–effect relationship can exist apart from the short-term elements in the analysis of economic phenomena. Of course, profits can be made by concentrating on short-term trading. Thus, making a profit, despite taking a different action principle, can complicate equity market analyses.

Sraffa standardized economic profitability. In the absence of any particular entry barrier, an industry that is earning excessive profits will be subject to an influx of capital; hence, its profits will eventually approach the mean. Therefore, the conventional thinking is that assuming a uniform rate of profit will allow for this type of capital movement over the long term. Shiozawa, on the other hand, views the rate of profit as a nationwide mark-up rate in the short run. In the following discussion, which is premised on the framework of trade analysis pioneered by Shiozawa, we will consider corporate competitiveness as we reinterpret it based on Sraffa's long-period analysis. In this study, the meaning of "long period" can vary, depending on the characteristics of the subject being analyzed. When trade issues are being analyzed, "long period" refers to the period of time during which the relevant facet of the production possibility set remains unchanged, despite changes in the production technology coefficient and other conditions, so that international normal prices remain unchanged as well. Each facet of the production possibility set incorporates a pattern for the international division of labor. If the pattern of the international division of labor changes dramatically over the course of five or ten years, then this period of 5 or 10 years will be deemed a "long period" when trade issues are the focus. Also, in the case of the aforementioned long-period equity transactions, a "long period" could be 3–5 years and sometimes even more than 5 years.

9 Elements of Corporate Competitiveness: Part 1

The task of maintaining employment while wielding the authority to allocate labor should not be relegated to the market. It is essential to have a setup in which the government, an affiliated body, someone in a corporate leadership position, or someone in a corporate division is responsible for preserving jobs while watching the market movements. In other words, employment must be preserved not only through corporate competitiveness but also through government policies. Corporate competitiveness comprises several elements.[3]

The following discussion focuses on several forms of corporate competitiveness and trade secrets, with particular reference to the competitiveness of Japanese corporations. We first examine cases where the well-being of employees is an element in competitiveness. In one study, a questionnaire survey was conducted in 1991. In response to whether to place priority on paying shareholder dividends or on preserving jobs during a recession, Japanese corporate managements responded that they would place more importance on jobs, which is contrary to Anglo-Saxon corporate managements, who prioritized dividends (Yoshimori 1993). Japan's economy was subsequently under pressure from a long-term recession, and for a while, it was fashionable for Japanese corporations to "Westernize." However, as we examined a report released a few years ago that gave examples of thriving Japanese small- and medium-sized enterprises, we found that many of these companies have adopted the strategy of not producing enough to meet surplus demand as a way of prioritizing jobs in the long term (Sakamoto 2008). This is because if the company were to expand its production capacity in response to a temporary surge in demand, it would be forced to cut jobs later on when that demand moderated. The outcome of such strategic hiring shows that these companies have been able to sustain steady growth over an extended period. Referring to effective demand at the corporate level, in contrast to the neoclassical corporate principle of "selling as much as the market will bear," Shiozawa (2014) referred to the principle of corporative behavior of "producing as much as the market will bear" as the "Sraffa principle." Taking this further, the previously mentioned companies are viewed as having competitive strategies that take a long-term viewpoint and responding to demand by merely producing what they can with their normal production systems. If most companies were to adopt such a strategy, it would only be a coincidence that the labor force would be completely absorbed by the production systems in an autonomous market.

10 Elements of Corporate Competitiveness: Part 2

The next symbol of Japanese corporate competitiveness is the so-called customization or responding to customers' needs. For example, more than 80% of Japan's computer software industry is involved in developing proprietary software or customizing basic software for corporate clients (Koike 2015). Offering such painstaking customization increases the likelihood of repeated business. Here is another example from a survey conducted by the author. This case concerns a medium-sized enterprise in Toyama Prefecture that is mainly involved in manufacturing equipment for automating automobile production lines. Each piece of equipment costs between 200 million and 500 million yen. The company, of course, designs each piece of equipment after finding out what each client's plant needs, but there are also times when the company further revises its design during the assembly process to incorporate feedback from the production manager. This company was originally a subcontractor for a large corporation, but it groomed young design engineers so that it would be able to supply products of its own, and it is now delivering its equipment to automobile plants around the world. Such a setup is common in Japan. Building a framework that can respond to customers' detailed needs becomes the foundation for a company's competitiveness.

The true worth of competitiveness in this regard comes into question; that is, what the customer communicates during discussions is not what the customer actually needs. When this happens, an attempt at deciphering the customer's true desires from what the customer states will have an effect on long-term relations with the customer. This brings to mind a product development anecdote that involves Steve Jobs, the cofounder of Apple Inc. Jobs said that he paid close attention to user feedback on his company's products, but at the same time, he also believed that users had no idea what the next-generation product would be. If Japanese corporations are to remain competitive in the future, they will need to be like Jobs and listen to what their customers want not only at the design stage but also in their production lines, on the assembly floor, and even when the sales force gets complaints and then demonstrate their talent for ingenuity. In other words, to enhance national competitiveness, it is necessary not only to raise productivity but to respond to the specific needs of customers.

11 Developing Regional Competitiveness

The third element is the aspect of food and agriculture, which Keynes feared would threaten the global competitive environment. Japan has long been concerned about its low rate of food self-sufficiency, and the status quo is somehow being maintained because agricultural chemicals have made it possible to supply large, stable volumes of agricultural products in the same way as the supply of industrial products. As is well known, developing countries are currently imposing extremely strict standards on the level of residual agrochemicals. Japan's standards on residual agrochemical levels are sometimes stricter than, if not as strict as, those in Western countries. For agrochemicals that are clearly harmful, regularly eating foods containing these residual chemicals will cause them to accumulate within the human body, even if strict residual agrochemical standards have been established. Some people believe that pollution will be less of a problem if such harmful substances are diluted by dumping them into rivers and oceans. However, the continual discharge of pollutants will cause them to build up in fish and other marine life; these will eventually be consumed by human society.

Agricultural products that are produced using as few agricultural chemicals as possible are labor-intensive and more expensive. The more affluent classes can choose to purchase organic agricultural products; however, people who are living at a subsistence level tend to be unconcerned about agricultural chemicals. It will not be possible to provide cheap organic agricultural products to a broad range of consumers if the matter is left up to market forces. Most practicing farmers do not have the luxury to think about how they can carefully farm without using agricultural chemicals. A company operating in Yamaguchi Prefecture supplies low-chemical agricultural products to the appropriate distribution networks. The company does not take a cut from the distribution networks, as do corporate executives (see the listing for Akikawa Foods & Farms Co., Ltd., in any edition of the Japan Company Handbook (Toyo Keizai ed.). Many farmers who have reduced their use of agricultural chemicals are setting up joint stock companies. Regional associations and associations made up of neighbors, families, and related groups are issuing shares.

Japan's remote regions, where the birth rate is low and the population is aging, are being further plagued by the problem of depopulation. An example of the depopulation problem being tackled while promoting local agricultural products is Ehime Prefecture's Muchachaen—an organization that engages in both low-chemical mandarin orange cultivation and home health care (according to an interview survey). Ehime Prefecture has a warm climate and has long been engaged in mandarin orange cultivation. The area has been hit hard by low birth rate and an aging population. In depopulated areas, many areas that used to be farmland have been abandoned. However, just because the land is no longer being cultivated does not mean that the current owners will give it away. Muchachaen leases an abandoned farmland, and if the landowner wants to farm, the company subleases the land to young people from other areas for the collective production of low-chemical mandarin oranges. The company has also set up a home health-care system for elderly people in the community as a way of responding to local needs and stabilizing the local economy.

These examples illustrate the process of creating competitiveness in the food and agriculture industry, which Keynes believed should be performed domestically and not be subject to international competition. If international competition places too much pressure on current costs, then there is the danger that such start-ups may fail before they even get off the ground. Although Japanese residual agrochemical standards are currently being met, agricultural products that are not chemical-free still occupy the aisles of supermarkets, and safe agricultural products are not widely available at affordable prices. Moreover, relying on market forces does not automatically improve the situation. Therefore, steps should be taken to intervene at various levels, along with long-term value judgments regarding the future of society.

12 Conclusion

The RSte model is notable for the vital role it plays in determining international normal prices as well as link commodities. It has also laid a new foundation for the study of employment conditions in trade analyses. Therefore, this study reviewed the elements regarding the long-term competitiveness of corporations and semiautonomous bodies.

Parrinello clearly shows that a bottom line of national competitiveness is established, that is, the condition that the international profit rate must be higher than the self-sufficiency profit rate during complete specialization. However, the RSte model shows that the complete specialization point does not occur in a more general international economic environment. Moreover, when full employment is not guaranteed by trade in the country, Keynes proposed spending time and carefully ascertaining the section that should be brought up in the country. Thus, it is necessary to examine these elements in detail to enhance the analytical explanation of national competitiveness.

Finally, this study viewed the RSte model as a possible theoretical basis for the national self-sufficiency concept of Keynes. In this concept, Keynes indicated that some domestic industries should be preserved from a long-term viewpoint and not merely be regarded as a short-term cost consideration. For instance, in Japan, promoting organic agricultural products nationwide may not seem to be worth it from a cost perspective. However, Japan is in a special situation of having a low rate of food self-sufficiency. Spending a little more money to halt any further deterioration of its people's health should thus be regarded as a necessary long-term investment. It is a business that the nation should be involved in and in a form of the socialization of investment envisioned by Keynes. Shiozawa's new theory of international values notes that it can only be coincidental if market forces alone bring full employment to the global economy; it could become the theoretical basis for supporting policy proposals of this kind.

> ... the new economic modes, towards which we are blundering, are, in the essence of their nature, experiments. We have no clear idea laid up our minds beforehand of exactly what we want. We shall discover it as we move along, and we shall have to mould our material in accordance with our experience. J.M. Keynes (1933, 1982, p. 246)

Notes

1. However, it is not clear whether or not the domain (productivity facet) of international normal prices gradually diminishes as the number of goods and countries increases.
2. See Shiozawa (2016) for a discussion of his mark-up rate.
3. See Fujimoto and Shiozawa (2011–2012) for a more comprehensive and detailed discussion of the concept of competition.

Acknowledgment The author thanks Yoshinori Shiozawa and an anonymous referee for their suggestions and useful comments but retains sole responsibility for any error.

References

Fujimoto, T., & Shiozawa, Y. (2011–2012). Inter and Intra Company Competition in the Age of Global Competition: A Micro and Macro Interpretation of Ricardian Trade Theory. *Evolutionary and Institutional Economics Review, 8*(1), 1–37, *8*(2), 193–231.

Keynes, J. M. (1933). *"National self sufficiency" in (1982) The collected writings of John Maynard Keynes* (Vol. 21). Macmillan: Cambridge University Press.

Koike, K. (2015). Long term competitiveness, (in Japanese), Nikkei.

Krugman, P. (1994). Competitiveness: A Dangerous Obsession. *Foreign Affairs, 73*(2), 28–44.

Kurz, H. (2015). David Ricardo: On the art of "elucidating economic principles" in the face of a "labyrinth of difficulties". *European Journal History of Economic Thought, 22*(5), 818–851.

Mainwaring, L. (1974). A Neo-Ricardian analysis of international trade. In I. Steedman (Ed.), *Fundamental issues in Trade Theory*. London: Macmillan.

Parrinello, S. (2009). The notion of national competitiveness in a global economy. In J. Vint, S. Metcalfe, H. Kurz, P. Samuelson, & N. Salvadori (Eds.), *Economic theory and economic thought : Essays in honour of Ian Steedman*. London: Routledge.

Sakamoto, K. (2008–2013). Japan's most treasured companies, 1–4, (in Japanese), Asa Publishing.

Schefold, B. (1989). *"A digression on international trade" in Mr. Sraffa on joint production and other essays*. London: Unwin.

Shiozawa, Y. (2007). A new construction of Ricardian Trade Theory: A Many-country, Many-commodity case with intermediate goods and choice of production techniques. *Evolutionary and Institutional Economics Review, 3*(2), 141–187.

Shiozawa, Y. (2014). *A final solution of Ricardo problems on international values*. Tokyo: Iwanami-shoten (in Japanese).

Shiozawa, Y. (2016). *The revival of classical theory of value*. Paper presented at Japan Society of Political Economy Annual Conference2014, held on 25–26 October at Hannan University, Osaka. A revised version published as Chapter 8 in Yokokawa et al. eds. (2016) The Rejuvenation of the Political Economy, Routledge.

Toyo Keizai. (Ed.). (various editions). Japan Company Handbook, (in Japanese), Toyo Keizai Sinpousya.

Wasik, J. F. (2014). Keynes's way to wealth, McGraw-Hill global education holdings.

Yoshimori, K. (1993). The Outlook for the Japanese-Style Corporate Administration System. *Organizational Science*, *27*(2), 24–36. (in Japanese)

Part III
Re-Examining the History of International Trade Theory

An Origin of the Neoclassical Revolution: Mill's "Reversion" and Its Consequences

Yoshinori Shiozawa

Abstract The neoclassical revolution was a shift from economics of production to economics of exchange. The study shows from an internalist point of view that one of the origins of the neoclassical revolution can be traced back to young John Stuart Mill, who tried to sort out a problem left unresolved by David Ricardo. Due to a peculiar reason that I would later clarify, he was led toward examining a pure exchange economy. In this setting, Ricardo's cost of production theory of value was invalid. When Mills found the answer to this, he came to the following conclusion: "we must revert to a principle anterior to that of cost of production, and from which this last flows as a consequence,—namely, the principle of demand and supply" (On Laws of Interchange between Nations. First essay in J.S. Mill, Essays on some unsettled questions of political economy, 1844. Citation is made from Library of Economics and Liberty, 1844, I.19). This thesis caused a long-lasting and strong influence on the research programs in economics. The study describes how Mill's thesis profoundly influenced three founding fathers of British neoclassical economics, namely, Stanley Jevons, Francis Ysidro Edgeworth, and Alfred Marshall. Different alternatives were researched and discovered, but it was Alfred Marshall, with his concept of demand and supply functions, who paved the way for today's mainstream economics.

Keywords History of economic thought • Marginal revolution • Demand and supply analysis • Pure exchange economy • International trade

1 Introduction

In the history of economics, we can detect two major price theories: the classical theory of value and the neoclassical theory of value. The change from classical to neoclassical economics, the biggest shift in the history of economic theories, is known as the "neoclassical revolution." Numerous factors paved the way for the

Y. Shiozawa (✉)
Osaka City University, Osaka, Japan
e-mail: y@shiozawa.net

Y. Shiozawa et al. (eds.), *A New Construction of Ricardian Theory of International Values*, Evolutionary Economics and Social Complexity Science 7,
DOI 10.1007/978-981-10-0191-8_7

neoclassical revolution. Many explanations have been given from the externalist viewpoint, that is, explanations from social, political, and philosophical backgrounds. Here, I present an internalist explanation, that is, one using the logic of scientific or theoretical development. I argue that, when John Stuart Mill attempted to solve the international value problem, he was forced to revert from the classical principle of the cost determination of value to a special form of the law of demand and supply. This marked a crucial turning point. The aim of this chapter is to prove this contention.

The paper is composed as follows. Section 2 explains the nature of neoclassical revolution. Section 3 complements Sect. 2 by clarifying the essential differences of the two value theories. Section 4 presents an argument about the nature of the Ricardian revolution and considers the neoclassical revolution from a new perspective. Section 5 pinpoints the point of origin of the neoclassical revolution, and Sect. 6 explains why Mill's "solution" and thesis for reversion were misplaced. Section 7 builds a bridge between Mill and the three founding fathers of British neoclassical economics, that is, Jevons, Edgeworth, and Marshall, while Sects. 8, 9 and 10, respectively, provide individual accounts of them. I show that all three individuals were deeply influenced by Mill's misleading solutions. Section 11 shows possibilities of alternative way outs. Section 12 summarizes the paper and presents some directions for further investigation.

2 The Nature of the Neoclassical Revolution

I have not explained how Mill's "solution" was misdirected. This is a task addressed in Sect. 6. First, let us consider the consequences of Mill's "solution." My bold conjecture is that this was the very point of the shift from classical economics to neoclassical economics or, put differently, the conversion from plutology (the economics of production) to catallactics (the economics of exchange), which must be the main feature of the neoclassical revolution. I owe this characterization of the neoclassical revolution to J. R. Hicks (1974, 1976), who explains as follows:

> while the classics looked at the economic system primarily from the production angle, the catallactists looked at it primarily from the side of exchange. It was possible, they found, to construct a 'vision' of economic life out of the theory of exchange, as the classics had done out of the social product. It was quite a different vision. (Hicks 1976, p. 212)[1]

Hicks defined plutology and catallactics as a difference in the focus of attention. This is an important difference, which contrasts two attitudes but might induce

[1] Hicks preferred to use the term "catallacticist" instead of "marginalist." I use the expression "neoclassical revolution" to indicate that the revolution comprises much wider changes of economic thinking.

a misunderstanding in that plutology or the economics of production does not consider exchange. This is a pure misapprehension.

The main difference between the economics of production and that of exchange lies in understanding how the values of goods are determined. As I argue in the next section, the core of economics is the theory of value. According to the classical theory of value, which is best represented by the Ricardian theory, values are determined by production conditions. The neoclassical theory of value assumes that the theory of value is possible without any reference to production conditions. Pedagogically, the theory of value starts from a pure exchange situation. As an inevitable consequence of this abstraction, psychological factors become dominant. The theory depends too much on the individual agent's ability to deal with total economic processes. The necessity arises to assume infinite rationality for agents.

The neoclassical theory of value may be a universal theory that is valid for any economic era and situation, but it is not a good theory for a capitalist economy in which producers play predominant roles. The classical theory of value presupposes capitalist production or a market economy whose principal players are industrial firms that produce commodities (goods and services) by employing workers and machines.[2] The neoclassical revolution presented not only a change in the focus of attention but also a change in the logic and structure of the theory of value.

When we talk about the neoclassical revolution, we have to take into account that, at the time of the neoclassical revolution, economics was not yet a unified discipline with the same research framework for economists in different countries. Indeed, the economics of French-speaking countries and that of German-speaking countries have very different histories to that of English-speaking countries. In this chapter, I examine the origin of neoclassical economics in only the UK. In the emergence of neoclassical economics across continental Europe, the neoclassical revolution took very different courses, and we have to observe it from totally different angles in each country, even if we admit that the neoclassical revolution was an episode of simultaneous discoveries.

The emergence of neoclassical economics is most often examined and argued with reference to the UK. However, as I indicate above, most explanations have been given from externalist views. In this chapter, I conjecture that the neoclassical revolution was led mainly by the internal logic of economics. This is totally an internalist view.

This internalist view is necessary because of the peculiarity of the neoclassical revolution in the UK. If the neoclassical revolution was a shift from the economics of production to that of exchange, we have to admit that this reversed the order of historical development. In a very ancient time when trade was undertaken only between communities, exchange rates might have had no close relation to the cost of production. Exchanged goods were surpluses of communities' necessities. We can also refer to the fact that mercantilist capitalism preceded industrial capitalism. The peculiarity of the neoclassical revolution is that it came after the industrial

[2]For a more detailed account of the classical theory of value, see Shiozawa (2016a).

revolution. The industrial revolution was the most conspicuous event to have taken place in the time of classical political economy. Why did neoclassical economics, which came after the industrial revolution, ignore this extraordinary phenomenon? It must have been apparent even then that a tremendous increase in commercial activities was supported by a rapid increase of production volumes. The externalist view does not explain this most conspicuous fact.

This chapter conjectures that the neoclassical revolution was a logical process in the development of economic theories. We contend that the groundwork for the neoclassical revolution was prepared by Mill's study, in which he aimed to solve an unsettled problem left by Ricardo. This was the question of the theory of value in international trade. As this is a conjecture that came to my mind recently (in the last 4 or 5 years), and as the stakes are extraordinarily high, I do not claim that this chapter provides sufficient evidence to prove my conjecture. This would require enormous work in the history of economics and would need the input of others to attempt to verify or falsify my conjecture. This section is simply a rough description of what might have occurred.

Mill did not abandon the very core of the classical value theory, but he was obliged to make a structural reform to its logic. There were three major drawbacks. First, labor was the most important category of commodities that did not obey the law of the cost of production (Mill 1848, III.2.19). The second involved joint cost cases (Mill 1848, III.16). International values formed the third category (Mill 1848, III.16.14). Mill had to concede, if reluctantly, that the law of demand and supply was an anterior and more fundamental law than the cost of production theory of value was. This was the first consequence of Mill's "solution." Ricardo once declared that "[t]he opinion that the price of commodities depends solely on the proportion of supply to demand, or demand to supply, has become almost an axiom in political economy, and has been the source of much error in that science" (Ricardo: Sraffa, p. 382; Library, 30.3). Mill knew this and had to reconcile Ricardo's theory of value and his new "solution." What Mill did was to admit the law of demand and supply as more fundamental and logically anterior to the cost of production theory of value. Mill stopped at this point, for this was the maxim possible concession he could make. However, economists after Mill did not feel an obligation to do so. They sought to make the theory more coherent and unified. Here arises the second consequence of Mill's "solution."

If the law of demand and supply were more fundamental, it would be natural to apply this law uniformly and universally. My suspicion is that the English founders of neoclassical economics were influenced by Mill's "solution" and conclusion. Jevons, Marshall, and Edgeworth showed more or less indicative evidence for this suspicion. As Jevons is thought to be the person to have led the marginalist revolution, an examination of Jevons will inevitably be longer than those of Marshall and Edgeworth.

I do not deny that other factors intervened in the arrival of neoclassical economics, including the spread of optimization techniques (in mathematics), the tradition of utilitarianism (in philosophy), the deep-rooted tradition of demand and supply thinking (in economics), the popularization of the concept of functions, and, finally, the general decline of Ricardian economics. In addition, as Mirowsky (1989) argued, the spread of "energy" concept might be crucial to the arrival of neoclassical economics.

All these factors indeed must have worked. However, we should ask how directly these factors worked to form exchange economics in place of production economics. Their influences are indirect, whereas the impact of Mill's "solution" was, I believe, more direct. Above all, Mill was logically forced to accept that the law of demand and supply was anterior and more fundamental than the cost of production theory of value was. If we admit that the cost of production theory of value was the center core of classical economics, then it was this internal logic that drove the economics of production to the economics of exchange. All other factors are external to the logic of economics. I agree that the change of society or sciences drove economics in a new direction. However, if we accept that economics is an independent science, then we should search for the internal logic that produced the neoclassical revolution. As it is called a "revolution," it was really a fundamental change in the logical structure of economics. It would be strange if there were no internalist explanation. My conjecture is one of the scarce internalist explanations in the literature.[3]

As far as I know, no one has ever claimed that Mill's "solution," or his situation setting, was one of the key factors that underlay the arrival of neoclassical economics. I believe this conjecture deserves scrutiny.

3 Essential Differences Between Classical and Neoclassical Economics

The theory of value is the core of economic theory. The theory of value exposes in a most abstract form how an economy works. It exhibits the vision of a specific theory. There are many different aspects between classical and neoclassical economics, but the essential differences appear in theories of value. Consequently, in this section, I discuss the differences between the two value theories and contrast them.

When we compare the classical and neoclassical theories of value, the first requirement is to define the classical theory of value, because there is no unified understanding of what it is. On the other hand, there is no serious confusion about the neoclassical theory of value. The neoclassical theory of value comprises various strands of economic thought from Marshall and Walras to Arrow and Debreu. However, these strands have a common core: they are all theories of *prices*. They

[3]Hicks (1976) asked how we could explain the rise of catallactics. He examined possible explanations, such as the reaction to socialism and the real-world changes, but he denied these were major factors and advanced a thesis in which the main appeal of catallactics lay in its intellectual quality. Walras made it possible to consider the economic system as a system of interactive markets and Menger as adjustments of means to ends. Hicks considered that Jevons did not complete his system. In this sense, Hicks attempted to understand the neoclassical revolution on an internalist standpoint, but he did not mention the points that I raise in this chapter.

place paramount importance on the function of prices. Prices permit the whole complex economy to work effectively and efficiently. The core of these theories is composed of two parts: the utility theory of value and demand and supply equilibrium. No specific structure is required as the theory holds for all situations.

However, the classical theory of value has no such unified theoretical core. In the era of classical economics, laws of demand and supply existed. Utility was an important component for many economists. Therefore, the main components of neoclassical economics already existed in the time of classical political economy. I do not attempt to unify the various strands into a single framework of classical economics. If I were to do that, it would result in total confusion. Instead, I pick the most typical classical theory and define it as the classical theory of value. I have chosen Ricardo as the representative of classical economics. Many people would agree with me. Indeed, Ricardo's theory of value is a culmination of various classical theories. However, there is still much misunderstanding about what Ricardo's theory of value is.

Many economists might consider the labor theory of value as Ricardo's theory of value. As Marx adopted this at the core of his theory, many Marxists would take it for granted, but my idea is different. My definition of Ricardo's theory of value, and consequently, of the classical theory of value, is the cost of production theory. This claim has textual evidence. Ricardo himself added a note in the third edition of his *Principles* explicitly claiming this. He wrote:

> Mr. Malthus appears to think that it is a part of my doctrine, that the cost and value of a thing should be the same—it is, if he means by cost, "cost of production" including profits. (Ricardo: Sraffa, p. 47; Library, I., n.7)

The idea is similar to what Marx later argued in terms of production price. There is not sufficient space here to argue this interpretation. Refer to Takenaga (2016) for a more detailed textual examination. As for the possibility of developing Ricardo's idea as a modern theory, see Shiozawa (2016a, 2017a).

The essential feature of the cost of production theory of value is its objective character compared to the subjective character of the neoclassical theory. By "objective," we mean that the value of a commodity is determined by social conditions. The core contention of the cost of production theory of value is that prices are primarily determined by the cost of production, including profit. In more modern terminology, prices are determined by the full-cost principle. Mathematically expressed, the price vector $\mathbf{p}$ is given by a system of simultaneous equations:

$$(I + M)\{w\mathbf{a}_0 + A\mathbf{p}\} = \mathbf{p}. \tag{1}$$

We skip the notations of symbols used here, because they are not used hereafter. Those interested in this expression can consult my paper (Shiozawa 2016a, equation 6).

Many factors are involved in the determination of the cost of production. First, there are production techniques. As a part of knowledge, production techniques have a subjective aspect, but can be specified by input–output relations, which are measured objectively. When all inputs that are necessary for the production of a unit of product are known, the cost can be calculated objectively.[4] The expenditure for an input is the value multiplied by the quantity of the input. If we know the value of all inputs, including wages, we can calculate the total expenditure. It is noteworthy that the expenditure is a totally observable quantity that has little relevance to toil and pain, even if human work is involved.[5] The cost is the sum of all expenditure necessary for production multiplied by a markup factor. The markup factor is 1 plus the markup rate. The total expenditure is the sum of all expenditure, and the full cost of the product is the total expenditure times the markup factor, or 1 plus the markup rate. The markup rate is fixed by custom or by calculation for the moment of consideration. We can argue that the markup rate changes with changes of market conditions, for example, with changes of competitive conditions.[6]

However, to determine the cost of production for all products in an economy is not a simple operation, because costs depend on values. Here, we have a logical cycle. To define the cost of a commodity, we should know the value of all inputs. At the time of Ricardo, the system of mathematical equations was not widely known and we can easily conceive how difficult it was for Ricardo to build a cost of production theory of value. This partly explains why Ricardo often talked about cases in which the product is made only by labor. He could include indirect labor but he could not provide a precise formula, as Eq. (1).

More difficult was the question of the choice of production techniques. Imagine a situation in which there are two production techniques for the same product. In that case, the superiority of a production technique would depend on values. When all input prices are known, it is a simple question of adding them up. However, for the theory of value, values have to be determined by costs and the costs of production depend on values. How does a system of production techniques come to be chosen in such a way that all chosen techniques have the least production cost among production techniques for the same product?

The substitution theorem, which later became the non-substitution theorem, was discovered and proved in the mid-twentieth century by Paul Samuelson. Samuelson proved only the two-commodity case while later, Tjalling Koopmans proved the three-commodity case and Kenneth Arrow proved the general case (Koopmans 1951). Confusion about the naming of the theorem shows that Samuelson did not understand the real meaning of the theorem. I prefer to call this theorem the "minimal price theorem," because doing so emphasizes the existence of the system of production techniques that informs the minimal price system given the wage

[4]For simplicity, we assume that production is linear, that is, inputs and outputs are directly proportional.

[5]There is no necessity to ask if it is proportional to toil and pain (or real cost after Viner).

[6]For a short account, refer to the section *How are markup rates determined?* in Shiozawa (2016a). The appendix in Shiozawa (2014) gives a more detailed analysis. Matrix M in Eq. (1) is the diagonal matrix whose jth diagonal is the markup rate m_j for product j.

rate.[7] In mathematical expression, the theorem assures for a given wage rate w the existence of a price vector p that satisfies the system of inequalities:

$$(I + M)\{w\mathbf{a}_0 + A\mathbf{p}\} \geqq \mathbf{p} \tag{2}$$

for all rows, each of which corresponds to a production technique, and for any given product equality holds at least for the row that corresponds to a production technique which produces the product. Readers of this book will easily notice that the fundamental theorem of the new theory of international values (Chap. 1, Theorem 3.4, Eq. iii) is a simple generalization of (2).[8] Sraffa (1960) made no remarks on this theorem, although he explicitly treated the question in Part III, Switch in Method of Production.[9]

As I argued in Shiozawa (2016a, 2017a), the classical theory of value took a complete form only in the latter half of the twentieth century. Evidently, this is in contradiction to Mill's contention that the classical theory of value was completed by 1848 (Mill: Library, III.1.2). Although Ricardo could not develop a theory of simultaneous equations or know the minimal price theorem, he understood by his deep insight that the value of a commodity stays constant regardless of sudden changes in the demand or supply of the commodity and that this occurred in the presence of a plurality of production techniques. He knew that "prices always vary in the market, and in the first instance, through the comparative state of demand and supply" (Ricardo: Sraffa, p. 119; Library, 6.28,). However, he knew also that the market price returns to the natural value, that is, to the cost of production, after the first disturbance settles down.

In Ricardo's theory of value, demand and supply play no role except that they are only disturbing factors. The values are determined by production conditions together with normal profit margins. If demand were to change, production would change correspondingly. If the production adjustment were to proceed, the market price would return to the natural value.[10] In this way, supply and demand have little to do with the determination of values. This abstraction of demand and supply relations from the theory of value is remarkable, because it recognizes that the essential

[7]It is widely reported in the literature that the minimal price or non-substitution theorem holds only when there are no fixed capital goods. This is a serious misunderstanding because the theorem is valid in the situation in which durable capital goods retain their efficiency within the depreciation period and are discarded thereafter. See Shiozawa (1975).

[8]In Chap. 1 of this book, A is the matrix of net production coefficients modified to an equivalent system. In this chapter, A denotes the matrix of input coefficients. Equation (iii) can be rewritten in this chapter's notation exactly in the same form as (2) if vector $\mathbf{w}$ were replaced by scalar w.

[9]Sraffa might have been aware of this theorem in the form of what he dubbed "Borkiewicz's dictum." See Gehrke and Kurz (2006).

[10]This became much more conspicuous because production adjustment speed became faster in many industries. Since the twentieth century, modern industrial firms have had less necessity to appeal to price adjustment in order to adapt to demand changes.

feature of a modern economy is the primal independence of values and quantities. Values are determined by production conditions that are mainly technological. Quantities are regulated by the amount of the demand that is expressed in the market at a given value of the product. This was the real substance of the Ricardian revolution. In an industrial society, Ricardo observed that it is the conditions of production that determine value relations. In fact, if the demand for a product were to change for any reason, the volume of production could change and be adjusted to the changed demand. As long as the cost of production remains constant, there is no reason that prices should change.

Few economists, in Ricardo's time and after, understood this. Ricardo must have had ideas for a kind of quantity adjustment. In fact, he argued as follows:

> the proportion between supply and demand may, indeed, for a time, affect the market value of a commodity, until it is supplied in greater or less abundance, according as the demand may have increased or diminished; but this effect will be only of temporary duration (Ricardo: Sraffa, p. 382; Library 30.1).

It is clear that Ricardo assumed that the supply of a commodity or its production would soon be adjusted to the demand for the commodity. If not, there would be no reason for the price to return to the original (natural) price. In this sense, Ricardo had in his mind a quantity adjustment mechanism although it was quite different from what we usually consider this term to mean. The classical theory of value presupposes that supply adjusts to changes in demand. This is a typical situation in the capitalist economy. Few economists in Ricardo's time understood that behind his theory of value lay a special supply assumption. It was not stated explicitly, but this principle of effective demand at the firm level is a twinned theory to Ricardo's price theory (Shiozawa 2016a).

Here, it is necessary to note that even after Ricardo, few people, including economists, understood this as the core of the Ricardian theory of value. I argue this point in the next section. The view that prices regulate discrepancies of supply and demand was common knowledge. Ricardo objected to this common sense. Mill was a good and loyal interpreter of Ricardo but he was the first to notice that the cost of production theory of value did not hold in the case of international trade. Probably after a long reflection, Mill claimed, albeit cautiously, that "we must revert to a principle anterior to that of cost of production, and from which this last flows as a consequence,—namely, the principle of demand and supply" (Mill 1844, Library, I.19; a similar expression appears in Mill 1848, Library III.16.5). After Mill, all three founding fathers of neoclassical economics (i.e., Jevons, Marshal, and Edgeworth) accepted Mill's contention and began to argue on the basis of demand and supply.

My main contention in this chapter is that Mill's "solution" was in fact misdirected, and his conclusion that recommended the reversion to the law of demand and supply wrongly guided economics toward the neoclassical revolution.

Mill started a tradition to examine the interior vertex of the production possibility frontier (PPF). This is the point that I name the Mill–Jones point. The tradition continued until very recently, as we observe in Sect. 6. I myself have fallen victim to this tradition (Shiozawa 2007). This paper was intended to be a successor of an earlier paper (Shiozawa 1985) and reflects that for more than 20 years, I continuously attempted to find a theorem that might provide sufficient conditions for the existence of interior vertices in models of input trade. Shiozawa (2007) was written when I found such a theorem (Theorem 3.4), although I cannot state that it was a very insightful one. Soon after, I was freed from my fixed idea that I have to search interior vertices, and my obsession fell away. I finally came to understand that Mill in the mid-nineteenth century and Ronald Jones in the mid-twentieth century were completely misguided by their obsession with price adjustment. A naive simple method to rebuild the international value theory would be to focus not on the interior vertex but on the points in the interior of facets. When I acknowledged this, I started to write papers that would come to comprise Shiozawa (2014). The new theory of international values was an extension of the cost of production theory of value. For me, it meant a revival of the classical theory of value.

After the 1960s, long discussions were held on the nature of neoclassical economics. Many economists revealed that microeconomics comprises vital flaws. Examples comprise the Sonnenschein–Mantel–Debreu theorem (inadequacy of representative agents), ubiquity of bounded rationality (impossibility of optimization), nonexistence of auctioneer (ubiquity of bilateral exchanges), capital reverse and reswitching (nonexistence of scalar quantity of capital), inconsistency of the neoclassical theory of firms (irrelevance of marginal productivity theory), and many other miscellaneous flaws (Beinhocker 2006; Keen 2011; Shiozawa 2016b). All these flaws come from the basic structure of neoclassical economics, which relies on individual agents' ability to adjust everything instantaneously. Such adjustments might be possible for a small economy composed of two or three people and a few kinds of goods. This is why neoclassical economics textbooks prefer to talk about Robinson Crusoe. All these defects result from the basic characteristics of neoclassical economics, that is, it is the economics of exchange. In contrast to the economics of production, the economics of exchange is intended to be a universal theory of economic activities. This abstract nature of neoclassical economics caused it to rely overly on individual human ability, because it could not assume any social structure that supports the economic activities of individuals. This is why in neoclassical microeconomics, omnipotent and omniscient human agents are generally assumed. Ricardo's cost of production theory of value assumes a concrete economic structure. There is no need to assume such agents. The prize at stake of the option between the two value theories is extremely large.

Mainstream macroeconomics now normally examines whether a theory has a micro foundation. However, this in no way guarantees that it is rational. Some economists have pointed out that the macroeconomics of these 30 years was spectacularly useless if not positively harmful. Minor inventions to adjust and modify macroeconomic models cannot save this state of affairs, because it is based on fundamentally flawed microeconomics. Insofar as microeconomics remains intact, micro-founded macroeconomics remains a fictitious entity ungrounded in any reality. It is now time to change our paradigm.

This paradigm change necessitates a new theory of value. Fortunately, it exists already. It is the classical theory of value. We can start from this theory. The classical theory of value once had a grave weak point. It lacked the theory of international values. Mill attempted to address this problem and he conceded to the law of demand and supply, because he could not build a theory of international values as an extension of Ricardo's theory of domestic values. The question of international trade was too important to neglect as an exception. As we see in Sect. 7 and sections that follow, this paved the way for neoclassical economics. However, currently, the theoretical situation has changed. We have a theory of international values that is an extension of the classical theory of value. The classical theory of value once had to cede its place of the economics orthodoxy to neoclassical economics, but now can resume its place.

In the following sections, we examine how neoclassical economics emerged from classical economics in the UK. As I have cautioned, I will not consider the neoclassical revolution in continental Europe or any other place. Before embarking on a concrete examination of this topic, it is useful to situate the neoclassical revolution in a wider historical context. Thus, we can link individual theorists' worm's-eye view contributions using a bird's-eye view of the history of economic theories.

4 The Neoclassical Revolution in a Wider Historical Context

The main feature of the neoclassical revolution was a change from plutology to catallactics or a change from the economics of production to the economics of exchange. However, this revolution was not a simple revelation of a new truth. Compared to the classical theory of value, the neoclassical theory of value was a retrogression to an old common sense. Almost one and a half centuries have passed since the arrival of neoclassical economics. During that time, it has been refined via the introduction of many components. First is the concept of marginal utility, followed by marginal products, preference order, the isoquant curve, decreasing returns to scale, the law of diminishing returns (due to input substitution), smooth production functions, constant elasticity functions, representative agents, fixed point theorem, (unbounded) rationality, rational expectations, stochastic dynamics, and many others. Does this prove there was a tremendous development of economics? Or was this an accumulation of rubbish in an attempt to rescue and beautify the vital flaws of neoclassical economics? The neoclassical revolution must be situated in a wider historical context. For this purpose, let us compare the history of economics with the history of science.

In the history of science, many wrong ideas were accepted as truth for a long period. Many school textbooks tell only how true theories were discovered by great scientists. However, as Thomas Kuhn (1962, Chapter 1 in particular) emphasized, true history is more sinuous and full of deviations. Science developed in the wrong direction for a prolonged period. Wrong ideas persisted and even developed in some cases. The progress of science is not necessarily "development by accumulation."

Once an intellectual tradition is established, regardless whether it is true or not, it continues in some cases for a long time before it comes to be understood as wrong.

For example, the phlogiston theory in chemistry survived for more than a century until Lavoisier established the oxidation theory of combustion. The aether hypothesis on the propagation of light was accepted for more than three centuries until Einstein announced his special relativity theory. Sometimes, an erroneous hypothesis has come to be strengthened by some of the most proficient scholars. In the case of the aether hypothesis, Newton played such a role by assuming an "aethereal medium" to explain refraction and diffraction.

At the time of a scientific revolution, something different to normal development occurs. Kuhn called it a paradigm change in which many parts of an old theory are thrown out simultaneously. However, the neoclassical revolution in economics is much more peculiar than common science revolutions. My view on the history of economics is quite inverted from the common understanding. In my view, there was once an essentially correct theory but it succumbed to a new theory. The latter was proliferated for more than one and a half centuries, but now, the old theory is being resurrected as a new hope. Even the history of science does not contain many similar stories. The only story I can cite is the heliocentric theory. It is commonly known that geocentrism occupied the mainstream of astronomy for a long time and was repudiated by Copernicus and others. The Copernican revolution paved the way for Newtonian physics. However, heliocentrism was not a new idea in a strict sense. We knew that a Greek philosopher named Aristarchus of Samos preached heliocentrism in the Alexandrian age before Claudius Ptolemy wrote or edited *Almagest*. The Ptolemaic system remained unchallenged for more than 1000 years until the arrival of the modern science age. Unfortunately, Aristarchus's paper was lost in the history of time, and only the knowledge was relayed that an idiosyncratic scholar named Aristarchus had advanced a theory that the earth revolves around the sun.

In considering this story, Ricardo is comparable to Aristarchus of Samos. Since the time of Mill until today, neoclassical economics (geocentricism in my comparison) occupied the mainstream of economics while Ricardo (Aristarchus in my comparison) was a subject for only historians of economic theory. In economics, we are now expecting what is akin to a Copernican revolution.

Some readers might object to my comparison by pointing out that Ricardo was much more important and influential an economist than Aristarchus of Samos was an astronomer. Indeed, Ricardo was once the most revered economist in England. Keynes wrote in his *General Theory* that "Ricardo conquered England as completely as the Holy Inquisition conquered Spain" (Keynes 1936, p. 32). Of course, this is Keynesian rhetoric. It does not prove how Ricardo was accepted in the nineteenth century. Ricardo might have had strong influence on economic policies, but policies and theories are different entities. I doubt that Ricardo's theory of value was really understood by a wide range of his contemporary economists.

If Ricardo was established as a firm orthodoxy, it is incomprehensible why there was a flood opposition to Ricardo's theory just after his death. A second-class economist, like Robert Torrens and members of the Political Economy Club, although it was founded by James Mill, a close collaborator of Ricardo, could claim that all Ricardo's great principles had been abandoned as erroneous by 1831 (Dobb 1973, Chapter 4). In England, few economists understood Ricardo's theory of value. Those who did might include James Mill, John Ramsey McCulloch, and John Stuart Mill. Ricardo's contemporary economists, such as Thomas Malthus and Jean-Baptiste Say, never agreed with or did not understand Ricardo's theory of value. Nassau Senior, whom I mention briefly in Sect. 11, was fundamentally opposed to Ricardo's theory of value (Bowley 1937[2010], pp. 17–19), although his theory is difficult to understand as an integral whole. Ricardo's objectivist theory never really infiltrated France (Faccarello 2014). The case of the German-speaking world is not very different from that of France. Major interpreters did not go further than the works of J. B. Say (Gehrke 2014). This fact partly explains why Karl Heinrich Rau preceded the Marshallian cross for many years.[11] Marx praised Ricardo as his precursor and as a superb analyst of the capitalist system but never understood (perhaps intentionally) Ricardo's cost of production theory of value. If Dobb's (1973) estimate is correct, Ricardo's theory of value was prestigious for only about 10 years. Even counting all the years after the publication of Ricardo's *Principles* until the arrival of the marginal revolution, the sum is only about 50 years. The period of neoclassical dominance is now about three times as long as the latter interpretation of the Ricardian years. Thus, Ricardo was an ephemeral phenomenon.

Aristarchus of Samos was once a famous mathematician and astronomer, but he was almost completely forgotten with the arrival of Claudius Ptolemy, and thereafter, Aristarchus was cited only as the creator of a strange doctrine claimed before the Ptolemaic system. If we observe Ricardo from afar, Ricardo is quite similar to Aristarchus. Ricardo is famous for an extreme case of classical political economy, but few economists have studied his theory of value seriously.

Keynes's (1936) parable of the Holy Inquisition is indicative. He mentions only the lack of an aggregate demand concept in Ricardo. Keynes was not fair to cite Ricardo in this way. Keynes's point had been to make a clear contrast between Say and Ricardo on the one hand and Malthus on the other.[12]

Keynes was correct when he stated that considerations on effective demand had been lacking in economists before him. If Keynes had really understood Ricardo's theory of value and if he had asked advice from P. Sraffa, he could have built his theory of effective demand in a completely different way. As I claim elsewhere, the theory of effective demand can and should be constructed on the basis of the classical theory of value (Shiozawa 2016a, Sect. 7; 2017a, Sects. 5 and 6). Keynes

[11] Blaug (2001, p. 159) reported that "the first appearance of subjective value theory and a demand and supply diagram-with price on the vertical axis as in Marshall-was in the fourth 1841 edition of Rau's Grundsätze der Volkswirtschaftslehre (1826)."

[12] Keynes cited the following economists on the side of Malthus: K. Marx, Silvio Gesell, and Major Douglas, who remained in the underworld.

missed his opportunity to define his concept of effective demand on a firmer basis if he had accepted Ricardo's theory of value. The Cambridge tradition did not permit him to do that. At the beginning of the twentieth century, Ricardo's idea had been almost completely wiped out, even in Cambridge, England.

Ricardo lived in reality in "the underworld" for a long time and remained a holder of a curious theory, as heliocentrism was before Copernicus. My aim in this chapter is to show how this occurred despite the appraised authority of Mill. The young Mill attempted to understand Ricardo's problem and to provide a "solution" for it. His solution was the real bifurcation point between classical and neoclassical economics. In the following sections, I show how Mill's solution was misguided and how his solution determined the path of economics after him.

Let me add a few more words on the Ptolemaic system. I remarked that there are parallels between Aristarchus of Samos and heliocentrism on the one hand and Ricardo and his cost of production theory of value on the other hand. This parable inevitably implies that neoclassical economics is compared to geocentrism. In this comparison, they have many common points. We know that the Ptolemaic geocentric system is a finely constructed precise science. Even after the publication of Copernicus's book, *On the Revolutions of the Heavenly Spheres*, in 1543, many eminent astronomers supported the geocentric system. Tycho Brahe (1546–1601), a Danish noble who could construct precision machinery for celestial observation, was among them. Aided first by the Danish King Frederick II and then by the Holy Roman Emperor Rudolf II, he was able to build observatories in Denmark and, later, in Prague. During his lifetime, he was the most precise and comprehensive observer of the celestial world. This reputation attracted Johannes Kepler to work as an assistant to Brahe for a short time. At this time, many astronomers supported the heliocentric system, including Kepler, but Brahe continued to retain the geocentric system. With small variation of the Ptolemaic system, Brahe could coherently predict the planetary movements within the observational error range. The heliocentric system became more precise than the Ptolemaic system only when Kepler discovered his three laws of planetary motion.

The Ptolemaic system was a complex system according to which planets move in a circular epicycle whose center moves around on a deferent, a circle with its center near to the earth (the offset is called eccentric). Planets move at a constant angular velocity viewed from an equant point that is also displaced from the center of the deferent. By this complex configuration, Ptolemy could explain the motions of planets, including retrograde motion and change of brightness. To improve the precision of prediction, Ptolemy and later astronomers added epicycles to epicycles. In this way, the Ptolemaic system became increasingly complex each time a new adjustment was added. In time, 70 cycles and spheres were employed.

The Ptolemaic system and its history are quite similar to neoclassical economics. Neoclassical economics can generate various models and can adapt them to any observed data. However, it is less precise than the Ptolemaic system. The theoretical components of the neoclassical system, such as marginal utility, marginal products, preference order, isoquant curves, and decreasing returns to scale, are similar to the deferent, eccentric, epicycle, equant, and other principles of the Ptolemaic system.

The former set of components was invented in order to keep the neoclassical system coherent without enquiring about the deep meaning of these abstract entities. Over the course of one and half centuries, neoclassical economics has accumulated too many irrelevant theoretical components. Now, many economists acknowledge that the neoclassical system has become too complicated and apologetic and contains no deep insights. The key concepts of the Ptolemaic system, such as the principles of the deferent, eccentric, and epicycle, lost value just after the Newtonian revolution. Similarly, after the Copernican-like revolution in economics, the key concepts of, for example, marginal utility, marginal product, preference order, and isoquant curves, will lose their meaning and will be discarded.

5 Mill's Unintended "Solution"

There is no reason to doubt Mill's good faith when he wanted to advance Ricardo's theory one step forward. Ricardo explained how trade is possible with gains for both countries when one country is more productive in absolute terms than the other country in two industries. In Ricardo's explanation, prices in international trade were simply assumed to be given and different from those of both countries when they produced without foreign trade. Ricardo's system comprised three theories of domestic values: prices of reproducible goods, prices of nonproducible goods, and the theory of rents. In chapter 7 on "Foreign Trade" in his *Principles*, Ricardo added a fourth element: prices in international trade.

Ricardo knew well that "[t]he same rule which regulates the relative value of commodities in one country, does not regulate the relative value of commodities exchanged between two or more countries" (Ricardo: Sraffa 1951, p. 133; Library, 7.10). However, Ricardo did not or could not produce any theory concerning how international values are determined.

As we see later in this section, the famous illustration using four magic numbers is a repetition of the logic that Viner (1937, p. 440; Library) named the "18th century rule." Viner was referring to Martyn's reasoning, but it is doubtful if it is appropriate to call the reasoning the "18th century rule," because Martyn's pamphlets (1701; 1720) were not widely distributed and remained obscure until the nineteenth century, when it was republished by McCulloch (1856). It is true that Adam Smith (Smith 1776; Library IV.2.11–12) made a similar argument to Martyn's, but there is no evidence that Smith had read Martyn's pamphlet.[13]

Another point that requires an adjustment to the traditional explanation is that Ricardo had discovered the rule to determine the specialization pattern by taking the ratios of two coefficients and comparing them. In view of Ricardo's footnote (Sraffa 1951, p. 136; Library 7.17 footnote 20), it is possible that Ricardo had

[13]See Maneschi (2002) for a history of acceptance of the pamphlets. The author of the pamphlets was not known until Christine McLeod (1983) established the author's identity.

compared the two ratios. However, it is doubtful whether Ricardo considered the task to determine the specialization pattern an essential part of his trade theory. He provided an example in which we can easily determine it, but Ricardo himself was thinking of a situation in which countries were not completely specialized. As he discussed in a later part of the chapter, Ricardo was considering the case in which many commodities are traded simultaneously. We find no mention of how to deal with these cases.

Despite all these ambiguities, Ricardo's example was ingenious. Paul Samuelson rightly named it Ricardo's four magic numbers. Mill started to study Ricardo from childhood. He was particularly interested in problems that he believed Ricardo had left behind. One such problem was to determine the terms of trade. The terms of trade are an expression to indicate the relative prices between exports and imports. In the two-commodity case, the terms of trade are the same as the relative price or the value of the two goods in the international market.

As Yukizawa (1974), Ruffin (2002) and Maneschi (2004) clarified,[14] Ricardo simply assumed that international exchange values were different from those of domestic exchanges. However, he gave no hint about how these values were determined and how they were related to each other. He simply remarked that "the same rule which regulates the relative value of commodities in one country, does not regulate the relative value of commodities exchanged between two or more countries" (Sraffa 133; Library 7.10).

Mill found that the terms of trade are not determined in Ricardo's *Principles*. Mill's judgment was correct. It is reasonable and justifiable that Mill set the problem in search of the logic that determines the relative value of commodities. Mill might have attempted to generalize Ricardo's cost of production theory of value into this international exchange situation. On this point, we have no record that tells how Mill thought about this. Even if he wanted to do make this generalization, it would be difficult to undertake because, in the time of Mill, there were no theories of convex polytopes and linear inequalities.[15]

Without any suitable mathematical tools, Mill and his followers were obliged to work on a simple two-country, two-commodity economy. If it were possible to imagine a case of many commodities, general analysis of such a case would have been very difficult in the middle of the nineteenth century. In fact, as we see later in Sect. 11, two or three economists wanted to move beyond the two-commodity case, but they could not elucidate why the case of two commodities and the cases of three or more commodities were different and where the difference arose.

The trouble with the two-country, two-commodity case lies in the fact that on the production possibility set (PPS), there is only one point at which both countries obtain gains from trade. It was a situation of complete specialization. Geometrically,

[14]For a detailed history on this connection, see Tabuchi (2017a).

[15]A true solution requires these theories, as revealed by Shiozawa (2014).

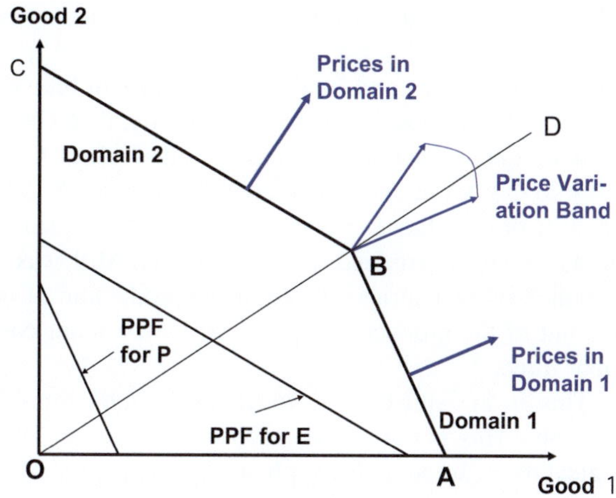

Fig. 1 Production possibility frontier of a two-country, two-commodity economy (*bold lines* in the first quadrant)

it is an internal vertex (extreme point)[16] of the PPS and a unique point in the positive quadrant (or first quadrant). Mill and his followers did not recognize that this was a singular circumstance that appears only for the two-country, two-commodity case. We can easily explain how Mill was forced to consider an economic situation that he had no particular intention to examine. Figure 1 shows a standard world PPF for the two-county, two-good case.

The PPF is the set of maximal points of the PPS. The frontier of the two-country, two-commodity economy has two segments: AB and BC. This feature or configuration does not change when the coefficients change. Let us omit two points at the coordinates, because they are degenerated cases. Any point between A and B has a price vector that is identical to the price vector in country P. Any point between B and C has a price vector that is identical to the price vector in country E. In both cases, one country enjoys no gains from trade, because the prices are the same as those of a closed economy for one country.[17] Mill excluded these cases because he considered that trade should be beneficial for both countries. That situation occurs only at point B. At that point, both countries P and E can enjoy gains from trade. Mill considered this to be the situation to examine and asked how the terms of trade are determined at that point. The young Mill did not know how grave a meaning this simple reasoning would have for the destiny of economics.

Point B is a point of complete specialization. A country has only one commodity that is competitive in the world market. In this situation, when the labor input coefficients are fixed and each country has a determined quantity of labor or

[16]The word "internal" here means "in the interior of the positive orthant." Vertices are always on the boundary of a polytope and never in its interior.

[17]From Fig. 1 itself, it is not clear that the wage rates of both countries are determined uniquely at the same time as prices. The international value system comprises wages of all countries and is uniquely determined when world demand lies in the interior of a facet or on an open domain of the frontier. See Theorem 3.4, Chap. 1 of this book.

labor power, the quantity of the product a country can produce competitively is determined. In Mill's example (Mill 1844, I.10–15; 1848, III.18.6–17.), Germany specializes in linen and England specializes in broadcloth (Mill 1844, I.10; 1848, III.18.7). If the model is interpreted strictly, then Germany employs all workers to produce linen and England employs all workers to produce broadcloth although Ricardo did not imagine such an extreme case. As labor power is determined, the quantity of linen that Germany produces and the quantity of broadcloth that England produces are determined. In other words, Mill was led to observe an economy in which two countries engage in the production of one commodity and the total amount of the product is determined by the input coefficients and the size of the labor force.

This is the situation in which Mill found himself in his international trade example. This is but a pure exchange economy. The above story is of course a logical reconstruction. No explicit words appear in Mill (1844, 1848).[18] Nominally, we have production in Mill's trade economy. However, the amount of production is predetermined, and a country can obtain the other commodity only by the exchange of its product. If we consider the situation in which two sides engage in negotiation, it is the same situation in which two sides negotiate with each commodity in their backyards.

Thus, Mill was guided by his good faith from Ricardo's production economy to a pure exchange economy, which by their characters is a totally different situation to what Ricardo examined.

How Mill solved the determination problem of the terms of trade is not of primary importance. The most important fact is that Mill displaced the problem from production economics to exchange economics. He did so unconsciously. I contend that this was the very point at which the economics of production of the classical political economy were replaced by the economics of exchange in the neoclassical tradition. After Mill, most trade theorists continued to examine the situation that Mill set and could not escape from it except for a few exceptional cases.

6 How Was Mill's Solution Misplaced?

Now it is time to examine Mill's "solution." Was it a really solution to Ricardo's problem of constructing a theory of international values? Was Mill's solution a unique possible solution to the theory of international values? By no means. Mill's solution was an observation of an exceptional case. Mill and many other economists in international trade theory mistook this exceptional case as a representative situation of international trade. Let me explain first using Fig. 2.

Figure 2 presents the world PPF in the case of two countries and three commodities. It is called the minimal model of Ricardian trade theory. I justify this name soon.

[18]Mill (1844, I.10, 11, 15) and Mill (1848, III.18.8) may have been a poor account of the logic here explained.

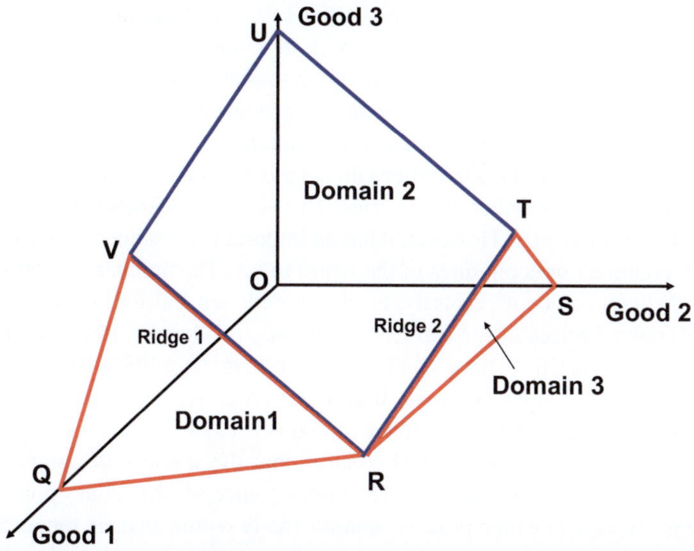

Fig. 2 A minimal model of Ricardian trade theory (two countries and three goods economy)

The world PPF in Fig. 2 consists of three facets (open domains), eight ridges, and six vertices. The first observation is that there is no vertex in the interior of the positive orthant. No point like B in Fig. 1 exists. We call such an internal vertex (or extreme point) the Mill–Jones point.[19] A formal definition of the Mill–Jones point is any vertex of the world PPS that lies in the interior of the positive orthant. Let us call such a point the PPF interior vertex. All coordinates of such a point must be positive. It is an extreme point of the world PPS (we omit the modifier "world" hereafter).

As we have observed, no Mill–Jones point exists in Fig. 2. One may consider that this is a mere exception, but it is not. No trade model has the Mill–Jones point if the number of commodities exceeds the number of countries. A vertex represents a situation of complete specialization. This means that every country has only one commodity that is competitive in the world. In the positive orthant, all commodities must be produced by at least one country. Then, there is one-valued mapping from the set of countries onto the set of commodities. Therefore, if a Mill–Jones point exists, the number of countries is equal to or greater than the number of commodities. In the Fig. 2 case, the number of commodities is greater than the number of countries, and it is impossible to have an internal vertex. In general, if the number of commodities is greater than the number of countries, no Mill–Jones point exists.

[19]Some call this point the "Ricardo point" or "Ricardo's Limbo point." Because these are not suitable names, I do not adopt them. The reason for my naming is given later.

There are vertices at the boundary of the PPS. At these points, one or more coordinates vanish. This means that one or more commodities are not produced at all. These are degenerate cases, and we do not consider them, because they represent a situation in which some commodities are not produced at all. We cannot state that there are three produced commodities in this situation.

Mill–Jones points or interior vertices do not exist when the number of commodities is greater than the number of countries. This simple proposition went unnoticed nearly 150 years after Mill. However, it has an important consequence. We can count about 200 countries or economies in the world today. The number of commodities, although difficult to count precisely, easily exceeds ten million. For example, the reference book of prices of the former Soviet Union's Gosplan (the State Planning Committee) is reported to have contained more than 20 million items. We could assume that any developed economy trades and produces more products than that number. On the other hand, the number of countries and economic zones is of the order of 200. Then, we could safely assume that the number of commodities is bigger than the number of countries. The nonexistence of Mill–Jones points means that there is no point at which prices can move freely within an open range. Scholars after Mill continued to attempt to determine prices on the assumption that they have some margin of free movement, but there is no such possibility when the number of commodities is bigger than the number of countries.

What happens at a point other than the Mill–Jones point? If we consider Fig. 2 again, we easily know that most points of the PPF lie in either of three domains. In Domains 1 and 3, the prices are proportional to those of a country. Mill excluded this situation and considered that if trade were to continue, both countries must have certain gains from trade. In the two-country, two-commodity case, such a point—point B of Fig. 1—is unique. In the case of the two-country, three-commodity case, there is a wide Domain 2. Any point in Domain 2 has a system of international values that remains constant as long as the point stays in Domain 2. This system of international values is different from either of the two countries' domestic values, and the two countries (and employed workers in both countries) can enjoy gains from trade.

There is a possibility that world production or demand comes on one of two interior ridges, RV or RT. We can exclude the points on the border of the frontier, as they are degenerate points. At a point on RV, for example, the prices can vary but they must remain perpendicular to the ridge RV. The degree of freedom of those prices is one-dimensional. Moreover, price changes have no effects on moving world demand and production. The price variation changes the exchange ratio between good 1 and a combined set of goods 2 and 3 but has no effect on changing demand from a point on RV to another point on RV. In fact, ridge RV is parallel to ridge TU and to the plane supported by axes OS and OU. All points on RV have the same quantity of good 1. The same explanation holds for ridge RT. Then, price adjustment does not work at any point on the frontier.

Domains 1 and 3 are cases that are sometimes studied as big country cases. Although Domain 2 in Fig. 2 is in a similar situation to point B of Fig. 1, they present very different characteristics. At point B, prices move freely within a certain range (the degree of freedom is the same as the number of commodities minus 1), and world production is fixed. At any point in Domain 2, the price system remains constant, and world production can change freely in the domain. These characteristics are quite similar to the classical value theory. There is only one value system, and supplies can be adjusted to any effective demand as long as it stays in Domain 2.

Now, the whole picture becomes clear. Mill examined a two-country, two-commodity case believing that the model provides a representative situation, thereby falling into an unexpected trap. He did not imagine that the situation changes drastically and essentially when the number of commodities is bigger than the number of countries. If we were to admit that the number of commodities exceeds the number of countries, the minimal model of international trade should be a two-country, three-commodity case. This is the reason I call Fig. 2 the minimal model of Ricardian trade theory.

Mill started a tradition to consider a complete specialization case, which meant as a logical consequence examining a pure exchange economy, because in such a case, the product and volume of each country's production would be determined and the countries (or representative agents) would negotiate the exchange ratios between commodities. This tradition was adopted by economists after Mill and remained in place even in the middle of the twentieth century.

In the 1950s, there was a kind of resurrection of Ricardian trade theory. Following the works of Frank D. Graham, Lionel McKenzie and Ronald Jones extended the Ricardian model from a two-country, two-commodity case to multi-country, multi-commodity cases. This was a great step forward in the Ricardian tradition, even though they conserved the main problem setting. McKenzie (1954a, b, 1955) wrote three papers on the Ricardian trade model and Jones (1961) followed him. Their works marked a new era as their analyses were based on the new mathematical tool developed by activity analysis, which was a topical trend at that time. Curiously, McKenzie dealt with cases in which the number of countries was equal to or greater than the number of commodities. Jones (1961) discovered a famous formula that I explain briefly hereafter. The formula provides a necessary and sufficient condition for realizing a pattern of full specialization ("efficient assignment" in Jones's terminology). This formula proves that there is at most one internal vertex in the N-country, N-commodity case. In addition, Jones considers unequal cases, that is, the case in which the number of countries is not equal to the number of commodities. However, as his definition of "class" (Jones 1961, p. 164) shows, he is interested in cases in which the number of countries is greater than the number of commodities. Why did McKenzie and Jones examine such singular perverse cases, whereas common sense or simple observation tells us that the number of commodities is far greater than the number of countries?

Jones's formula is expressed in the form of next theorem.

[Jones's Theorem] (Jones's formula)[20]

Let a_{ij} be labor input coefficients for country i to produce commodity j. Suppose there are N countries and N commodities. Then, the following two conditions are equivalent:

(1) Positive wage rates $w_1, w_2, \ldots, w_N$ exist, which satisfy

$$
\begin{aligned}
w_1 \cdot a_{11} &< w_2 \cdot a_{21}, w_3 \cdot a_{31}, \cdots, w_N \cdot a_{N1};\\
w2 \cdot a_{22} &< w_1 \cdot a_{12}, w_3 \cdot a_{32}, \cdots, w_N \cdot a_{N2};\\
&\cdots\\
w_N \cdot a_{NN} &< w_1 \cdot a_{1N}, w_3 \cdot a_{3N}, \cdots, w_{N-1} \cdot a_{(N-1)N}.
\end{aligned}
$$

(2) For any permutation τ of N indexes that is different to identity,

$$a_{11} \cdot a_{22} \cdot \cdots \cdot a_{NN} < a_{1\tau(1)} \cdot a_{2\tau(2)} \cdot \cdots \cdot a_{N\tau(N)}.$$

If one of two equivalent conditions is satisfied, each country i can specialize in the production of commodity i competitively, and this is the unique efficient pattern of specialization.

The second condition can be translated as follows: the left-side product is the strict minimum among permutation products in the form of the right side. Condition (1) is equivalent to the existence of the internal vertex or positive extreme point.

This is a beautiful generalization of Mill's formula (a comparison of two ratios from four magic numbers) for comparative advantage. In Mill's case, country 1 specializes in commodity 1, and country 2 specializes in commodity 2 when two ratios satisfy the following relationship:

$$a_{11}/a_{12} < a_{21}/a_{22}. \tag{3}$$

Here, a_{ij} are labor input coefficients. The condition (3) is equivalent to

$$a_{11}/a_{21} < a_{12}/a_{22}. \tag{4}$$

These two conditions are in turn equivalent to

$$a_{11} \cdot \mathrm{a}_{22} < a_{12} \cdot a_{21}. \tag{5}$$

[20]Jones (1961) effectively proved the proposition (1) ⇒ (2), but it is doubtful if Jones proved the inverse proposition (2) ⇒ (1). See Shiozawa (2015, Sect. 10).

Jones's theorem tells us that the above equivalent conditions (3), (4), and (5) are also equivalent to *the existence* of positive numbers w_1 and w_2, which satisfies the conditions:

$$w_1 \cdot a_{11} < w_2 \cdot a_{21} \text{ and } w_2 \cdot a_{22} < w_1 \cdot a_{12}. \tag{6}$$

This equivalence theorem can be demonstrated as follows. If positive numbers w_1 and w_2 that satisfy (6) exist, by multiplying both sides of the two inequalities, and eliminating $w_1 \cdot w_2$ from both sides, we obtain (5). The converse also holds. In fact, if we assume that condition (4) holds, we can take positive numbers w_1 and w_2 in such a way that

$$a_{11}/a_{21} < \mathrm{w}_2/\ \mathrm{w}_1 < a_{12}/a_{22}.$$

Then, we can derive the first inequality from the left inequality of (6) and the second inequality from the right inequality of (6).

The left inequality of (6) means that the cost of production of commodity 1 in country 1 is cheaper than the cost of production of commodity 1 in country 2. In other words, the production of commodity 1 in country 1 is more competitive than that in country 2. In the same way, the right inequality of (6) implies that the production of commodity 2 in country 2 is more competitive than that in country 1. With wage rates w_1 and w_2 that satisfy (6), we obtain an economy in which country 1 specializes in commodity 1 and country 2 specializes in commodity 2. Thus, the abovementioned equivalence theorem implies that the real cost and money cost approaches are in fact equivalent in the two-country, two-commodity case.

Jones's theorem gives a sufficient condition for the existence of a Mill–Jones point.[21] In addition, the theorem implies that if Mill–Jones points exist at all, they are unique. Thus, we can observe keen interest in Mill–Jones points at the time of Ronald Jones. However, as we have observed, no such point exists when the number of commodities N is bigger than the number of countries M. If $M = N$, there is at most one Mill–Jones point. If $M > N$, many Mill–Jones points may exist. In fact, there are generally as many Mill–Jones points as the number of "classes." If Jones knew all these facts, why did he not consider the nonexistence case? It must have been easy for him to reason that an internal vertex did not exist in the case of $M < N$. This strange fact proves how the tradition Mill started was strong and binding.

Mill began a well-established tradition that he may not have intended to. Mill might have been guided to this tradition simply by the easiness of examination, but the stakes were large. This simple examination marked the start of new economics, which later came to be named neoclassical economics. Jones was a deep thinker but was fully immersed in the tradition of neoclassical economics. He was too preoccupied by the price adjustment framework to examine the most common cases.

[21] For the existence of an internal vertex, it is necessary and sufficient that the permutation products have a unique strict minimum.

The cases $M < N$ were not suitable for the price adjustment paradigm and were excluded from examination. In the time of classical political economy, this paradigm was not such an established tradition. Mill, intentionally or not, set a problem that had not been solved by Ricardo and, thus, started the new economics of exchange.

7 Mill's Solution and After

Mill posed his question. How it was solved is not important. What matters is how he posed it. We know his solution well. The international values come to an intermediate point of two extremes that are values of each of two countries. Mill showed how, at a value, two countries would exchange their products for the products of the other. Existence is not discussed in a modern way. In Sects. 6, 7 and 8, which were added in the third edition, Mill started discussing the question of multiple equilibria, but he gave no sufficient condition for a unique solution. These questions were studied by Alfred Marshall, and we provide a short comment in the section on him (i.e., Sect. 10).

Despite the inadequacy of Mill's solution, I have no intention of criticizing the lack of precision of his solution. At the time of Mill, that is, in the middle of the nineteenth century, mathematics itself was still intuitive rather than logical. Even many years later, Jevons and Walras were satisfied by counting the numbers of equations and unknowns. What is more important and crucial to the history of economics is what Mill believed to have established in his *Principles*. It is sometimes difficult to distinguish the influence of his solution from the unsettled trade question and other examinations he made in the course of his economics formation. Although I admit that the following may not be an exclusive result of Mill's solution, we can observe that Mill made a tremendous step forward toward the economics of exchange in the course of considering and examining international value problems. The following is a list of propositions that Mill affirmed in the course of this examination:

(1) It is not the absolute advantage of an industry but its comparative advantage that determines the pattern of trade and specialization (Mill 1844, I.2, I.9; 1848, III.18.60).
(2) The law of demand and supply is anterior and more fundamental to the law of cost of production (Mill 1844, I.19; 1848, III.16.5, III.18.4).
(3) Price regulates itself in such a way that demand will be equal to supply (Mill 1844, I.23; 1848, III.18.15, a citation from Mill 1844 I.19).
(4) Supply and demand are but another expression of reciprocal demand (Mill 1848, III.18.24).
(5) Demand depends on value (or price) (Mill 1844, I.27; 1848 III.18.10).
(6) The equation or equilibrium of international demand is the equality of exports and imports (Mill 1844, I.52, I.72; 1848, III.20.5).
(7) Each country completely specializes in the production of one commodity (Mill 1844, I.6; 1848, III.18.4).

(8) Two-country, two-commodity analyses can be generalized without fundamental change of logic (Mill 1848, III.18.17, III.18.20; no explicit mentions in Mill 1844).

Mill believed that proposition (1) was what Ricardo wrote in his "Foreign Trade" chapter. However, as we now know it, Ricardo did not explain it in this way, at least in his four magic numbers (see Yukizawa 1974; Maneschi 2004; Faccarello 2015; Tabuchi 2017a, b).[22] In practice, this was Mill's invention, and yet, his idea still influences international trade thinking. Proposition (2) became a kind of manifesto for the neoclassical revolution. In *Principles*, Mill advanced this statement in chapter 16, Book III, Some Peculiar Cases of Value, but it is my guess that he put it there to attenuate too drastic an announcement for many Ricardian loyalists, including Mill himself. However, what Mill prepared in the international value chapter was not this manifesto alone.

Proposition (3) was the real content of proposition (2). In *Principles*, there was no explanation of the principle of demand and supply, but in *Essays*, Mill explained that the principle of demand and supply meant that price or exchangeable value depended on demand and supply (Mill 1844, I.23–24). In *Principles*, Mill gave no such clue before the international value chapter, except for two minor comments in a part that was omitted in the third edition in which he spoke about the adjustment of wages (note #84) and the adjustment of prices of two jointly produced products (Mill 1848, III.16.15). On the contrary, Mill emphasized the opposite causal relationship at the beginning of Book III:

> [T]he value of things which can be increased in quantity at pleasure, does not depend (except accidentally, and during the time necessary for production to adjust itself,) upon demand and supply; on the contrary, demand and supply depend upon it. (Mill 1848, III.3.7)

The notion that value depends on supply and demand already had appeared in *Essays* (Mill 1844, I.23). The essays were written in 1829–1830 (Mill 1844, P. 1), but an examination of how gains from trade were divided between trading countries went back to the early 1820s when Mill was still a teenager. Mill knew very well that the principle of the cost of production was the core of Ricardo's whole doctrine. Mill discovered by his study on international trade questions that this central dogma had a deep crack. Knowing this from his young days, Mill remained at surface a loyal adherent of Ricardo's theory. He proposed that we should revert to a more general and fundamental law of demand and supply. In *Principles*, this proposal was placed only after the main argument on values was completed and just before the two important International Trade chapters. Despite this careful presentation of Mill's new doctrine, students after him have found that this is the key point that may have revolutionized the whole system of economics.

---

[22]Ricardo's footnote on the shoemaker and hatmaker (Ricardo; Sraffa 1951, p. 138, Library 7.30 footnote 20) is more ambiguous. He might have been thinking as we normally do by taking two ratios of two different industries for both countries.

What Mill named peculiar cases might have influenced to some extent the adoption of the new doctrine. For example, Alfred Marshall studied the cases of joint cost of production in his early days when he started to study economics as his major research field. However, the importance of international values is without comparison. As we observe in the Sects. 8, 9 and 10 hereafter, the three pioneers of British neoclassical economics picked up the international trade problem as one of their major research targets.

The international trade situation as set by Mill was more influential than we can imagine in the formulation of neoclassical economics. Mill might have not been well aware of the effects of complete specialization, even though he clearly stated this point as proposition (7). In a two-country, two-commodity case, what each country produces and how much it produces are determined as shown in Sect. 5. This introduces a pure exchange situation. It is evident that this setting paved the way for the economics of exchange.

In addition, this situation helped to establish the symmetry of demand and supply. In this two-person, two-good pure exchange economy, supply is in fact the demand for another good. As Mill put it, supply and demand were in fact an expression of reciprocal demand. This significantly helped to formulate the supply curve, because supply was a symmetrical counterpart to demand. The notion of the demand curve or function was easily formulated by utility maximization. The supply curve or function, involving the intervention of production, was much more difficult to formulate. The marginal principle was first introduced for utility, but its introduction in production was delayed for many years. We know how Marshall struggled to harmonize the supply curve with production conditions. He invented the notions of internal and external economies. He barely succeeded in explaining the increasing supply curve by excluding economies of scale within a firm. Such a complication did not exist in the case of reciprocal demand.

When we acknowledge that demand depends on prices, it is easy to understand that supply depends on prices in the case of reciprocal demand. This acknowledgement established a symmetric framework of supply and demand. Although it seems that Mill did not arrive at the notion of the demand and supply function with prices as independent variables (Yoshii 2017), there was only one step involved in doing so. What was lacking for Mill was the mathematical concept of a function in general. In the time of Mill, the idea of a function in general was not very common. Mathematicians might have used such an abstract concept, but for nonspecialists, a function was something expressed by algebraic expressions.[23] It must have been difficult for Mill to conceive the relationship between demand and price as a function.

[23]This explains in part why A. A. Cournot (1838) introduced the concept of the demand function as early as 1838. Cournot was trained as a mathematician, In addition, mathematics education was much more developed in France than it was in England in the first half of the nineteenth century.

Mill's concept of the equilibrium of international demand was special. It referred to the trade balance, and in this sense, equilibrium did not represent the equality of demand and supply as it does in the modern sense. It is possible that Mill must have been thinking of a kind of cybernetic process.[24] However, *Principles* wiped out some traditional misconceptions. As Mill put it, many people assumed that value depended on the proportion between demand and supply. He emphasized that it is not the proportion between demand and supply but their equality (Mill 1848, III.3.5). However, this equilibrium concept might have induced a grave error when Mill examined the case of more than two countries and two commodities. Mill suggested that these cases could be treated just the same as the two-country, two-commodity case (Mill 1848, III.18.17), but it seems that he did not reflect on these questions deeply. Mill claimed that the introduction of a third country or commodity would not alter the theory. He is correct in one sense. In his understanding, the total value of exports and imports must be equal. He was preoccupied with this equality or the trade balance and did not consider how specialization was defined in these situations. It is difficult to suppose Mill had any concrete idea in mind, because this question remained a difficult one even in the middle of the twentieth century. Curiously, Mill did not mention explicitly that his two-by-two case analysis could be generalized to the case of many countries or commodities in the first *Essay* (Mill 1844). It is probable that he was aware of this difficulty when he was working on specialization and had forgotten it some 10 years later when he started to write *Principles*.

As a simple conclusion of this chapter, we could state that Mill's international value chapter (Mill 1848, III.18) prepared much of the neoclassical framework and marked a clear shift from the economics of production to that of exchange. In Chapter 3 of Book 3, Mill (1848, III.3.7) stated clearly that the value of things did not depend on demand and supply, but demand and supply depended on the value of things. In Chapter 18 of the same book, Mill set up a situation in which the exchange value adjusted itself so that demand and supply were equal. In Chapter 3, the mechanism that brings the supply equal to the demand was the change of production volume. In Chapter 18, the production was put aside and a pure exchange situation was introduced. In this state, two parties have a certain quantity of each of two commodities. In the course of long explanations, a clear change of adjustment mechanism occurred and causal orientation was reversed. Without mentioning this change of adjustment mechanisms, and without investigating why this change of adjustment occurred, Mill declared that the law of demand and supply was more general than that of the cost of production. As we explain in Sect. 6, this was a grave oversight. We observe in the following three sections how this oversight guided latecomers to the construction of the economics of exchange. More precisely, in

---

[24]See my argument on Marshall in Sect. 10. See also Yoshii (2017, Sect. 2).

this economics, exchange plays the major role whereas production is interpreted as simply a variant of exchange (production as exchange with nature).

8 William Stanley Jevons

In the case of Jevons, my judgment is symptomatological. There is no textual evidence that Jevons was influenced by Mill's solution. Despite a lack of direct evidence and Jevons's apparent hostility and confrontational attitude toward Mill, it is highly probable that Jevons unconsciously accepted Mill's fundamental framework on international trade.

I start my argument by citing two interesting papers: Donzelli (2007) and Aldrich (2000). Donzelli claims that Jevons did not develop the law of demand and supply while Aldrich (2000) debated why the Jevonian revolution did not take place in international trade theory unlike the case of domestic theory.

Donzelli (2007) questioned as follows:

> One of the most surprising features of Chapter 4 of TPE [*Theory of Political Economy*] is that, in spite of the reiterated emphasis laid by Jevons on the allegedly fundamental role played by the so-called "laws of supply and demand" in his theory of exchange, no formal demand-and-supply analysis is actually employed by the author in deriving such theory nor, in spite of what Jevons himself occasionally claims, can be deduced from the formal statement of the theory, as can be found in TPE. (Donzelli 2007, p. 2)

It is true that the expressions "demand function" and "supply function" do not appear in Jevons (1871) (see also White 1989; Nakano 2009). In the preface to the second edition, Jevons referred to what we call the "demand function" when he talked about Cournot, but he cited it as "a function of the price, or D = f(p)." Even in his mathematical theory (Jevons 1874), no such expression appears. Jevons was trained as a natural scientist, was particularly good at chemistry, and had a concept of function in general. He used such terms as "function of utility," but he never used the concept of the "demand function" or "supply function" in his principal book as his own concept. What does this mean?

A simple explanation is that Jevons did not think in terms of demand and supply functions. Then, what did Jevons mean by the laws of supply and demand?[25] In the preface to the second edition of his *Theory*, Jevons cited Lardner (1850) in which he found a mathematically and graphically treated example of the laws of supply and demand. Lardner's Chapter 13, as indicated by Jevons, is an account of receipts and

[25]Expressions law/laws of demand and supply did not appear in Jevons (1871). By contrast, the expression "law of supply and demand" appeared twice in chapter 5 (V.47. V.48), and the expression "laws of supply and demand" appeared 18 times in 17 paragraphs, including the prefaces for the first and second editions.

profits.[26] In a diagram Lardner (1850, p. 249) discussed how tariffs influence profits and argued that there is a maximum point of profit at which tariffs continuously increase from very low values to higher values. This must be a good example of the mathematical treatment of profit analysis, and it must be true that Jevons derived the "idea of investigating Economics mathematically" (Jevons 1871, PS.8). However, it is difficult for us to observe that this is related to the law of demand and supply. What Lardner argued is in fact a calculation of profit maximization.

In the body of Jevons's *Theory*, the expression "laws of supply and demand" appears about ten times, but none provides a detailed explanation of the laws. In many instances, it is indicated that the laws of supply and demand are the consequences of the law or theory of exchange (Jevons 1871, Library I.2, I.24, IV.39). The most detailed account of the laws appears in Chapter 1 Introduction. In the second appearance, Jevons explained: "The ordinary laws of supply and demand treat entirely of quantities of commodity demanded or supplied, and express the manner in which the quantities vary in connection with the price" (Jevons 1871, I.5). This explanation of the law of demand and supply does not seem to differ much from the commonly accepted notion of the law. However, we might infer that the main focus shifts between Mill's and Jevons's interpretations of the law of demand and supply. By these expressions, Mill mainly referred to the equality of demand and supply (see Yoshii 2017, Chap. 8 of this book). In the case of Jevons, he may have implied the equality of supply and demand, but he seems to have been more concerned with the form of how demand and supply change with price changes. In addition, there seems be a discrepancy between what Jevons understood by demand change and what we might imagine by the same expression. When we consider the demand function (or curve), we normally think of a function between demand and price. This kind of conception is very rare in Jevons. Instead, he spoke about the utility function (or curve). When he spoke about demand changes, he claimed with no preliminary explanation that the final degree of utility (or derivative of the total utility function) decreases as the quantity of the commodity increases. Although there is no such expression, I have an impression that Jevons considered the laws of supply and demand to be equivalent to the diminishing law of marginal utility.

Jevons's account of exchange is a strange one. He emphasized that total utility must be the maximum after the exchange for both traders. He wrote the corresponding equation as follows:

$$\varphi_1\,(a-x)\,/\,\psi_1(y) = y/x = \varphi_2(x)/\psi_2\,(b-y)\,. \tag{7}$$

Jevons explained that the quantities exchanged, x and y, must satisfy these two equations, which he defined as "whenever two commodities are exchanged with each other, and *more or less can be given or received in infinitely small quantities*" (Jevons's emphasis. Jevons 1871, IV. 36). Jevons assumed that the total utility and consequently the final utility are separable by each commodity. Functions

[26] Jevons cited the wrong pages.

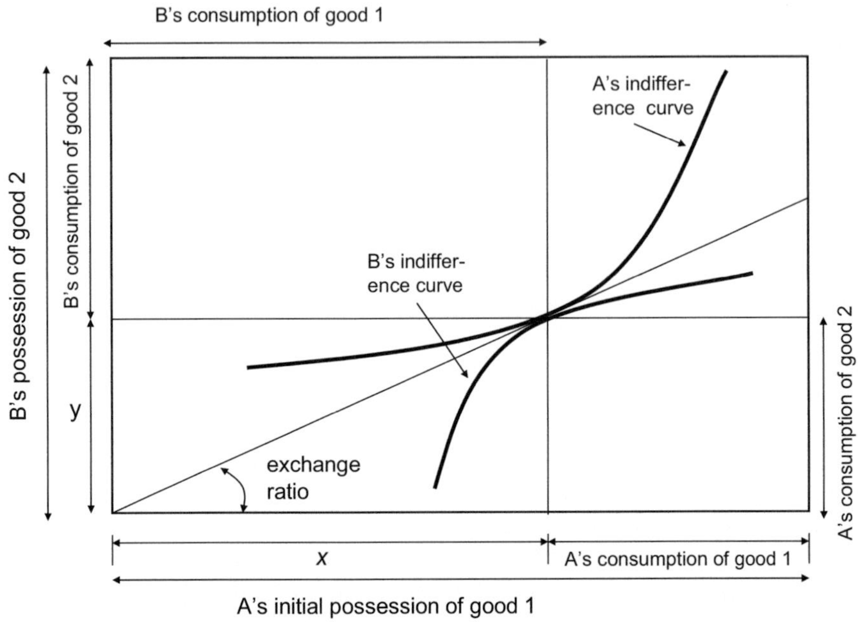

Fig. 3 Jevons explained by an Edgeworth box diagram

$\varphi_1(s)$ and $\psi_1(t)$ signify trader A's final utilities when A possesses s units of corn and t units of beef. Functions $\varphi_2(u)$ and $\psi_2(v)$ signify trader B's final utilities when B possesses u units of corn and v units of beef. Even if the utility functions were not separable, the equations would hold when we replace $\varphi_1(a-x)/\psi_1(y)$ and $\varphi_2(x)/\psi_2(b-y)$ by $\partial\Phi/\partial s/\partial\Phi/\partial t$ and $\partial\Psi/\partial u/\partial\Psi/\partial v$, evaluated at $(a-x, y)$ and $(x, b-y)$, respectively.

If we were to use an Edgeworth box diagram, we could express these relationships as in Fig. 3. The only difference between Jevons and Edgeworth lies in whether they admitted the middle side, y/x, was equal to the left- and right-hand sides. Jevons claimed that the middle side equals either side. This is a simple consequence of the law of indifference.[27] We might have a drawn-out argument about this law, but this is not the right place to do so. Edgeworth, on the other hand, would not accept this law. Consequently, we could state that the difference between Jevons and Edgeworth is whether or not they accepted this law. Except for this single point, Jevons could be said to be the forerunner of Edgeworth. We discuss this point in the next section.

Contrary to well-established belief, the laws of demand and supply do not require demand and supply functions. We could argue that this implies an intermediate characteristic of Jevons's economics. A common understanding is that Jevons failed to grasp these crucial concepts. Another strand of understanding is to situate Jevons

[27]The whole of Chapter 4 of Jevons (1871) delves into the explanations of this law.

on the path to a deeper understanding of the exchange process (Fonseca and Ussher 2002; Nakano 2009). I am inclined toward the first understanding and I explain the reason for this soon hereafter.

If we were to admit the law of indifference, Eq. (7) would have a good chance of having a unique solution. The cases of nonexistence of solutions are not excluded. When solutions exist, they might not be unique, but in general, they form a separate discrete set. If we were not to ask how we obtain one of these solutions, Jevons's argument could be estimated quite satisfactorily. There would be no need to examine the exchange process by demand and supply curves.

Fonseca and Ussher (2002) and Nakano (2009) considered that there must be a route other than the path comprising demand and supply functions. They are correct to consider that demand and supply functions are artifacts with little empirical support and no deep theoretical basis. I do not investigate this question but those who doubt my view are requested to read Erik Beinhocker (2006), Steve Keen (2011), and my papers (Shiozawa 1999, 2016b). However, I consider that Jevons (1871) had another aspect to which Fonseca and Ussher (2002) and Nakano (2009) did not pay due attention. This is a point observed by Aldrich (2000), who claimed that Jevons's Theory of Exchange chapter was not a simple theory of (domestic) exchange, but a unified theory of both domestic and international trade.

Aldrich (2000) is correct. Jevons introduced in chapter IV, the Theory of Exchange, the term "trading body," which refers to any trader who may be either "a single individual in one case" or all "inhabitants of a continent" (Jevons 1871, IV.19). Jevons justified this singular terminology because "the principles of exchange are the same in nature, however wide or narrow may be the market considered" (Jevons 1871, IV.20). Jevons added that "our laws of Economics will be theoretically true in the case of individuals, and practically true in the case of large aggregates; but the general principles will be the same, whatever the extent of the trading body considered. We shall be justified, then, in using the expression with the utmost generality" (Jevons 1871, IV.20). Thus, Jevons's theory of exchange was intended from the start to be a unified theory that is applicable to both domestic exchange and international trade. Aldrich emphasized this possibility as follows:

> As in the aggregative neoclassical analysis of the 1930s, international trade was conceived as exchange between countries, each with its own preferences and production possibilities. Jevons developed marginal conditions for consumption and production and extended the theory of comparative advantage in a way that did not become established in the literature until the 1930s (Aldrich 2000, p. 65).

Aldrich (2000) asked why the Jevonian revolution did not take place in international trade theory unlike the case of domestic theory, on which Jevons's agenda setting, mode of argument, and analytical tools made a tremendous impact. In my judgment, this lack of a Jevonian revolution in international trade theory is instead a natural course of economics, because Jevons did not contribute any substantial development to international trade theory. Except for mathematical formulations, Jevons and Mill treated the problem in the same framework: a pure exchange economy.[28] Moreover, Jevons could not deal with any topic that was specific to international trade situations. For example, he could not devise any account of the large disparity in real wages between developed countries and the many colonized underdeveloped countries that existed at the time. Moreover, Jevons could not explain how international specialization occurred.[29] This was inevitable, in my opinion, because Jevons did not understand the essential difference between exchange in a country and exchange between countries, and this is why he thought he could have unified exchange within a country and between counties.

Aldrich (2000) posed an interesting question, but it was ill posed, or at least, it was shallow. If we were to move to the depth of the problem, we should ask why Jevons came to deal with his pure exchange economy or an economy in which there are two people and two commodities. I cite Jevons's own proclamation:

> The keystone of the whole Theory of Exchange, and of the principal problems of Economics, lies in this proposition—*The ratio of exchange of any two commodities will be the reciprocal of the ratio of the final degrees of utility of the quantities of commodity available for consumption after the exchange is completed.* (italics by Jevons himself, Jevons 1871, IV.29)

This is a manifesto of Jevons's marginal utility doctrine.

> Imagine that there is one trading body possessing only corn, and another possessing only beef. It is certain that, under these circumstances, a portion of the corn may be given in exchange for a portion of the beef with a considerable increase of utility. How are we to determine at what point the exchange will cease to be beneficial? (Jevons 1871, IV.30)

Jevons's economy was an exchange economy with two people and two commodities. Where does this setting come from? When we consider exchange, we ordinarily think of a two-person, two-good situation. Do we do so because this is logically the minimal? We should reflect whether this kind of setting was common before Jevons. Did Adam Smith or David Ricardo start discussing exchange from this setting? In fact, they did not.[30] Present-day economists are accustomed to consider this abstract setting, but it is actually a custom created by the neoclassical revolution. Smith, Ricardo, and other classical economists considered the existing economy, which we could call industrial capitalism. This is a system of labor division in which tools and machines play an important role and exchanges are monetary. An economy of two people and two goods might exist when two people meet by chance in the middle of

---

[28] Aldrich (2000, p. 74) claims that Edgeworth (1881) "showed how Mill's theory could be deduced from Jevons's theory of exchange."

[29] In chapter 5 of the *Theory*, in which Jevons explained labor and production, he dealt with international trade under the headline of Various Cases of the Theory. He argued correctly when trade ("foreign commerce" in Jevons's words) was excluded. This is the case in which $\omega_2/\omega_1 = \mu_2/\mu_1$ (Jevons's substitutes for Ricardo's four magic numbers), but, in order to study specialization, he had to analyze the case in which the two ratios were not equal (Jevons 1871, IV.42). See also footnote 27 on Turgot.

[30] Ricardo (1817) and James Mill (1822) referred to a two-country, two-commodity exchange, but they did not consider it as an abstract economy isolated from other part of the entire economy.

a desert, say, one with water and the other with food.[31] This is the world that Jevons wanted to analyze. It is a possible situation but not a very important one for real life. It is true that Smith talked of the story of a deer and two beavers. This was but a simple illustration and not the target of analysis. What Smith wanted to describe and analyze was the emerging industrial capitalism, which is why he started to talk about pin making.

Jevons set up an abstract economy comprising two traders and two goods. The traders could have their own utility but no other information nor social institutions. This is a very simple situation that can be imagined and is adapted for pure analysis. How did Jevons come to devise this setting? Was it the result of logical deduction? If we consider this in an abstract way, exchange must comprise at least two people and two different goods. Then, the two-person, two-good economy must be the minimal setting for analysis without any other givens except that the people involved in trade have their preferences and they do not appeal to violence. We are accustomed to think in this way. However, Jevons himself did not think in this way. He wrote, for example, "It is also essential that the ratio of exchange between any two persons should be known to all the others" (Jevons 1871, IV.16).[32] If this is a minimal prerequisite for a market to work, Jevons implicitly violated his own situational setting.

My hypothesis is as follows. Mill's trade theory led many economists to focus on the abstract setting of a two-country, two-commodity exchange and this triggered the emergence of the theory of the pure exchange economy.

This hypothesis is informed by several pieces of circumstantial evidence.

1. Mill's *Principle* (1848) was the most influential economics textbook in the third quarter of the nineteenth century in the UK.
2. Mill advocated a reversion from Ricardo's cost of production theory of value to more the fundamental law of demand and supply.
3. The structure of Mill's economics is composed of two theories of value, one domestic and the other international, and despite the claims of Mill,[33] these theories of value were not sufficiently unified.

---

[31] Turgot, in his unfinished paper *Value and Money* (1769), examined just this kind of situation. See Turgot (2011, pp. 173–178). Murray Rothbard considered that this was the first Crusoe economics (ibid, p. xiii). It is not certain if Jevons knew Turgot's *Value and Money*. In Jevons (1871), Turgot's name appears only as the author of Vie de Condorcet in the bibliography. In Jevons (1875), Turgot appeared once each in Chapters IV and VII, but there was no mention on Turgot's theory of exchange. In the nineteenth century one of first references to Robinson and Friday appears in Bastia's Chap. 6 Property and Plunder of his Selected Essays (Bastiat 1845, 6.63). I owe this insight to a hint by Giulio Palermo. Economic Sophisms (Bastiat 1848) contains three sophisms on the same theme.

[32] Turgot (1769) did not make such an assumption in the two-person exchange case.

[33] Mill (1848) claimed that "there is nothing in the laws of value which remains ... to clear up" (Mill 1848, III.1.2). It is a mystery why Mill inserted this self-betraying proclamation.

4. It was normal for ambitious economists in the 1860s to want to build a unified theory of economics (or a value theory), and Mill's example provided an almost unique important case from which to start their attempts.[34]
5. International trade was the predominant situation, which was too important to be treated as an exception and required a new value theory.
6. Jevons, despite his generally hostile, confrontational attitude to Mill, adopted a particularly conciliatory outlook to Mill's trade theory.[35]

What is important here is not how Jevons himself thought consciously. Jevons made a great leap forward in economics: he changed economics from plutology (the economics of production) to catallactics (the economics of exchange). His *Theory of Political Economy* proves this. Characteristically, production is discussed in chapter 5, after chapter 4, which deals with the theory of exchange. Chapter 5 is titled "Theory of Labour," while there is no chapter with "production" in the heading. In the preface to the first edition, Jevons spoke of exchange but did not use the word "production." Jevons started his revolution unconsciously. He did not make his catallactic revolution on the basis of deep study of the economics of production. Even when he spoke about production, it was a production that did not suggests industrial capitalism. The workers he referred to were not employed workers but were better interpreted as self-employed.

On the conscious level, Jevons was decisively influenced by two ideas and he was aware of this fact. One was the idea to apply mathematics to society. Another was the idea of utility. The two were combined to coin a new idea, "final utility" or marginal utility in more standard terminology. Some economists have praised this for being particularly revolutionary. However, I do not think so. If the idea of "utility function" might be new, the conception that utility determines the prices of goods was as old as economic thinking. Classical economics emerged from the *rejection* of this "common sense." Jevons did not understand this crucial point.[36] Except for his adoption of mathematical formulations, as Marshall (1872) pointed out, Jevons's *Theory of Political Economy* contained no substantially new propositions.

Jevons started a revolution, or more precisely, a counter-revolution, but he was not aware of it. Then, from where did this change come? If my conjecture is correct, Jevons came to know about Mill's international values chapter, and he came up with the idea to construct the whole of economics on the basis of the international trade situation. If this is in fact the case, it is natural that Jevons emphasized there is no difference of logic between domestic and international economics, because in his understanding, his new theory was a generalization of international exchange.

[34]Mill (1848) cited the "joint cost of production" case as a reason to revert from the cost of production theory to the more fundamental law of demand and supply (Mill 1848, III.16), but this is not comparable to the international trade case in terms of importance.

[35]Jevons praised Mill's theory of international trade as "ingenious" and "nearly always true" (Jevons 1871, IV.100). In addition, Jevons mentioned that he thoroughly concurred with his citation from Mill (1848): "Almost every speculation respecting the economical interests of a society thus constituted, implies some theory of Value" (Library IV.2).

[36]It might have been possible because Jevons was an amateur economist.

Aldrich (2000) had a good insight to debate the fundamental nature of the Jevonian revolution, but his understanding was completely wrong. Jevons did not succeed in constructing a theory of international values. He claimed to have done so, but that does not mean he actually did. He might have built a theory of pure exchange, which might be applicable to long-distance trade between different communities, but it is not a theory applicable to an industrial economy.[37] The lack of such notions as specialization and comparative costs is evidence of Jevons's failure. In opposition to Aldrich's expectation, it was inevitable that the neoclassical revolution in the theory of international trade did not take place until long after Jevons.

Aldrich (2000) assumed that this delayed revolution arrived in the 1930s, but this is not the place to trace the theoretical development of the "modern theory" of international trade (Chipman 1965).[38] We are more concerned with the neoclassical revolution, and for this purpose, we have to turn our attention toward two other eminent names: Edgeworth and Marshall. In Sect. 9, we discuss Edgeworth, and in Sect. 10, we discuss Marshall. Both of them were heavily influenced by Mill's international value chapter but in very different ways.

9 Francis Ysidro Edgeworth

Edgeworth was born 3 years after Marshall and died 2 years after him. Although Edgeworth expressed an intellectual obligation to Marshall, I discuss Edgeworth before Marshall because (1) they stood at opposite extremes, and (2) Edgeworth has more in common with Jevons than does Marshall.

Figure 3 in the previous section provided an illustration of Jevons's theory of exchange. The situations Jevons and Edgeworth set up are quite similar, that is, both are two-person, two-commodity exchange cases. The similarities between Jevons and Edgeworth's ideas are apparent if we use what became known as an Edgeworth box diagram.[39] Of course, there are also some differences. Most importantly, Edgeworth did not admit that y/x (the middle side of (7)) is equal to $\varphi_1(a - x)/\psi_1(y) = \varphi_2(x)/\psi_2(b - y)$. In the Edgeworth box diagram (see Fig. 3), this difference is reflected in whether the common tangent line passes through the origin. Jevons considered that it does while Edgeworth considered that it does not necessarily do so.

[37] Jevon's main contention that "value depends entirely on utility" (Jevons 1871, I.2) does not apply to an industrial economy. If Jevons had claimed that the final utilities of two newly purchased goods are the same, he would have been correct, but this does not mean that final utility determines the values of goods. If goods are produced as much as they are demanded at the price set by producers, it is this price that determines the actual price. Final utility selects those who estimate it higher than the final utility of the payable money in exchange for the product. This does not determine the price but who buys the product at that price. More over, Jevons was inconsistent with himself. See Jevons (1871, IV.131).

[38] As for international trade theory in 1930s, Tabuchi (2017a, b) gives an interesting account.

[39] Creedy (1992) emphasized the same point.

This difference reflected the two economists' views on the exchange process. Jevons's exchange is *mediated* by prices but not by money. In other words, when exchange occurs, Jevons assumed that two people change their possessions in such a way that the value one receives is equal to the value the other renounces at mutually agreed prices. Edgeworth was a much more radical thinker of exchange. He aimed to analyze a pure exchange situation before the notion of price is established between people. Edgeworth assumed an extremely pure exchange economy. Such notions as prices were invented a very long time back and were institutionalized after the repeated experience of people. By rejecting these preconceptions, Edgeworth aimed to place his theory of exchange on a firm basis.

I am not certain that he succeeded in this project. *Mathematical Psychics* (Edgeworth 1881) is one of Edgeworth's earliest works and decidedly his major work. In this book, Edgeworth defined the field of competition as an exchange economy among an indefinite number of individuals. The field of competition is defined as perfect when four conditions of free contract are satisfied. In this field of competition, a *settlement* is "a contract which cannot be varied with the consent of all the parties to it,"[40] a *final settlement* is "a settlement which cannot be varied by recontract within the field of competition," and finally, "[c]ontract is indeterminate when there are an indefinite number of final settlements" (Edgeworth 1881, p. 19). The intent of introducing these concepts, after Edgeworth, is to investigate "How far contract is indeterminate" (ibid. p. 20).

To begin with, Edgeworth started by examining Jevons's two-trader, two-commodity case. As Edgeworth did not admit Jevons's law of indifference, the common tangent curve expressed by $\varphi 1(a-x)/\psi 1(y) = \varphi 2(x)/\psi 2(b-y)$ is not necessarily equal to y/x.

Lacking the middle side of (7), Edgeworth's condition does not determine an exchange rate, whereas Jevons's conditions normally determine a solution. In the later analysis, Edgeworth proceeded to examine the cases of "several persons and several variables" (Edgeworth 1881, p. 26). Jevons could not deal with these cases, but the extension was not difficult for Edgeworth, because he knew how to use Lagrange multipliers. Edgeworth even introduced if vaguely the concept of Pareto efficiency using the term "relative maximum" (ibid., p. 23).

Mathematical Psychics is full of sinuous arguments, and it is difficult to grasp Edgeworth's real contention. It seems, however, that Edgeworth was unsatisfied by the treatment of exchange process by Jevons, Marshall, and Walras. I cite only those names that Edgeworth used explicitly. These authors assumed there was a price prevailing in a market and appealed to the concept of aggregate quantity demand. For Edgeworth, such treatment did not consist of a general theory of exchange in view of haggling deployed in the market. He explained the superiority of his method by stating that his theory was applicable to cases of imperfect competition.

Edgeworth's idea was buried soon after him and "remained dormant until it was resurrected by Martin Shubik (1959) as the theory of 'core'" (Fonseca n.d.

[40]I question the use of "with the consent of all parties" and wonder whether it should be "without the consent of all parties." Another possible solution is to interpret "cannot" as a typo of "can".

(A) Edgeworth's "Indeterminacy of Contract," 1. Introduction). The literature on this topic exploded in the 1960s, and Edgeworth's three conjectures were proved (Debreu and Scarf, 1963). Edgeworth's conjectures were now theorems. Citing this fact, many historians of economic analysis now remark on Edgeworth's method. Some scholars, like Fonseca and Ussher (2002), Donzelli (2007, 2009) and Nakano (2009), claimed there was another route of development other than that of Marshall and Walras or economics based on the concept of demand and supply functions.

I heartily admit this possibility, but I doubt if it could be realistic economics. The concept of "core" assumes that a solution is not blocked by any coalition of the set of traders in the market. Let there be a set of N traders. Then, the number of possible coalitions grows rapidly by the order of 2^N. This signals an explosion of information and communication. Edgeworth's set of final settlements might shrink to Walrasian equilibrium without checking all coalitions. Suppose that the interventions of a third party are sufficient for the shrinking. Even in such a case, the extent of information that should be communicated between traders is exorbitant. It would not provide an exchange system that is allowed to run an economy as big as a small nation. It would be better to interpret Edgeworth's idea as an illustration of a multipartite bargaining process. In this regard, I believe that Marshall's treatment is much more effective and realistic, although I do not adhere to any demand and supply cross-point parable. I pick up the Marshall case in the next section and discuss problems of his formula. However, our main concern is not the theory's possible future. Instead, our main objective is to know where Edgeworth's idea came from.

In *Mathematical Psychics*, we find no substantial argument on Mill's theory of value.[41] little mention on Mill (only two paragraphs). In a later long article in the *Economic Journal*, of which Edgeworth was the editor, Edgeworth (1894–1895, part 1) repeatedly mentioned Mill and his international value chapter, calling it a "great chapter" (three times), a "stupendous chapter" (twice), and a "classical chapter" (once). The reason Edgeworth did not debate Mill's international value chapter in *Mathematical Psychics* is unknown. Edgeworth might have obtained his idea from Jevons (1871) or Marshall (1879a, b). Although Edgeworth's main example was a two-trader, two-good case, he knew how to deal with the many-trader, many-good case appropriately. Despite these facts, I believe I can claim that Edgeworth was also influenced by Mill's international value chapter at a deep level.

The first observation is that Edgeworth considered a pure exchange economy without any production. Such a situation exists in Mill (1848) only in the international value chapter. Second, when Edgeworth reflected on the theory of international values (Edgeworth 1894–95), he considered a two-country, three-commodity case. In such a case, he had two options: (1) a complete specialization case in which a country produces two commodities and another country produces another third commodity and (2) a partially specialized case in which two countries have one commonly produced commodity. In the second case, as Edgeworth himself

[41]The name "Mill" appears 31 times if I am right, but Edgeworth mentions on *Political Economy* (Mill 1848) only three times (p. 118, p. 119, and p. 126). There is no mention on Mill's *Essays* (1844).

discussed in Part 3, the relative prices of all commodities are determined through this common good. Why did Edgeworth prioritize the first option? Is this not a symptom of the influence of Mill's theory of international trade? Third, Edgeworth's emphasis on indeterminacy is unique among economists. If we were to search for such a case in the texts before Edgeworth (1881), Mill (1844, first essay) and Mill (1848, Book III. Chapter 18) would emerge as probable sources of inspiration.

As mentioned above, Edgeworth was doubly influenced by Jevons and Marshall, and while it is difficult to find direct evidence, we can at least state that he was not out of Mill's field of problem setting. Using Mill's problem setting, Edgeworth developed a theory of exchange that led toward the extreme opposite to that of Mill. If we were to combine this fact with Mill's direct influence on Marshall, we could claim that Mill's proposal to revert to a more fundamental and anterior law compelled the generation after him to a widely variant economics but within a definite range of direction.

10 Alfred Marshall

The case of Marshall is much easier to analyze than that of Jevons and Edgeworth, because we have clear evidence. In fact, Marshall (1879a, b) was never officially published at the time but was distributed privately by H. Sidgwick. The date of 1879 represents this year of private publishing. The work formed part of a two-volume book that was originally planned to be *Outline of Political Economy*, which Marshall started to compose between 1872 and 1874 (Whitaker 1975, I. p. 260). Jevons (1871) was published before this. If Marshall's recollection is correct, he was not very impressed by Jevons's book (Pigou 1925, appendix to review of Jevons 1871). It would require scrupulous study of Marshall's theoretical development to discern the influences of Mill on different themes, but in this chapter, I concentrate on examining two pure theories, which were intended to comprise a unified book.

Although Marshall has devised some contrivances, his *Pure Theory of Foreign Trade* is a kind of mathematical annotation of Mill's international value chapter. Of course, Marshall is much clearer than Mill and more successful at providing a proof of uniqueness (Proposition VIII). From a modern mathematical viewpoint, his explanation does not pay attention to the existence problem, even though a sufficient condition is already given by Mill. As many economists mention, Marshall started to consider the stability and instability of equilibrium. This can be connected at least to the cobweb theory of the twentieth century. It is still questionable if his treatment of stability is really dynamic, but our main point of investigation does not lie herein.

Our main interest is in how Marshall was influenced by Mill's treatment and argument of international trade. The direct influence of Mill' international value chapter on Marshall's *Pure Theory* is so evident that we have no necessity to prove it. The question is how Marshall reformulated Mill's problem. A conspicuous fact is that Marshall invented the concept of the demand function. Simply speaking, Jevons had no concept of the demand function (see Sect. 8 of this chapter). Edgeworth, having read Marshall's *Pure Theory*, criticized that concept. Consequently, Marshall was unique among the British founders of the neoclassical school, who put the demand function at the core of economic analysis.[42] A more subtle question is why Marshall did not use the expression "supply curve." In the foreign trade case in Mill's setting, there is no difference between the demand curve and the supply curve. As both came to be called the "offer curve" by later authors, the same relationship is called demand or supply depending on differences in points of view. In the case of *The Pure Theory of Domestic Values*, the companion volume of the *Outline*, the two expressions of the demand curve and the supply curve appear almost as frequently as each other. The natural question is, then, as follows: did Marshall have an opportunity to invent the demand and supply curves in circumstances other than his examination of Mill's international value chapter?

We have no direct evidence to prove this. Instead, we have some indirect evidence that Marshall obtained the idea of the demand and supply function in the course of studying the theory of international values. There are three circumstances for this thesis, although each is closely connected with each other.

The first circumstance is the special composition of *Outline of Political Economy*. As indicated above, this unfinished book contains two major parts: "The Pure Theory of Foreign Trade" and "The Pure Theory of Domestic Values." The contrast between domestic and international economy continues to this day. The particularity of *Outline of Political Economy* is that the volume on foreign trade precedes that on domestic values. This is quite a strange structure in the present day. Normally, we first study the theory of domestic economy and then proceed to the international economy. Marshall took the opposite approach and we should ask why.

This strange composition seems to reflect Marshall's process of theory construction. At the beginning of the *Domestic Values*, Marshall criticized Mill's usage of the term "theory of value." In Marshall's idea, "theory of value" must be a generic term that should include both domestic and international values. He is correct but why did he put the theory of international values before the theory of domestic values? Marshall justified this by stating that "[t]he apparatus of diagrams which was best adapted for the investigation of the latter will not be of service here [in the theory of domestic values]" (*Domestic Values*, p. 1). The "apparatus of diagrams" is apparently the diagram in which two demand curves (or, in more modern terms, two offer curves) cross. This provides a clue that the so-called Marshallian cross came first to Marshall's mind when he considered international values and the domestic version was completed thereafter.[43]

---

[42]The work of Marshall (1879a, b) contained the term "demand curve" but not "demand function." However, we assume that "demand curve" is employed to express the demand function in the two-goods case. As Marshall adopted an exchange between a good and money, any function could be called, simply, a "curve." Yoshii (2017, Chap. 8 in this volume) examines Jenkins' contribution to the formation of the demand function concept.

[43]Sraffa's (1926, p. 535) criticism of Marshall's price theory starts by pointing out its fundamental symmetry. He also pointed out that the notion of constant cost was "the most dangerous enemy of the symmetry of demand and supply. (Sraffa 1926, p. 541, footnote 1)

The second circumstance is the argument and reason that Marshall gave as to why the theory of domestic values is more difficult than that of international values. Most present-day economists think in the opposite way, but Marshall thought differently. In the theory of international values, the demand curve of one country is the supply curve of the other country. In the theory of domestic values, Marshall warned that there is no such symmetry between the demand and supply curve. It was Marshall who established the logical symmetry of demand and supply curves. On this particular point, he is correct. Marshall's situation setting in *Pure Theory of Foreign Trade* is quite ambiguous, because he contended that he considered a case of many commodities. However, it is clear that he was thinking of a situation in which a pattern of specialization already was determined. If the bundle of commodities that is produced in a nation has predetermined proportions, the amount of production of these bundles is determined uniquely. Then, Marshall in fact was considering a pure exchange economy. In the theory of domestic values, he adopted money as the commodity to be exchanged against a commodity to be considered. This presupposes an economy that Edgeworth would have preferred to avoid. Even if Marshall had assumed commodity money, the definition of the supply curve must have been no easy attempt.

We could conjecture that this introduction of the supply curve is Marshall's major contribution and the second crucial bifurcation point between classical and neoclassical economics.[44] Marshall might have assumed that he had defined the supply curve in a satisfactory way. On the other hand, he might have been dissatisfied by his clumsy and in fact confused definition of the supply curve. We cannot tell firmly which was correct. If it were the latter case, this might have been the reason why Marshall did not complete *Outline of Political Economy*. If it were the former case, then Marshall in fact would have been trapped in a logical error. Another possibility is that Marshall wandered between these two states of mind.

Marshall's definition of the supply curve is based on what he calls "the great central law of economic science" (*Domestic Values*, p. 3; Whitacker, 1975 Volume 2, p. 188) Citing a long part of an article[45] he published in 1867, Marshall explained as follows:

> This law is [1] that "producers, each governed under the sway of free competition by calculations of his own interest, will endeavour so to regulate the amount of any commodity which is produced for a given market during a given period, [2] that this amount shall be just capable on the average of finding purchasers during this period at a remunerative price. [3] A remunerative price is to be interpreted to be a price which shall be just equal to the sum of the exchange or economic measures of those efforts and sacrifices which are required for the production of the commodity when the amount in question is produced. [4] These economic measures are the expenses which must be incurred by a person who would

[44]The first bifurcation point was, as I explained above, Mill's choice as the standard situation of international trade situation and the Mill–Jones point of the two-country, two-commodity economy.

[45]"Mill's Theory of Value" appeared in *Fortnightly Review*, April 1876. It was reprinted in Pigou (1925, Part II, Chapter 3, pp. 119–133).

purchase the performance of these efforts and sacrifices." (Marshall 1925, pp. 126–127; Whitacker 1975 Volume 2, p. 188; the square brackets are my own)

Recall that the cited part is Marshall's summary of what he considered as the central truth of Mill's theory of value. When we compare the above citation to the original text, we observe that "the central law of economic science" expressed an essence of Mill's *Principles of Political Economy*. Of course, Marshall made some changes, such as a proposal to replace the "cost of production" with an expression like "expenses of production," which, Marshall believed, is less prone to misunderstanding.

The question we have to examine here is the fact that Marshall used the term "law" in the singular form. In the citation, many different propositions are in fact made. Is this a single law? Or, is it a composition of several different laws? Part [3] explains the meaning of the term "remunerable prices." Part [4] is related to the correction or more exact interpretation of the cost of production. For a long time, the cost of production was commonly interpreted in real terms as "those efforts and sacrifices which are required for the production of the commodity when the amount in question is produced" (latter half of Part [3]). Marshall corrected this common interpretation. It should be understood as the sum of expenses required for production.[46] However, this central law comprises at least two different ideas. Part [1] states that producers regulate the amount of their production for a given market during a given period. Part [2] states that the produced goods are sold at remunerable prices.

If we were to understand this law as a law of classical economics, Part [2] would mean that commodities were sold at remunerable prices on average. The classical theory of value assumes that the value is determined by the cost of production (in modern terms, the "full cost"). It could fluctuate by the discrepancies of demand and supply, but it returns to the natural price or value sooner or later. Producers assume the quantity of demand that will be sold at this natural price and regulate the quantity of production accordingly. The adjustment process might be complicated but on average, the price of a product moves around the natural price and production is regulated so as not to increase the discrepancy between demand and supply significantly. This is the central truth of the political economy that Marshall interpreted. However, is this the same law that Marshall formulated by his demand and supply curves?

Mill characterized the law of demand and supply as "a law of value anterior to cost of production, and more fundamental" (Mill 1848, III.16.5).

By the law of demand and supply, many economists now consider that a producer chooses an amount of his/her production for any price given in the market. However, the abovementioned central truth tells of no such thing. The fundamental assumption is that prices fluctuate around the natural price. If the market price were to move

[46]In spite of this remark, the real cost interpretation continued until the 1930s at the time of Viner. See Tabuchi (2017a, b, Chap. 9 in this volume).

far from the natural price, producers would not know and could not calculate the optimal quantity at that market price. They would increase their production only if the market price were higher than the natural price and vice versa. This is a similar process that Leijonhufvud (1968) called the *cybernetic approach* in macroeconomics. If the cost of production is constant, the producer cannot know any definite quantity to produce that is best for him or her. I consider this point soon hereafter.

A misinterpretation must have occurred in Marshall, because he induced the existence of the supply curve from this central law.[47] The formal definition of the supply curve is given as follows:

> The Supply curve SS' for a commodity in a market is such that if any point P_2 be taken on it, and P_2M_2 drawn perpendicular to Ox, P_2M_2 represents the price per unit at which a supply of the commodity of which the amount is represented by OM_2 can be remuneratively produced and brought into the market in each year (or other given period). (Marshall 1879b, p. 5; Whitaker 1975, Volume 2, p. 192)

In order for this definition to be valid, we need an important condition: the law of decreasing returns. Marshall might have known this fact. But it is not certain. Just before giving the formal definition, Marshall illustrated how to draw Supply Curve by assuming that "every increase in the amount supplied involves a more than proportional increase in the expense of producing it." (Marshall 1879b, p. 5 Whitaker 1975, Volume 2, p. 191) However, after the formal definition, he also spoke about the case of increasing returns to scale. In contrast to a "raw commodity" that we could assume obeys the law of decreasing returns, Marshall explained that, in the cases of most manufactured commodities, the law of increasing returns holds (Marshall 1879b, p. 6; Whitaker 1975, Volume 2, p. 192). How can we define the supply curve in these cases? We cannot.

Let me explain this using more modern terms. It is assumed that producers regulate the amount of their product for a given period of time and at a given price system. This means that they prefer that particular amount of production in comparison to all other amounts. How can this happen? The only possible situation is that they maximize their profit at the particular amount of production. The concept of the supply curve implies this. Then, for profit maximization to hold, production must be at a point of decreasing returns to scale. In fact, let $f(x)$ be the total cost function when amount x is produced, and assume that it is continuously differentiable. Profit is given by

$$px - f(x) \tag{8}$$

under the assumption that all products are sold at price p. If profit is at a maximum at point x, expression (8) must have a derivative of 0. In other words,

$$p–f'(x) = 0. \tag{9}$$

[47] Yoshii (2017) examined the same point from a different angle.

If the product has a positive price p, the marginal cost $m(x) = f'(x)$ must be positive. If the firm is operating with profit, then this means that the firm is operating at decreasing return to scale. By definition, decreasing returns to scale means that the average cost is increasing. This is easy to confirm. In fact, let $a(x)$ be the average cost, that is, the total expenses divided by the amount of the product, and take a derivative of function $a(x)$. Then,

$$a'(x) = \{f(x)/x\}' = \{f'(x)x - f(x)\}/x^2 = \{p - a(x)\}/x > 0,$$

when profit is positive and x satisfies (9).

The contrapositive of the above observation is as follows: when returns to scale are increasing or constant, the supply curve as a function of the product price has no proper meaning. It is not clear if Marshall was well aware of this fact. In *The Pure Theory of Domestic Values*, no explicit mention of this is found. Marshall cited cases of increasing returns but issued no associated warnings.

Another symptom that Marshall was confused is that he often identified the supply curve with the expense curve (or the total cost function). We can define the expense function without any condition, but the expense function and the (inverse function of the) supply function do not coincide unless the return to scale is decreasing.[48] If the average cost function is constant or decreasing, we cannot define the desired supply level of production.[49]

Marshall overlooked the difficulty of defining the supply curve. This point remained a source of trouble throughout Marshall's work. In *Principles of Economics*, he evaded the question by introducing the distinction of internal and external economy. This period was still at the early stage of giant firms. However, over time, it became clear that the majority of firms faced increasing returns to scale. Just after Marshall died, Piero Sraffa (1926) wrote a famous paper, *The Laws of Returns*, and pointed out that production was not limited by the increase of costs but by sales. This is a simple denial of not only the Marshallian cross framework but also the general equilibrium framework. However, at that time, neoclassical economics was too firmly established. No reformulation was made within neoclassical economics, and it now faces the same difficulty when Sraffa debated the law of returns.[50]

We can now understand the real nature of the problem. Marshall wanted to build a theory of domestic values on the same principles as the theory of international values. In *Pure Theory of Foreign Trade*, Marshall successfully reformulated Mill's trade theory to a more formal symmetric theory of reciprocal demand curves. Marshall knew there was no similar symmetry in the domestic economy, and he

[48]This misidentification partly explains why Marshall employed quantity instead of price as the horizontal axis, whereas he assumed that prices were given. The expense curve could be defined for any production quantity, provided that production were possible.

[49]This is the major reason why neoclassical economics prefers to assume decreasing returns to scale even though it is rare to find such an industry or firms. See Shiozawa (1999, 2004, 2016b).

[50]Joan Robinson's theory of imperfect competition was not a thorough breakthrough to amend this difficulty. For more details, see Shiozawa (2016b, Sects. 2 and 3).

introduced the concept of the supply curve. This was a natural reaction to the theoretical situation that Marshall faced. He did not know the real problem that the notion of a supply curve implies. He must have been driven by the desire to re-establish the symmetry of supply and demand. When viewed from present-day knowledge, Marshall's choice might not seem a special decision. However, if we were to know the real problem associated with the concept of the supply curve, we would understand how deeply Marshall was influenced by Mill's thesis on the necessity to revert to the law of supply and demand.

The third circumstance is the introduction of the concept of "consumers' rent" (or "consumer's surplus," although the latter term was not used in *The Pure Theory of Domestic Values*). If we read the definition of the term "consumers' rent" in Section 1 of Chapter 2 (Marshall 1879b, p. 20; Whitaker 1975 Volume 2, p. 213), or the mathematical footnote at the end of Section 3 of Chapter 2 (Marshall 1879b, p. 25; Whitaker 1975 Volume 2, p. 219), we can easily understand that it is an antecedent of the later "consumers' surplus." Why did Marshall feel it necessary to introduce this concept? A simple answer is to insert a theory in domestic economy that is comparable to that of the gains from trade. I have no space to argue this point in detail here. This must have been one of three pieces of circumstantial evidence that deeply influenced Marshall when he started to build his new theory of economics.

At the end of this discussion on Marshall, let us briefly compare Edgeworth and Marshall. In summary, Edgeworth was more inclined to a pure logic of exchange while Marshall was more realistic. Although Marshall's concept of the supply curve has a serious defect, his idealization is based on the monetary economy. On the other hand, Edgeworth wanted to examine the exchange economy which underlies the monetary economy. He was more loyal to the idea of catallactics. However, he could not really consider how the large network of exchanges actually could be organized. If we consider this point, money and price systems (including price-mediated exchange) are indispensable institutions that make modern large-scale economies work. As the trouble with neoclassical economics deepens, there are some reflective movements in search of alternatives. It is true that Edgeworth shows the extreme opposite to the Marshallian paradigm. However, the search for a pure theory of exchange is itself a product of Mill's misguided problem setting. Thus, it is necessary to return to alternative ways out which can replace Mill's solution.

11 Alternatives to Mill's Solution: Senior, Mangoldt, and Sidgwick

We have traced the history how Mill's solution paved the way for the neoclassical revolution. It might seem that the force that led this history was overwhelming, and there were no alternatives to this trend. However, we could ask if there was a way out. Were there no trials in the direction of a new theory? The answer is yes. There were some attempts which failed. McKenzie's Princeton teacher, Frank D. Graham

(1948), was the most famous dissident to the mainstream tradition in international trade. Graham was the most remarkable theorist who thought on the basis of the Ricardo's theory of value, but he came long after the neoclassical revolution.[51] Even before or at the time of the neoclassical revolution, there were three notable exceptions: Nassau Senior (1830), Hans von Mangoldt (1863) and Henry Sidgwick ([1883] 1901). The latter two cases are reported in Edgeworth's survey (Edgeworth 1894–1895, Part 3).

I do not explain this in detail. Edgeworth (1894–1895 Part 3) gave a good and concise account of Sidgwick and Mangoldt (Edgeworth 1894–1895, Part 3, I.(4) and II.(3), respectively). The influence of Mangoldt (1863), writing in German, on the English world was limited although he was a somewhat forgotten economist even in the German world (Schneider 1960). Jevons's (1871) second edition includes Mangoldt's book in the appendix, but he made no explicit mention of him in the preface to the second edition, where Jevons attempted to explore all-important precursors in mathematical economics. Marshall once cited Mangoldt's name in *Principles*, but no mention is made of him in Marshall's correspondence (Schneider 1960). Without Edgeworth's subsection, Mangoldt might have been forgotten almost entirely. It is possible that Mangoldt was the first person to acknowledge clearly that the existence of a commodity produced in common fixes the international prices for a wide range of demand and production. Although there is now an English translation of Appendix II (Chipman 1975), I move straight on to introduce Sidgwick's argument:

> [L]et us suppose that there is at least one other commodity—say corn— which is produced both in England and in Spain. According to Mill's general theory of value, discussed in the preceding chapter, the relative values of cloth and corn in England must be determined by their comparative costs of production; and, again, the relative values of wine and corn in Spain must be determined in the same way. (Sidgwick, third edition [1891], p. 213, Book II. chapter III, section 2.)

As Sidgwick assumed that England exports cloth to Spain and Spain exports wine to England, this is a two-country, multi-commodity economy.[52] If a third commodity exists, which is competitively produced in both countries, the (relative) wages of both countries are determined by the corn-producing industry. Then, through this common commodity the prices of all commodities are determined.

Sidgwick's contention can be paraphrased as follows. In a two-country, multi-commodity economy, suppose there is a commodity that is produced in two countries, E and S. Let $w\mathrm{E}$ and $w\mathrm{S}$ be the wage rates of countries E and S, respectively. If the transportation cost is negligible, we could write a price equality condition for the common product as follows:

$$(1 + \pi\mathrm{E})\, w\mathrm{E}/c\mathrm{E} = (1 + \pi\mathrm{S})\, w\mathrm{S}/c\mathrm{S}, \tag{10}$$

[51]For a short account of Graham' work, see Sato (2017, Chap.10, Sect 3 in this volume). See also Gomes (1990), pp. 91–98.

[52]In Sidgwick's expression, many commodities other than corn might exist. In this sense, he considered the two-country, multi-commodity economy, but we could represent it as a two-country, three-commodity economy. See Fig. 2.

Here, πE, πS and cE, cS are the standard profit rate and labor productivity for each country, respectively. If we suppose that the profit rates of the two countries do not differ much, we have approximately

$$wE/wS = cE/cS. \tag{11}$$

In other words, the wage rate ratio is proportional to the labor productivity ratio. If the wage rate of a country is determined, the prices of all other commodities that are produced in the country are also determined. As the wage rate ratio is determined by Eq. (11), the prices of all commodities are determined. The cost of production theory of value holds.

The difference between Mill and Sidgwick lies in whether they assume a commonly produced good. Graham called this state in which common products exit *linked competition*. Thus, the simple existence of a commonly produced commodity resurrects the cost of production theory of value à la Ricardo.

Sidgwick first published his *Principles* in 1883, a few years after he privately published Marshall's *Pure Theory of Foreign Trade*. A curious question arises. Did Sidgwick have this idea when he published Marshall's *Pure Theory*? Sidgwick used the book as a text for a lecture in Cambridge. Did he obtain the idea exposed in his *Principles* in the course of his teaching *Pure Theory*? A second interesting question is how Marshall reacted to Sidgwick's opinion. We know that Marshall did not complete his *Outline*, which comprises two *Pure Theory* volumes (Marshall 1879a, b). Is this why he did not complete it? If so, then, why did he continue to employ the idea of demand and supply curves at the core center of his theory?

Naturally, the debate that followed Sidgwick ([1883] 1901) was concerned with the existence of common commodity. In the third edition (1901, p. 213, footnote 1) Sidgwick cited Charles Bastable, who had criticized this point. Bastable's critique was more apologetic than analytic, because he appealed to the authority of Mill. Citing Bastable, Edgeworth (1894-95, Part 3) also argued against Sidgwick, but I do not consider it sufficiently persuasive. In the subsection presenting Mangoldt, Edgeworth questioned whether it is coincidence that two countries have the same cost of production for a product. However, he neglected the possibility of a mechanism that could make them coincide, like the old price–specie flow mechanism or the present-day fluctuating exchange rate regime. Of course, all of these economists including Mill, Marshall, Edgeworth, and Bastable concentrated on the terms of trade and did not analyze how demand and production might come to be equalized. For this analysis, we should know the world PPS. If Sidgwick and Edgeworth knew the shape of the PPS, for example, if they had a rough idea of Fig. 2, the course of the argument might have been very different. If they knew Fig. 2 and the reason why Mill–Jones points do not appear, they might have taken a different historical path.

After all, Sidgwick's contention was a complete refutation of Mill's solution and all other followers. This totally contradicts what Edgeworth considered the fundamental principle of international trade. This principle, according to him, implies in its negative form that the value of articles in the international market is not proportional to the cost (Edgeworth 1894–1895, Part 1, p. 36) and this was the starting point of all economics of exchange programs. Edgeworth and Marshall were divided in their orientations but they could not accept Sidgwick, because doing so would have implied that all their contributions to economics would fall down. The only defense possible for Edgeworth was to point out the implausibility of the existence of such a commonly produced commodity. In the face of Sidgwick and Mangoldt, Edgeworth obstinately attempted to defend the framework of the exchange economy. This was a natural result of the revolution of economics, and Edgeworth was too deeply absorbed in this revolution to convert to a new (or more classical) interpretation.

An interesting question is how Mangoldt and Sidgwick reacted to their own discoveries. If they had considered the true significance of their economy setup, they might have had a chance to open a path to a totally different economics than that based on demand and supply functions.

The case of Nassau W. Senior (1830) is more difficult to interpret, but may be more interesting as an alternative to Mill's solution. Senior left a series of lectures that contain an interesting style of argument. He was more interested in the particular question of how the value of money was determined. Consequently, he did not argue explicitly on the theory of international values as it is commonly understood, nor did he not write a textbook explaining his system of economics.[53] Thus, we have to guess through his explanations on other topics what theory he might have held.

His lecture on the value of money is commonly understood as the cost of production theory. I cannot tell if it is essentially different from that of Hume and other scholars' price–specie flow mechanisms. The only statement I can make here is that Senior had an image that the same commodities were produced and exported and these common commodities helped to determine the wage rate disparity between countries. Let me cite a short paragraph from Senior:

> Or to use a still more concise expression, that labour in England is eight times as productive of exportable commodities as in Hindostan, and labour in North America is one-fourth more productive of exportable commodities than in England. (Senior 1830, p. 12)

Senior pointed that the price of exported commodities depends on "the amount of the wages which has been paid, and the time for which they have been advanced" (ibid., p. 14). If we combine these propositions, we can make a similar equation to (10), and the wage rate disparity follows from (11). If the relative wage rates are known, it is possible to know the prices of all other commodities, because the same wage rates are paid in the same country.

By Senior's explicit reference to labor productivity and wage rate and his firm framework of the cost of production theory of value, he had the opportunity to

[53] Although Senior talked a lot about wages and profits, his *Political Economy* (1850) contains few discussions on how prices are determined.

develop a totally different theory of international values. In fact, in view of the new theory of international values presented in Chap. 1 in this volume, almost all Senior‘s arguments can be translated into the new theory.[54]

Senior was 18 years younger than Ricardo and 15 years older than Mill. Senior's lectures in Oxford were delivered in the 1820s, and some of them were published between 1828 and 1830. Mill had a chance to know Senior's cost of production theory of international values before the publication of Mill (1844, 1848). Indeed, as Bowley (1937) put it, Senior, Torrens, and Mill exchanged opinions on the terms of trade:

> The famous controversy between Senior and Torrens on the terms of trade, which was taken up again by J.S. Mill, turned exactly on this question of the relevance of an analysis confined to two commodities and two countries to the real world. (Bowley 1937 [2010], p. 225)[55]

If Mill had reflected more closely on his solution and studied Senior, he might not have advanced his famous thesis that lent appeal to the reversion to the more fundamental law of demand and supply.

12 What Was the Neoclassical Revolution? Implications for Future Research

This paper showed an origin of the neoclassical revolution. It goes back to the young Mill, when he attempted to solve an unsettled problem. This was a question left by Ricardo in the field of international trade.

---

[54]Later in 1843, Senior published in Edinburgh Review an anonymous article titled "Free Trade and Retaliation." Senior criticized Torrens that the latter's claim was valid only when "each country possesses, against the other, a strict monopoly;—a monopoly unaffected by the existence of any third market or of any third commodity, capable of serving as medium of exchange." Only in such case, "[t]he prices of the two commodities in question would be governed, not by the general and permanent regulator of price, cost of production, but by the occasional and disturbing causes, demand and supply." (Senior 1843, p. 36)" Similar claim appears in p. 42. This implies that, in a more general case, the value of a commodity is decided by the cost of production. In fact, he claimed that "So far as the price of a commodity is not affected by any natural or artificial monopoly, it coincides with the cost of production to the producer. . . . That this is true with respect to domestic commerce, is obvious; it appears to us obvious, that it is equally true with respect to international commerce. (Senior 1843, p. 37) Thus by 1843, just before the publication of Mill (1844), Senior had a clear critical view of Mill's "solution."

[55]Bowley (1937, chap. 6, p. 201) pointed out the possibility of two different ways of treating international trade: one investigates comparative physical costs, and the other analyzes in monetary terms. Bowley placed Ricardo, Mill, Taussig, Marshall, and Haberler in the first group and Senior and Ohlin in the second group. My understanding is 90 degrees different from her classification. By their theories of value, Senior, Mangoldt, and Sidgwick are much closer to Ricardo, because all three considered that the cost of production theory of value was applicable to international trade as well as to domestic exchange.

The question Mill posed to himself was how the advantages of trade were divided among trading nations. He intended to solve this problem by providing a theory of international values. The simplest setting was a two-country, two-commodity case. Nobody doubted that within this setting lay a deep trap. Mill simply excluded the situations in which one country could not enjoy gains from trade. By this pure inference, he was led to examine a pure exchange economy. As a result of his examination, a conclusion came to his mind In the form of his famous thesis: "we must revert to a principle anterior to that of cost of production, and from which this last flows as a consequence,—namely, the principle of demand and supply" (Mill 1844, I.19).

This thesis had a deep, strong influence on the research programs of economics. In the UK, three founding fathers, namely, Jevons, Edgeworth, and Marshall, were deeply influenced by the thesis and setting, although they worked in different ways and constructed their own economics. Among the three, Alfred Marshall with his demand and supply functions paved the way for today's mainstream economics.

The neoclassical revolution in the UK was a shift from the economics of production to the economics of exchange. Nearly 150 years later, economics is in trouble. Many economists and noneconomists now recognize that economics requires a change. What is the remedy? A series of modifications will not be helpful. Instead, a fundamental redesign is required. The history of the Copernican revolution is illustrative. Neoclassical economics is a kind of Ptolemaic geocentric system. It developed significantly and explained a lot, but it is essentially wrong. It requires a Copernican-type revolution.

This chapter described a new story. A question that was not hitherto asked is as follows: why was the story not conceived in the past? A possible answer seems to lie in Takashi Negishi's postcard to the author.[56] It reads:

> In our country, international economists are not interested in the history of economic doctrines and historians of economic thought know little about international economics. (Postcard dated July 8, 2011; translated from Japanese by the author)

Negishi was speaking about the intellectual situation in Japan, but a similar situation is observed in many other countries. If there was a tipping point of economics in international trade theory, then a coincidence of history and theory was necessary. I am a theoretician rather than a historian. Yasuaki Tsukamoto taught me that history is a strong tool to persuade people to adopt a new theory. He made me consider the meaning of the Mill's solution from a historical perspective. Thus, he helped me to identify the point at which classical economics turned to neoclassical economics.

Acknowledgment An earlier version of this paper was first read at a conference jointly organized by the Southwest branch of the Japanese Society for the History of Economic Thought and the Kyushu branch of the Japan Association for Evolutionary Economics held on June 21, 2014.

[56] I reproduce this sentence with Negishi's permission (postcard dated December 19, 2014).

The original English version, titled *On Ricardo's Two Rectification Problems*, was completed in September 2014. It was then divided into two parts. The first half is to be published under the same title as Shiozawa (2017a). The second half of the original version became the first draft of this chapter. The idea for this chapter was reported with the title *Why did John Mill retreat?/An Internalist View on an Origin of the Neoclassical Revolution* as part of a special session for the annual conference of the Japanese Society for the History of Economic Thought held in Hikone, Shiga Prefecture, on May 31, 2015. The original version was completely revised owing to preparatory discussions for the special session organized by Satoko Nakano. I thank Masashi Izumo, Akinori Isogai, and Satoko Nakano for giving me the chance to present my work. I give special thanks to Satoshi Yoshii, who gave me the difficult task of responding to his comments and advice on this paper. In addition, I owe much to members of the Workshop on the Theory of International Values for their valuable comments and discussions. Robin Edward Jarvis encouraged me with his short review that had deep understanding of the first version of this paper.

References

Aldrich, J. (2000). The Jevonian revolution in international trade theory. *Journal of History of Economic Thought, 22*(1), 65–84.

Bastiat, F. (1845). *Economic sophisms. Library of economics and liberty*. Irvington-on-Hudson: Foundation for Economic Education.

Bastiat, F. (1848). *Selected essays on political economy. Library of economics and liberty*. Irvington-on-Hudson: Foundation for Economic Education.

Beinhocker, E. D. (2006: Paperback Ed. 2007). *The origin of wealth: evolution and complexity and the radical remaking of economics*. (Paperback Edition Subtitle: The radical remaking of economics and what it means for business and society). Boston: Harvard Business Review Press.

Blaug, M. (2001). No history of ideas, please, we're economists. *Journal of Economics Perspective, 15*(1), 145–164.

Bowley, M. (1937: Paperback 2010). *Nassau senior and classical economics*. London: Routledge.

Chipman, J. S. (1965). A survey of the theory of international trade: part 1, the classical theory. *Econometrica, 33*(3), 477–519.

Chipman, J. S. (1975). On the equation of international demand, Hans von Mangoldt, with introductory note. *Journal of International Economics, 5*, 55–97.

Cournot, A. A. (1838). *Recherches sur les principes mathématiques de la théorie des richesses*. Paris: L. Hachette

Creedy, J. (1992). Jevons's complex cases in the theory of exchange. *Journal of the History of Economic Thought, 14*(1), 55–69.

Debreu, G., & Scarf, H. (1963). A limit theorem on the core of an economy. *International Economic Review, 4*(3), 235–246.

Dobb, M. (1973). *Theories of value and distribution since Adam Smith*. Cambridge: Cambridge University Press.

Donzelli, F. (2007). *Jevons, Jenkin, and Walras on demand-and-supply analysis in the theory of exchange* (Working paper). Dipartimento do scienze economia aziendale e statistiche, University of Milano.

Donzelli, F. (2009). *Edgeworth vs. Walras on Equilibrium and Disequilibrium* (Working Paper no. 2008.47). Departimento di of Economy and Statistics, State University of Milano.

Edgeworth, F. Y. (1881). *Mathematical psychics: An essay on the application of mathematics to the moral sciences*. Reprinted in F. Y. Edgeworth's Mathematical psychics and further papers on political economy. Edited by Peter Newman, Oxford Unviersity Press, 2003.

Edgeworth, F. Y. (1894–95). The theory of international values. (in three parts) *Economic Journal* Part 1 *4*(13), 35–50, Part 2 *4*(15), 424–443, Part 3 *4*(16), 606–638.

Faccarello, G. (2014). 'Nous marchons sus un autre terrain.' The Reception of Ricardo in the French Language: Episodes from a Complex History. In Faccarello & Izumo (Eds.), *The reception of David Ricardo in contemporary Europe and Japan* (pp. 10–75). London: Routledge.

Faccarello, G. (2015). Comparative advantage. In H. D. Kurz & N. Salvadori (Eds.), *The elgar companion to David Ricardo* (pp. 69–77). Cheltenham: Edward Elgar.

Fonseca, G. L. (n.d.). Edgeworthian exchange http://www.hetwebsite.net/het/essays/edgew/edgecont.htm

Fonseca, G. L., & Ussher, L. J. (2002). Microfoundation and the marginalist revolution. Paper presented at the History of Economic Society Conference, University of California, Davis, July 6.

Gehrke, Ch. (2014). The reception and further elaboration of Ricardo's theory of value and distribution. In Faccarello and Izumo (Eds.) *The Reception of David Ricardo in Continental Europe and Japan (Chapter 2)*. London: Routledge.

Gehrke, C., & Kurz, H. (2006). Sraffa on von Bortkiewicz: Reconstructing the classical theory of value and distribution. *History of Political Economy, 38*(1), 91–149.

Gomes, L. (1990). *Neoclassical international economics: A historical survey*. London: MacMillan Press.

Graham, F. D. (1948). *The theory of international values*. Princeton: Princeton University Press.

Hicks, J. R. (1974). The scope and status of welfare economics. *Oxford Economic Papers (New Series), 27*(3), 307–326.

Hicks, J. R. (1976). Revolution in economics. In S. J. Latsis (Ed.), *Method and appraisal in economics* (pp. 207–218). Cambridge: Cambridge University Press.

Jevons, W. S. (1871; 1879, 1888). *The theory of political economy* (1st Ed. 1871, 2nd Ed. 1879, 3rd Ed. 1888) Available at Library of Economics and Liberty.

Jevons, W. S. (1874). The progress of the mathematical theory of political economy. *Transactions of the Manchester Statistical Society, 1874–1875*, 1–19.

Jevons, W. S. (1875; 1876). *Money and the mechanism of exchange*. Available at Library of Economics and Liberty.

Jones, R. (1961). Comparative advantage and the theory of tariffs: A multi-country, multi-commodity model. *Review of Economics Studies, 28*(3), 161–175.

Keen, S. (2011). *Debunking economics*. London: Zed books.

Keynes, J. M. (1936[1973]). *The General Theory of Employment, Interest and Money*. Volume VII in the *Collected Writings of John Maynard Keynes*, Cambridge University Press.

Koopmans, T. C. (Ed.). (1951). *Activity analysis of production and allocation*. New York: Wiley.

Kuhn, T. (1962). *The structure of scientific revolutions*. Chicago: University of Chicago Press.

Lardner, D. (1850). *Railway economy; a treatise of the new art of transport*. New York: Harrer and Brothers.

Leijonhufvud, A. (1968). *On Keynesian economics and the economics of Keynes: A study in monetary theory*. Oxford: Oxford University Press.

Maneschi, A. (2002). The tercentenary of Henry Martyn's considerations upon the East India trade. *Journal of the History of Economic Thought, 24*(2), 233–249.

Maneschi, A. (2004). The true meaning of David Ricardo's four magic numbers. *Journal of International Economics, 62*(2), 433–443.

Mangoldt, Hans von, (1863: 1963). *Grundriss der Volkswirthschaftslehre*. English translation of Appendix II in Chipman (1975).

Marshall, A. (1879a). *The pure theory of foreign trade*. Privately published and circulated by Sidgwick. Reprinted in Whitaker (1975) 2, 117–181.

Marshall, A. (1879b). *The pure theory of domestic values*. Privately published and circulated by Sidgwick. Reprinted in Whitaker (1975) 2, 186–236.

Martyn, H. (1701: 1720). *Considerations upon the East India Trade, anonymous pamphlet*. Reprinted with the title in 1720. McCullock (1856) included the pamphlet in his book.

McCulloch, J. R. (Ed.). (1856). Early english tracts on commerce. Reprinted by Cambridge University Press, Cambridge, 1954.

McKenzie, L. (1954a). On equilibrium in graham's model of world trade and other competitive systems. *Econometrica, 22*(2), 148–161.

McKenzie, L. (1954b). Specialization and efficiency in world production. *Review of Economic Studies, 21*(3), 165–180.

McKenzie, L. (1955). Specialization in production and the production possibility locus. *Review of Economic Studies, 23*, 56–64.

McLeod, C. (1983). Henry Martin and the authorship of 'Considerations upon the East-India Trade.' *Historical Research, 56*(139), 222–229.

Mill, J. (1822). *Elements of political economy*. Boston: Gould and Lincoln.

Mill, J. S. (1844). On Laws of Interchange between Nations. First essay in J.S. Mill, *Essays on Some Unsettled Questions of Political Economy*, 1844. Citations are made from Library of Economics and Liberty.

Mill, J. S. (1848). Principles of political economy. Citations are made from Library of Economics and Liberty.

Mirowsky, Ph. (1989; 1991). *More heat than light, economics as social physics, physics as natures's economics*. Cambridge: Cambridge University Press. Paperback 1991.

Nakano, S. (2009). Jevons's market view through the dynamic trajectories of bilateral exchanges. In A. Ikeo & H. Kurz (Eds.) *A history of economic theory/essays in honour of Takashi Negishi*, Chapter 10, pp. 169–201. London: Routledge.

Pigou, A. C. (1925). Memorials of Alfred Marshall. MacMillan.

Ricardo, (1817:1821). *On the principles of political economy and taxation*, Edited by Sraffa (1951) Volume I of The Works and Correspondences of David Ricardo, Cambridge University Press, 1951. Cited as "Sraffa 1951". "Library" stands for texts and paragraph numbers in the Library of Economics and Liberty.

Ruffin, R. (2002). David Ricardo's discovery of comparative advantage. *History of Political Economy, 34*(4), 727–748.

Sato, H. (2017). An overview of research into international 1 values in Japan. In Y. Shiozawa, T. Oka, & T. Tabuchi (Eds.), *A new construction of ricardian theory of international values (Chapter 10)*. Tokyo: Springer.

Schneider, E. (1960). Hans von Mangoldt on price theory: A contribution to the history of mathematical economics. *Enonometrica, 28*(2), 380–392.

Senior, N. W. (1830). *Three lectures on the cost of obtaining money*. Available at https://archive.org/stream/threelecturesont033052mbp#page/n0/mode/2up

Senior, N. (1843). Free trade and retaliation, *Edinburgh Review, 78*(157): 1–47. (July, 1843) Anonymous.article. Available from Google Books: Art. 1.__1. of the following page: https://books.google.co.jp/books?id=WFA7AQAAMAAJ&printsec=frontcover&hl=ja&source=gbs_ge_summary_r&cad=0#v=onepage&q&f=false

Shiozawa, Y. (1975). *Durable capital gods and their valuation* (KIER Discussion Paper No. 8). Institute of Economic Research, Kyoto University. Uploaded in Research Gate.

Shiozawa, Y. (1985). International trade and the choice of techniques (in Japanese). *Keizaigaku Zasshi, 85*(6), 44–61.

Shiozawa, Y. (1999). Economics and accounting: A comparison between philosophical backgrounds of the two disciplines in view of complexity theory. *Accounting Auditing and Accountability Journal, 12*(1), 19–38.

Shiozawa, Y. (2004). Evolutional economics the 21st century. *Evolutionary and Institutional Economics Review, 1*(1), 5–47.

Shiozawa, Y. (2007). A new construction of Ricardian trade theory: A many-country, many-commodity case with intermediate goods and choice of techniques. *Evolutionary and Institutional Economics Review, 3*(2), 141–187.

Shiozawa, Y. (2014). A final solution of ricardo problem on international values, Iwanami Shoten (in Japanese).

Shiozawa, Y. (2015). International trade theory and exotic algebra. *Evolutionary and Institutional Economics Review, 12*(1), 177–212.

Shiozawa, Y. (2016a). The revival of classical theory of values. In Yokokawa, N. et al. (Eds.) *The Rejuvenation of Political Economy* (Chapter 8, pp. 151–172). Oxon/New York: Routledge.

Shiozawa, Y. (2016b). A guided tour of the backside of agent-based simulation. In H. Kita, K. Taniguchi, & Y. Nakajima (Eds.), *Realistic simulation of financial markets/analyzing market behavior by the third mode of science* (Chapter 1, pp. 3–50). Tokoyo: Springer.

Shiozawa, Y. (2017a). On Ricardo's two rectification problems. In S. Senga, M. Fujimoto & T. Tabuchi (Eds.) *Ricardo and international trade (Chapter 10)*. London: Routledge.

Shiozawa, Y. (2017b). The theory of international values: an overview. In Y. Shiozawa, T. Oka, & T. Tabuchi (Eds.), *A new construction of ricardian theory of international values (Chapter 1)*. Tokyo: Springer.

Shubik, M. (1959). Edgeworth market games. *Annals of Mathematics Study, 40*, 267–278.

Sidgwick, H. (1883; 1901). *Principles of political economy*. The third edition (1901) is available at http://ebook.lib.hku.hk/CADAL/B3139033X/

Smith, A. (1776). *An inquiry into the nature and causes of wealth of nations*. Citations are from Library of Economics and Liberty.

Sraffa, P. (1926). The laws of returns under competitive conditions. *Economic Journal, 36*(144), 535–550.

Sraffa, P. (1951). Volume I of The works and correspondences of David Ricardo, *On the Principles of Political Economy and Taxation*, Edited by Sraffa (1951) Cambridge University Press, 1951.

Sraffa, P. (1960). *Production of commodities by means of commodities*. Cambridge: Cambridge University Press.

Tabuchi, T. (2017a). Yukizawa's Interpretation of Ricardo's 'Theory of Comparative Costs'. In S. Senga, M. Fujimoto, & T. Tabuchi (Eds.), *Ricardo and international trade (Chapter 4)*. London: Routledge.

Tabuchi, T. (2017b). Comparative advantage in the light of the old value theories. In Y. Shiozawa, T. Oka, & T. Tabuchi (Eds.), *A new construction of ricardian theory of international values (Chapter 9)*. Tokyo: Springer.

Takenaga, S. (2016). Structure of the theory of value of David Ricardo, uploaded in Reserach Gate. doi:10.13140/RG.2.1.5068.2646.

Turgot, A. R. J. (1769). Value and Money in Turgot (2011) Chapter 8, pp. 163–182.

Turgot, A. R. J. (2011). *The Turgot Collection: Writings, Speeches, and Letters of Anne Robert Jacques Turgot, Baron de Laune*. Edited by David Gordon, Ludwig von Mises Institute.

Viner, J. (1937). *Studies in the theory of international trade*. Reprinted in 1955. London: George Allen & Unwin.

Whitaker, J. K. (Ed.) (1975). *The early economic writings of Alfred Marshall*, 1867–1890. Two volumes. London: McMillan for the Royal Economic Society.

White, M. V. (1989). Why are there no supply and demand curves in Jevons? *History of Political Economy, 21*(3), 425–456.

Yoshii, S. (2017). An extinction of adjustment time and an introduction of stability condition in economics through misunderstandings to J.S. Mill's law of supply and demand and international value theory, Chapter 8. In Y. Shiozawa & T. Oka (Eds.), A new construction of Ricardian theory of international values (Chapter 8). Tokyo: Springer.

Yukizawa, K. (1974). Original Meaning and Deformed Interpretation in Ricardo's 'Comparative Advantage Theory.' *Shogaku Ronsan* (Chuo University) 15(6): 25–21. English translation by Takenaga is available.

An Extinction of Adjustment Time and an Introduction of Stability Condition in Economics through Misunderstandings to J.S. Mill's Law of Supply and Demand and International Value Theory

Satoshi Yoshii

Abstract It is generally considered that the theoretical development in supply-demand equilibrium theory by J.S. Mill from supply and demand ratio theory by Adam Smith and David Ricardo is the most important event in the history of economics. Marshall (On Mr. Mill's theory of value. Fortnightly Review, April, reprinted in Pigou AC (ed) Memorials of Alfred Marshall, London, 1925, 1876) and Schwarz (The new political economy of J.S. Mill. Duke University Press, Durham, 1972) identify Mill's pricing model about absolutely limited commodities in quantity as a partial equilibrium theory. However, his pricing model is not governed by the laws of simultaneous determination between a quantity and a price at all. Followers have been misunderstanding the term "equation" that Mill used in the pricing. Mill's system is not supply-demand equilibrium theory, but rather sequential process model with time, like Robinson (Econ J 63(251):579–593, 1953) and Leijonhufvud (On keynesian economics and the economics of keynes. Oxford University Press, Oxford, 1968).

When Jenkin (North Br Rev 9:1–62, 1868; The graphic representation of the laws of supply and demand, and their application to labour. In: Sir Alexander Grant (ed) Recess studies. Edmonston and Douglas, Edinburgh, 1870, 151–185. Reprinted in series of reprints of scarce tracts in economic and political science, No. 9, The London School of Economics and Politcal Science, London, 1870) introduced the functions of supply and demand and an expression by a graph into the British economics for the first time, he misconstrued J.S. Mill's system as the view that was almost identical to present-day microeconomics model. Additionally, Marshall (Pure theory (foreign trade-domestic values). No.1 in series of reprints of scarce tracts in economic and political science. London, 1930, 1879), who was very

S. Yoshii (✉)
Nagoya University of Commerce and Business, Nagoya, Japan
e-mail: yoshii@nucba.ac.jp

Y. Shiozawa et al. (eds.), *A New Construction of Ricardian Theory of International Values*, Evolutionary Economics and Social Complexity Science 7,
DOI 10.1007/978-981-10-0191-8_8

influenced by Jenkin, misread Mill's reciprocal demand theory and introduced into economics the theme about the research of an equilibrium and the stability condition in his construction of pure theory.

Keywords Value theory • Equation • Cybernetic • Process analysis • Pure theory

1 Introduction

In the neoclassical economics, a price is considered as an important signal. Consumers choose the combination of goods to maximize their utility at the level that rate of substitution is equal to the price ratio of goods. Producers determine the optimal production at the level of price = marginal cost. In the market equilibrium, a price is determined by the intersection between the supply function and the demand function. If there is a discrepancy between the quantities of supply and the quantity of demand, their equalization is again recovered by the quick reaction of a price. That is, (1) the quantity of supply and the quantity of demand are expressed, respectively, as a function of price: the quantity is reflected directly by the price on one-on-one level; (2) there is a subjective desire or utility behind these functions, and utility is included in the determinant of the quantities; (3) a price and the quantity of exchange are simultaneously determined by balance of two functions; and (4) there is a stable equilibrium point to which an economical state should return. These are the features of supply and demand equilibrium model in microeconomics.

In the history of economics, classical economists had adopted the pricing by the production cost. Generally, it is said that the transformation is related to a rediscovery of the marginal utility in the 1870s (marginal revolution), but Hutchison (1972) states that classical economics had been collapsed itself before the rediscovery. As Schumpeter (1954) says "the scrapping of Ricardo's central concept" in J.S. Mill's Half-Way House (Part IV, Chap. 6, p.604), it is generally considered that the theoretical development in supply-demand equilibrium theory by J.S. Mill from supply and demand ratio theory by Adam Smith and David Ricardo is the most important event in the history of economics. Marshall (1876) and Schwarz (1972) identify Mill's pricing model about absolutely limited commodities in quantity as *a partial equilibrium theory*.

However, according to Yoshii (2014a, b, 2016), his pricing model is not governed by the laws of simultaneous determination between a quantity and a price at all. Followers have been misunderstanding the term "equation" that Mill used in the pricing. Mill's system is not supply-demand equilibrium theory, but rather sequential process model with time, like Robinson (1953) and Leijonhufvud (1968).

Then, why is J.S. Mill's system recognized as a supply-demand equilibrium theory? When F. Jenkin (1868, 1870) introduced the functions of supply and demand and an expression by a graph into the British economics for the first time, he misconstrued Mill's system as the view that was almost identical to present-day microeconomics model. Additionally, Marshall (1879), who was very influenced by Jenkin, misread Mill's reciprocal demand theory and introduced into economics the theme about the research of an equilibrium and that stability condition. Mill's system itself had a cause of inducing misunderstanding and transformation of a view on economics, but the misreading by Jenkin and Marshall was the turning point of the great transformation in economics. At the same time, we can say that the theoretical purity of economics advanced strongly, because Marshall defines pure theory by way of the misreading and by way of J.S. Mill's methodology itself. The purpose of this paper is to clarify the history of these points.

2 The True Meaning of J.S. Mill's Law of Supply and Demand

In this section, we consider the true meaning of J.S. Mill's law of supply and demand according to Yoshii (2014a, b, 2016). Mill classifies goods into three categories according to the difficulty of attainment.

Category 1

The quantity of goods cannot be increased at will, even if the labor and expense are input infinitely: Supply is constant at a certain time, e.g., wine made in special land, ancient sculptures, pictures by the old masters, rare books or coins, etc.

Category 2

The quantity of goods can be increased without increasing the cost per unit, if the labor and expense can be input: goods to which the law of constant return and constant-cost industry is applied.

Category 3

The quantity of goods can be increased with increasing the cost per unit, if the labor and expense can be input: goods to which the law of decreasing return and increasing-cost industry is applied.

Although the case of diminishing return is lightly treated in Ricardo, it is significantly different from Ricardo that J.S. Mill makes the category of goods for diminishing return. So to speak, the categories 1 and 2 express the extreme example of goods, and Mill himself considers the category 3 as the most important in economics (J.S. Mill 1848, vol. I, p.531). In addition, this classification of Mill is *unrelated* to the concept of time. Hollander treats category 1 as a case of short period pricing model (S. Hollander 1987, p.118), but he misreads it. It is a very important point that the classification is *unrelated* to a concept of time. And then, as for pricing method, the goods of category 1 are governed by law of supply and demand, and categories 2 and 3 are governed by production cost principle. Monopolistic goods[1] cannot increase in the quantity of supply because of artificial restrictions, so law of supply and demand governs the price.

We cannot understand essential point of the law of supply and demand, only if we consider concepts of supply and demand individually. Therefore, it is necessary to consider the adjustment process, by which the quantity of supply and the quantity of demand are equalized.

2.1 The Case of Excess Demand

In this case, all quantities are sold out, and so there are the consumers who are not able to purchase the goods. And then, "competition takes place on the side of the buyers, and the value rises" (J.S. Mill 1848, vol. I, p.534). The competition means that there are consumers, who purchase the goods even if a price is high than now, and suppliers understand such a situation and so they will make a price rise on the next market day. If a price is high than before, "additional sellers" and/or consumers who cannot purchase from restrictions of purchasing power may emerge more. Such a process is performed in the market day by day, and by increasing the quantity of supply and/or decreasing the quantity of demand, finally the quantity of demand and the quantity of supply are equalized. In the final day, namely, equalization day, there is a price for exchange, and the price becomes an exchangeable value of category 1.

2.2 The Case of Excess Supply

In this case, since the quantity brought to the market cannot be sold out all, there are unsold goods. And then, "The competition will now be on the side of the sellers" (Ibid., p.535). About unsold goods, the supplier who wants to sell even by lowering a price and so "an additional demand" (Ibid., p.535) is stirred up. Because new consumers are found and/or those who are already consumers purchase more (Ibid., p.535), unsold goods are sold out. If the quantity of supply decreases and so a price decline stops or returns to a former level by "the farmers withdraw their corn, and hold it back in hopes of a higher price" (Ibid., p.535) and/or "the operations of speculators who buy corn when it is cheap, and store it up to be brought out when more urgently wanted" (Ibid., p.535), unsold goods of "the supplier who did not want to sell even by lowering a price on a former market day" may be sold out on this market day. By the increased demand through a price decline and/or a price rising by withdrawal of supply, such a process is performed in the market day by

[1] Strictly speaking, monopolistic goods themselves may be classified into category 1, 2, and 3, respectively, but price itself is governed by a supply and demand principle at all.

day, and finally the quantity of demand and the quantity of supply are equalized. In the final day, namely, equalization day, there is a price for exchange, and the price becomes an exchangeable value of category 1.

By using a price determined by such procedures as a reference point like *Prospect Theory*, and by taking account of subjective evaluation (utility) and external factors (purchasing power in the society, social requirement, law, institution, foreign demand, etc.), suppliers will determine the quantity of supply, and consumers will determine the quantity of demand for the next market. In any case, what J.S. Mill put an importance to is equalization between the quantity of demand and the quantity of supply:

> we see that the idea of a ratio, as between demand and supply, is out of place, and has no concern in the matter: the proper mathematical analogy is that of an equation. Demand and supply, the quantity demanded and the quantity supplied, will be made equal. (Ibid., p.536)

Therefore, J.S. Mill uses the term "equation." However, it is a mistake to understand the term as the meaning of solving a formula like a present-day mathematical term, if Mill's procedures and the level of mathematics that economists treat and/or understand at that time are taken into consideration. That is, the view, that a price is determined on the intersection between a demand function and a supply function (if $D(p) = S(p)$, then p is determined), is misreading to Mill's system. If the term "equation" is understood as solving a formula, the evaluations, which Mill was a pioneer of partial equilibrium theory (Marshall 1876), which Mill advocated the partial equilibrium theory superior to Jevons (Schwarz 1972), could be true at all. However, the term "equation" has another meaning. It is a meaning showing the equivalence of different situations: the set of different facts, ideas, or people that all affect a situation and must be considered together (see Longman Dictionary of Contemporary English):

> Mill's equation, indeed, is not explicitly the same as any at which we have arrived above. His equation states that the quantity of a commodity given by A is equal to the quantity received by B. This seems at first sight to be a mere truism, for this equality must necessarily exist if any exchange takes place at all. (Jevons 1871, p.102)

Therefore, in this case, a mere relationship that the quantity of demand and the quantity of supply are equal is appropriate. Such an interpretation of "equation" is seen by J.M. Keynes (1936), too:

> The law, therefore, of values, as affected by demand and supply, is that they adjust themselves so as always to bring about an equation between demand and supply, by the increase of the one or the diminution of the other; the movement of price being only arrested when the quantity asked for at the current price, and the quantity offered at the current price, are equal. This point of exact equilibrium may be as momentary, but is nevertheless as real, as the level of the sea. (J.S. Mill 1869, p.636)

The term "equation" represents just a relationship (equivalence), and the true meaning of J.S. Mill's law of supply and demand is the law that "the quantity of demand becomes equal to the quantity of supply." Thus, although the demand-supply equilibrium model in present-day microeconomics puts great importance mainly on pricing and stability condition of equilibrium point, it differs from Mill's law. Mill thinks as follows: by a price fluctuated on market day by day or by interacting with the subjective factor of suppliers and consumers and social factors through a trial-and-error process, the quantity of demand becomes fortuitously and luckily equal to the quantity of supply. That is, he considers the adjustment process about whole economy. Moreover, because he states "the level of the sea," even if the quantity of demand and the quantity of supply are equalized, a price does not stop stably in an equilibrium point forever:

> The value always adjusts itself in such a manner, that the demand is equal to the supply. (J.S. Mill 1848, vol. I, p.574)
>
> If unequal at any moment, competition equalizes them, and the manner in which this is done is by an adjustment of the value. (Ibid., p.536)

It is natural that the price at that time is temporary, but this is called as equilibrium price because of equivalence between two kinds of quantities.

The quantity of demand itself fluctuates according to the factors like an individual taste and a situation (purchasing power is included), the market price at a certain time, social environment (social taste and purchasing power), a foreign demand, etc. That is, since the concept of time is not contained in J.S. Mill's goods classification, no situations other than a price may also be constant, and the quantity of demand may change because of these factors. Therefore, even when the price is the same about a certain commodity, a phenomenon, which the quantity of demand changes according to location and/or time, is also assumed in Mill's system. The definition of Mill's Homo economicus is consistent with this phenomenon too (see J.S. Mill 1843; 1844). Mill only says that one factor, which changes the quantity of demand during one market day and the other market day, is a price. In general equilibrium theory and partial equilibrium theory, if a price system is given, the function can uniquely, directly, and unambiguously derive the quantity of demand from a price.

Piecing together the consideration in this section, the existence of demand function itself may be acknowledged in J.S. Mill's system, but it cannot be acknowledged that Mill advocates the price theory a la neoclassical economics, which a price and an exchange quantity are simultaneously determined on the intersection between a demand function and a supply function. Mill does not describe stable point, so Mill's system is unrelated to the ways of thinking, such as Walrasian stability condition, Marshallian stability condition, and cobweb adjustment process.

Such Mill's method is closely similar to cybernetic method (Leijonhufvud 1968). According to Howitt (2002), cybernetic method explains the set of acts that transaction agents might adopt and the action rules (if-then rules) that they choose a specific act from the set, in a given situation. A price is determined by step-by-step process, not by the intersection between two functions or cobweb theorem. Cybernetic method is called as process analysis, period analysis, sequence analysis, and step-by-step analysis, and this method was an important analytical method also for A Treatise on Money (1930) by Keynes, R. G. Hawtrey, D. H. Robertson, the Swedish School, and the Austrian School. J.S. Mill's "law of demand and supply" is not a pioneer of partial equilibrium theory, but the process analysis, and it is similar to the system that Robinson advocates as follows:

> Any movement must take place through time, and the position at any moment of time depends upon what it has been in the past. The point is not merely that any adjustment takes a certain time to complete and that (as has always been admitted) events may occur meanwhile which alter the position, so that the equilibrium towards which the system is said to be tending itself moves before it can be reached. The point is that the very process of moving has an effect upon the destination of the movement, so that there is no such thing as a position of long-run equilibrium which exists independently of the course which the economy is following at a particular date. (Robinson 1953, p.590)

3 Most Misunderstood Application of the Law of Demand and Supply

> This law of International Values is but an extension of the more general law of Value, which we called the Equation of Supply and Demand. We have seen that the value of a commodity always so adjusts itself as to bring the demand to the exact level of the supply. But all trade, either between nations or individuals, is an interchange of commodities, in which the things that they respectively have to sell constitute also their means of purchase: the supply brought by the one constitutes his demand for what is brought by the other. So that supply and demand are but another expression for reciprocal demand: and to say that value will adjust itself so as to equalize demand with supply, is in fact to say that it will adjust itself so as to equalize the demand on one side with the demand on the other. (J. S. Mill 1848, vol. II, p.136)

About applying the law to international trade, there are two important points. Firstly, J.S. Mill considers international trade as the theory of exchange: catallactics. However, Mill states as follows:

> One eminent writer has proposed as a name for Political Economy, "Catallactics," or the science of exchanges: by others it has been called the Science of Values. If these denominations had appeared to me logically correct, I must have placed the discussion of the elementary laws of value at the commencement of our inquiry, instead of postponing it to the Third Part; and the possibility of so long deferring it is alone a sufficient proof that this view of the nature of Political Economy is too confined. (Ibid., vol. I, p.519)

He agrees surely that economics is not a "catallactics," but in international trade, his theory becomes a "catallactics." That is, a schism which thinks a price determined by a balance of supply and demand acquires Ricardo's comparative advantage theory internally, in spite of Ricardo's repugnance:

> When the trade is established between the two countries, the two commodities will exchange for each other at the same rate of interchange in both countries. (Ibid., vol. II, p.125)

> Since all trade is in reality barter, money being a mere instrument for exchanging things against one another, we will, for simplicity, begin by supposing the international trade to be in form, what it always is in reality, an actual trucking of one commodity against another. (Ibid., vol. II, p.124)

According to Yukizawa (1974), Ruffin (2002), Maneschi (2004), Aldrich (2004), Meoqui (2014), and Faccarello (2015), Ricardo never considers Chap. 7 of *Principle*

Table 1 The numerical examples by J.S. Mill

Exchange value	England	England	Germany	Germany
Linen/cloth	The quantity of supply of cloth	The quantity of demand of linen	The quantity of demand of cloth	The quantity of supply of linen
1.5	600	600 × 1.5	1200	1200 × 1.5
1.6	700	700 × 1.6	1100	1100 × 1.6
1.7	800	800 × 1.7	1000	1000 × 1.7
1.8	900	900 × 1.8	900	900 × 1.8
1.9	1000	1000 × 1.9	800	800 × 1.9
2.0	1100	1100 × 2.0	700	700 × 2.0

Mawatari (1997), p.257

(Ricardo 1817) as a barter economy. This fact relates to the problem called now as "the original meaning" and "the deformed interpretation" about comparative cost theory (see chapter 7 and 9 of this book.)

Secondly, therefore, J.S. Mill's international trade theory is often misunderstood as the closest to the view of neoclassical economics. Moreover, his trade theory is application of the law of supply and demand, but followers are premised on *erroneous interpretation* of the "equation":

> In common with most of my predecessors, I find it advisable, in these intricate investigations, to give distinctness and fixity to the conception by numerical examples. These examples must sometimes, as in the present case, be purely supposititious. I should have preferred real ones; but all that is essential is, that the numbers should be such as admit of being easily followed through the subsequent combinations into which they enter. (J. S. Mill 1848, vol. II, p.125)

Mill adopts the method of explaining trade theory by a numerical example. The numerical examples by Mill are Table 1.

Since a price corresponds strictly to the quantity of demand in Mill's numerical example respectively, the existence of a demand function may be permitted to Mill's system. However, in the theory of international value, equalization between the quantity of demand and the quantity of supply is explained by the *process analysis* over time, and it is the same as logic in the category 1 too. In fact, he considers trade theory as the adjustment process: "To procure the remaining 200, which she would have no means of doing but by bidding higher for them" (Ibid., vol. II, p.127–8). In addition, changes of the quantity of demand "depends on the nature of the particular commodity, and on the tastes of purchasers" (Ibid., vol. II, p.140):

> It may be considered, therefore, as established, that when two countries trade together in two commodities, the exchange value of these commodities relatively to each other will adjust itself to the inclinations and circumstances of the consumers on both sides, in such manner that the quantities required by each country, of the articles which it imports from its neighbour, shall be exactly sufficient to pay for one another. As the inclinations and circumstances of consumers cannot be reduced to any rule, so neither can the proportions in which the two commodities will be interchanged. We know that the limits within which the variation is confined, are the ratio between their costs of production in the one country, and the ratio between their costs of production in the other. [. . .] The ratios, therefore, in which the advantage of the trade may be divided between the two nations are various. The

> circumstances on which the proportionate share of each country more remotely depends, admit only of a very general indication. (Ibid., vol. II, p.128–9)

As he states that "As the inclinations and circumstances of consumers cannot be reduced to any rule, so neither can the proportions in which the two commodities will be interchanged," the stable point of an exchange value cannot be determined either, because of the subjective tastes of the consumers and/or social circumstances in each country. That is, there is no stable or same equilibrium point to which a price backs again. Probably, J.S. Mill's using the numerical example for explanation of trade theory may be evil for the followers to understand his *process analysis* correctly.

4 Jenkin's Misunderstanding: Extinction of Adjustment Time and Return to Same Equilibrium

Although category 1 is an extreme case and category 3 is most important for J.S. Mill, as Marshall (1876) pointed out, sight lines in academic community are directed to the theory of exchange in a market after the controversy between Mill (1869) and Thornton (1869). Fleeming Jenkin is inspired from the controversy and publishes "Trade Union" in 1868 and "The Graphic Representation of the Laws of Supply and Demand" in 1870. These are contained in the most seminal papers in the history of economics, and the influence to followers is also great:

> The graphic method has been applied, in a manner somewhat similar to that adopted in the present Chapter, by Dupuit in 1844; and, independently, by Fleeming Jenkin (Edinburgh Philosophical Transactions) in 1871. (Marshall 1961, p.476)
>
> he was the first Englishman to discuss, with nearly the same clearness as had Verri and Cournot, demand functions. (Schumpeter 1954, p.837)
>
> In the process of attacking the wages fund doctrine, Thornton and Longe drew attention to the possibility of perverse demand and supply functions in the labour market; inspired by this controversy, Fleeming Jenkin drew demand supply curves in a paper published in 1870-Cournot had done so as early as 1838, but he was almost unknown in England. (Blaug 1962, p.279)
>
> Jenkin, moreover, was the first Englishman to discuss demand functions and he introduced the diagrammatic method into English economic literature. (Brownlie and Prichard 1963, p.211)

Since Cournot's pioneering achievement (Cournot 1838) was buried for about 40 years and Jules Dupuit's graph (Dupuit 1844) was only a demand curve, Jenkin is considered as the first man that introduces the graph of supply and demand, which is called as Marshallian Cross Diagrams now, into economics in England. However, since Jevons and Jenkin had a private relationship, it seems that they argued privately about graphical expression.[2]

---

[2]In regard to this point, see Keynes (1972), p.138.

The main subject of "Trade Union" in 1868 is a consideration of the principles that affect a labor union, a right to organize, the actual state of a labor union, and a needful law (see Uemiya 1981), but Jenkin also attempts to express pricing model in J.S. Mill's category 1 by mathematics. He starts with the conceptual definition of supply and demand. In order to avoid an ambiguous expression, supply and demand are defined as the quantity. On the other hand, he also recognizes "the word demand, as popularly used, signifies a desire" and "the readiness to sell" (Jenkin 1868, p.15), namely, Malthusian intensity of the will. Jenkin defines supply and demand, respectively, by "the quantity" and "the will," and each becomes a pair: "the quantity of demand and the quantity of supply" and "a desire and the readiness to sell." Since pricing cannot be considered from the side of "the will," he adopts the method that, first of all, he considers pricing by the equalization between "the quantities" of supply and demand and then introduces the "will" into the model.

After defining supply and demand, he states as follows:

> We may now try to write the equation indicated by Mr. Mill. (Ibid., p.17)

If D is the demand, x is the variable price, and D diminishes along with a price increase, D = f (1/x). Here, "f is not a simple factor, but is a mere symbol," and D "is affected by no other circumstance, as assumption which on any given market-day may be true" (Ibid., p.17). S is defined as the quantity of supply at price x on the same market day; S increases as the price increases. Therefore, S = F (x), but S is affected by no other circumstance. "When D is equal to S, we have the equation f (1/x)=F (x), by which the price x could be calculated, and would be determined" (Ibid., p.17). The price determined in this way is "only one natural and invariable value or price for each article" (Ibid., p.17). After the birth of classical economics, there were many discussions about the concept of demand and supply and the relationship between a price and these concepts. Especially, classical economists define the supply as constant, and there is no much development in the consideration about the relationship between a price and supply. However, Jenkin's descriptions are the "moment" that a supply function appears suddenly in the history of economic thought.

And then, Jenkin introduces the "will" into the model. "If the desire for the article increases, the value tends to rise," and "the readiness to sell at a given price may diminish, and so diminish the quantity supplied" (Ibid., p.17). Therefore, each function becomes no longer a mere function of price; an equation is corrected: f (A + 1/x) = F (B + x). Here, A and B are "some unknown variable quantity," and "so long as A and B and f and F were all constant in value and form, x would remain constant, and would be fixed in terms of these magnitudes" (Ibid., p.17). If x becomes higher accidentally than a price determined by the equation, the excess supply D < S would occur. In this case, the free competition would lower the price and "would bring back x to its true market value" (Ibid., p.18). In the case of excess demand, a price rises and adjustment is finished.

Here, it is an important thing that Jenkin describes "bring back x to its true market value." He thinks an activity on a given market day and "natural and invariable value or price" determined by the equation on a certain day or time must be only one. Therefore, "bring back x to its true market value" means "back x to the same equilibrium point as before." Such a Jenkin's view is the same as the view of so-called microeconomics, and it is certain that Jenkin is one of the origins about view of microeconomics: (1) there is a subjective desire or utility behind a supply and demand functions; (2) a price and the quantity of exchange are simultaneously determined by the equation or intersection between two functions; and (3) when a price deviates from an equilibrium point, it converges to the same equilibrium point again; these are familiar propositions:

> Our equation thus expresses every relation between value, demand, and supply, which Mill states as expressing the law of value with respect to all commodities not susceptible of being multiplied at pleasure. (Ibid., p.18)

Is this Jenkin's statement correct? It is a very important point that J.S. Mill's classification about the goods is unrelated to a concept of time, and so category 1 is not a case of short period pricing model. Although Jenkin also considers the situations that the quantity of demand and the quantity of supply are affected by the factors other than a price, there is no adjustment process through trial and error with time like Mill, in Jenkin's model. That is, Jenkin considers the stability condition of a price, and the quantity of exchange and a price are simultaneously determined, so "the concept of adjustment time" is lacking absolutely. The reason why the time is extinct is that Jenkin grasps Mill's term "equation" as the meanings of "solving a formula." Mill thinks that a price also fluctuates and the prices are one of the mediums of the adjustment during the adjustment process through trial and error with time. When the quantity of demand and the quantity of supply become equal fortunately, fluctuations of a price can stop, and it is an equilibrium price. However, it is not a stable price. Jenkin perverts the view of classical economics, in which cybernetic method is adopted and which Robinson states its feature as "the very process of moving has an effect upon the destination of the movement," to the view of neoclassical economics. That is, Jenkin's paper in 1868 is the turning point of a "great transformation" about a view on economics.

Next, "The Graphic Representation of the Laws of Supply and Demand" in 1870 is considered. Although, needless to say, the first graphic representation is important, it is more important for a "great transformation" that Jenkin exterminates the concept of adjustment time in economics. J.S. Mill classifies the goods into three categories according to the difficulty of attainment, but Jenkin classifies the pricing method according to the time differential.

Short period is the period, in which the whole supply and a purchase fund do not change. However, the individual will influence price determination. A price is determined on the intersection between a demand curve and a supply curve.

Middle period is the period in which the whole supply and a purchase fund change. A price is determined on the intersection between a demand curve and a supply curve.

Long period is the period in which the concepts of the whole supply and a purchase fund are rejected, and "the price of the manufactured article is chiefly determined by the cost of its production, and the quantity manufactured is chiefly determined by the demand at that price" (Jenkin 1870, p.165).

Jenkin introduces the idea of the whole supply which J.S. Mill uses the term in a reply to Thornton (J.S. Mill 1869, p.641). The whole supply is a sales potential at the selling time (Jenkin 1870, p.76). In the demand phase, Jenkin introduces the idea of purchase fund. The purchase fund is a budget constraint, and a possibility that other goods are bought by its budget is excluded. Jenkin defines that *qw* is whole supply, a price is *pw* when supplier is willing to sell *qw*, a price is *pd* when consumer is willing to buy *qw*, a supplier doesn't sell goods at the price *ps* or lower, a consumer doesn't buy goods at the price *pf* or higher, *pe* is a market equilibrium price, and *qe* is an equilibrium quantity. And then, we can graph an exchange at short period as below (Fig. 1):[3]

Jenkin recognizes, actually, that the demand curve and the supply curve are unknown and that a price determined by such procedure is "the theoretical price" (Ibid., p.78). When the shape of the function changes by the nature of human side, it's drawn like Fig. 2. This is a situation of the increase or decrease in supply and demand by a change of propensity. In middle period, whole supply and purchase fund are changeable (Fig. 3).

Fig. 1 Short period 1

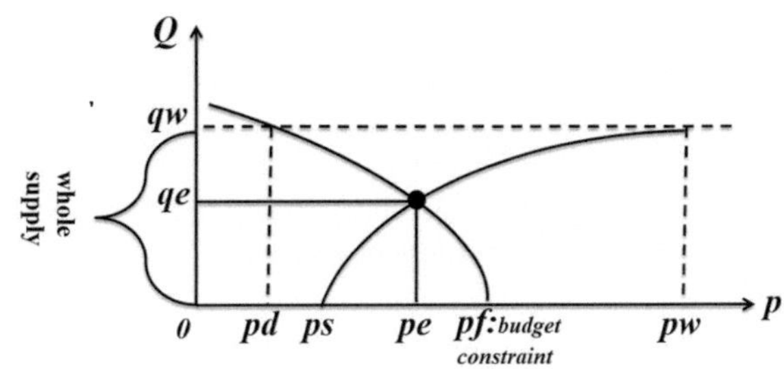

Fig. 2 Short period: the change of propensity

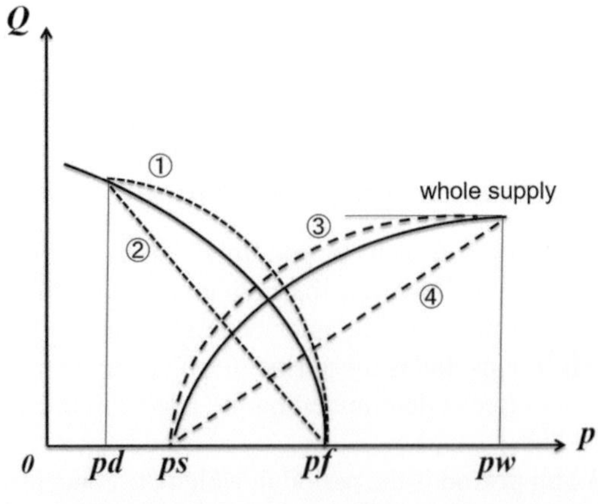

[3]Jenkin graphs numerical examples. Uemiya (1981) is used as reference in all grapshs.

Fig. 3 Middle period

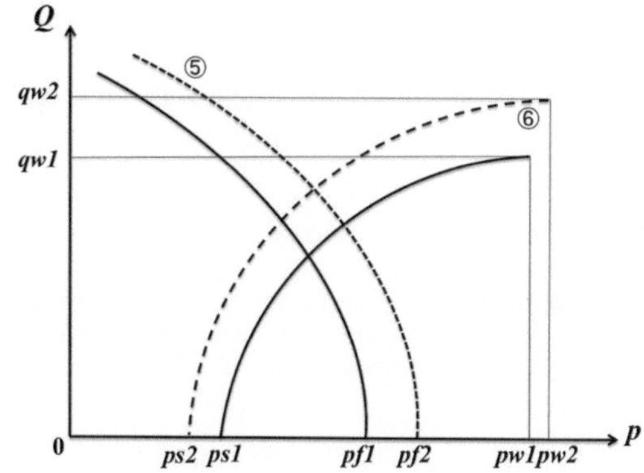

In any case, the historical process of economy that Robinson and Leijonhufvud point out can no longer be analyzed by graphic representation and the classification of the time concept. It is because an economic situation returns to an identical equilibrium point along the function. Jenkin's paper in 1870 plays a role of the reinforcement of a great transformation about a view on economics by the graphic representation.

5 Marshall's Misunderstanding: Extinction of Monetary Term and Introduction of Stability Condition

All influences that J.S. Mill and Jenkin give to Marshallian economics, especially *Principles of Economics* (1890), should be considered in another paper, but only the point, which Marshall may further accelerate the great transformation of the view on economics by Jenkin, is considered in this subsection. As discussed previously, "The law of demand and supply in the category 1" and "the numerical example in trade theory" by Mill mislead the followers, especially Marshall, to the view on economics that a price and the quantity are determined simultaneously by the intersection between two functions, through Jenkin as the medium[4]:

> H. C. Fleeming Jenkin (1833–85) was an economist of major importance, whose main papers belong chronologically to the previous period but who has been reserved for discussion here because these papers form an obvious stepping stone between J.S. Mill and Marshall. (Schumpeter 1954, p.838)

Marshall (1879) is the pamphlet that Marshall makes oneself clear about the issues of value theory. It's a characteristic that he makes international value theory a first issue before domestic value theory. Each country follows same law of demand, only reciprocal demand, in case of international value theory. Therefore,

[4]Leijonhufvud (1993) states that Marshall (1890) adopts the process analysis. Since Marshall (1876) evaluates Mill's value theory highly, there is no wonder that a similarity is observed in both theories. However, Marshall (1879) is the simultaneous determined system by two functions.

it's possible to graph the symmetrical demand function without thought. In case of domestic value theory, however, there are different laws in demand side and supply side. So, before considering the law that a supplier obeys, he can't graph.

He starts with defining "pure theory" in the first chapter of *The Pure Theory of Foreign Trade*:

> The function of a pure theory is to deduce definite conclusions from definite hypothetical premises. The premises should approximate as closely as possible to the facts with which the corresponding applied theory has to deal. But the terms used in the pure theory must be capable of exact interpretation, and the hypotheses on which it is based must be simple and easily handled. The pure theory of foreign trade satisfies these conditions. (Marshall 1879, p.1)

J.S. Mill says in trade theory, "we are now in the region of the most complicated questions which political economy affords; that the subject is one which cannot possibly be made elementary" (J. S. Mill 1848, vol. II, p.125), but Marshall replies, "The unavoidable difficulties of the subject are great: but students frequently fall into errors which they may easily avoid if they will resolve that when discussing the pure theory they will not speak of the imports or exports of a country as measured in terms of money" (Marshall 1879, p.3). That is, the consideration in terms of money is eliminated from pure theory in economics.

Additionally, "The only apparatus which Ricardo and Mill brought to bear on the problems of pure economic theory was that of arithmetical illustration. But this is inadequate to the work" (Ibid., p.4). So, Marshall thinks of better method as "mathematical methods" and of best method as "the method of diagrams" (Ibid., p.5). From such a viewpoint, Marshall draws the graph known as "offer curve" (Fig. 4).

Commercial terms are determined in offer curve by the intersection between the demand functions of two countries. All trustworthy data for drawing cannot be obtained, but "for the purposes of the pure theory we are at liberty to suppose that the curve is properly drawn throughout its entire length" (Ibid., p.7). Furthermore, Marshall states as follows:

> The terms in which the curves are described imply that there is no change in the circumstances. (Ibid., p.7)

Fig. 4 Offer curve

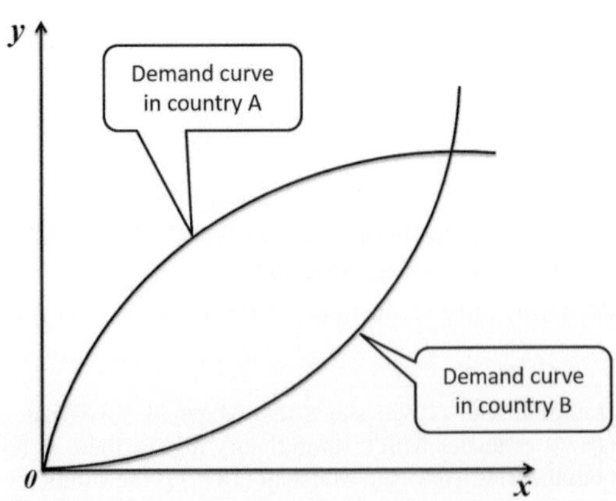

> we should aim at simplicity in our first approximations, in order that they may be easily manageable. Therefore, we are to neglect for the present all consideration of the disturbances arising from such variations; leaving account to be taken of them in the applications of the results of the pure theory to practical issues. (Ibid., p.8)

To be able to understand from these descriptions is that Marshall seeks "pure theory," that it has to get closer as much as possible to the facts with which applied theory has to deal, that variables used have to be reduced as much as possible in terms of mathematics, and that Marshall assumes ceteris paribus. These are subject to influences by J.S. Mill and/or Jenkin: Mill pursues unitary law of economics (J.S. Mill 1843, 1844). He gives realistic validity to a definition of Homo economicus in contradiction to Ricardo's hypothetico-deductive method (Ricardo 1817). James Mill and J.S. Mill make trade theory "catallactics" and eliminate monetary terms in it. Factors inducing fluctuations other than a price are eliminated in J.S. Mill's numerical example in trade theory. J.S. Mill considers the theory of value under the fixed condition about a value of other goods (J.S. Mill 1848, p.458). Jenkin eliminates the concept of an adjustment process over time by mathematization and drawing the graph.

Marshall points out that J.S. Mill's trade theory is almost entirely devoid of the stability condition of equilibrium, and then he proves stability conditions from 17 pages to 28 pages in Marshall (1879):

> The possibility of more than one position of equilibrium in such cases as this has been noticed by Mill. His treatment of the matter is certainly inadequate: for he has failed to discover the laws which determine whether any particular position of equilibrium is stable or unstable. It is, generally speaking, true of Mill as of Adam Smith, that much of his work which appears at first sight to contain error, proves itself on further investigation to be only incomplete or incompletely expressed. This is however one of the few instances in which careful study has failed to convince me that Mill's work is right as far as it goes. (Ibid., p.12)

However, in J.S. Mill's trade theory, the change of quantity of demand "depends on the nature of the particular commodity, and on the tastes of purchasers" (J. S. Mill 1848, vol. II, p.140), and so terms of trade are not stable. Mill's purpose is to explain the equalization *process* between the quantity of demand and the quantity of supply through time. This is the same as explanation in the category 1 and belongs to the *process analysis* in which an event is located at historical time. In addition, the equilibrium for Mill is meaning that the quantity of demand is equal to the quantity of supply and the price (terms of trade) at that time is an equilibrium price, but an equilibrium point is just only one reference point in the adjustment process over time. Therefore, as described above, Mill's equilibrium concept is *unrelated* to a stability condition. Such Marshall's interpretation on equilibrium may be the influence of "true market value" of Jenkin. Namely, it is considered that a certain equilibrium point is the same equilibrium point as before.

Marshall's view on economics is expressed in the following quotations:

> It has been remarked, that in economics every event causes permanent alterations in the conditions under which future events can occur. This is to some extent; the case in the physical world, but not to nearly so great an extent. The forces that act on a pendulum

> in any position are not to any appreciable extent dependent on the oscillations that the pendulum has already made. And there are many other classes of movement in the physical world, which are exact copies of movements that have gone before. But every movement that takes place in the moral world alters the magnitude if not the character of the forces that govern succeeding movements. And economic forces belong to the moral world in so far as they depend upon human habits and affections, upon man's knowledge and industrial skill. (Marshall 1879, p.26)

Human's subjectivity alters the magnitude, not the character of the forces. In Marshall, the alteration of magnitude means "some alteration in the shapes of the curves" (Ibid., p.26). Human's nature is considered in a framework of the stability condition of equilibrium. Therefore, once the shape of functions is given, a price and the quantity would be determined simultaneously and a stable equilibrium point would be calculated. Of course, disequilibrium price may exist, but, at any rate, the phenomena take place within the same view on economics. Marshall does not have the same view as Robinson. The facts that Marshall emphasizes the concept of stable equilibrium and considers it as the theme that economics should pursue have played a large role in great transformation of view on economics. Therefore, it can be said that the root of the acceleration of great transformation is Marshall's reading mistake about J.S. Mill's trade theory.

6 Concluding Remarks

J.S. Mill is apt to be misunderstood theoretically. That's also related to the academic background where he was put. Mill was on cordial terms with Ricardo. Although Mill himself tried to be a pious Ricardian, the status of Ricardian economics was lower around him, especially in "Political Economy Club." After Ricardo's death, Mill abandoned pure labor embodied value theory and the concept of invariable measure of value in 1823, and the concept of absolute value in 1825. Moreover, Mill thinks that a law of demand and supply is antecedent than a principle of cost of production. Thus, Mill experienced a tension between economic theory that he believed and a respect to Ricardo. What concepts among Ricardian economics should be survived? Probably, Mill worried about the reply to this problem most.

The key to reply it would be comparative advantage theory, because only the theory within Ricardo's one was accepted widely in academian such as Robert Torrens. How to leave this theory in posterity? This point might be a problem for J.S. Mill. The law of cost of production is not applicable to international value theory, and so a theory of value has to be reconstructed. If the law of supply and demand was anteceded, Mill might think that terms of trade could be determined. Therefore, a schism which thinks a price determined by a balance of supply and demand acquired Ricardo's comparative advantage theory internally, in spite of Ricardo's repugnance. Mill might be evaluated as eclecticism from such a reason. Moreover, his influence was so big for his authority in those days.

Just because J.S. Mill's system doesn't exclude the influence of demand to determine a price, his system isn't also an equilibrium theory like present-day microeconomics. Mill uses the term "equation" in a derivation of equilibrium. However, we can understand that it is a mistake to interpret the term as the meaning of solving a formula like a present-day mathematical term, if Mill's procedures are taken into consideration. The term "equation" has another meaning. It is a meaning showing the equivalence of different situations. That is, a mere relationship that the quantity of demand and the quantity of supply are equal is appropriate. Moreover, because he states "the level of the sea" about equilibrium point, even if the quantity of demand and the quantity of supply are equalized, a price does not stop stably in an equilibrium point forever.

In general, equilibrium theory and partial equilibrium theory, if a price system is given, the function can uniquely, directly, and unambiguously derive the quantity of demand from a price. The existence of demand function itself may be acknowledged in J.S. Mill's system, but it cannot be acknowledged that Mill advocates the price theory a la neoclassical economics, which a price and an exchange quantity are simultaneously determined on the intersection between a demand function and a supply function. Mill does not describe stable point, so Mill's system is unrelated to the ways of thinking, such as Walrasian stability condition, Marshallian stability condition, and cobweb adjustment process. Mill's method is similar to a method called now as process analysis, period analysis, sequence analysis, and step-by-step analysis. It's obvious that J.S. Mill's system is not a partial equilibrium theory.

About applying the law of supply and demand to international trade, the theory is often misunderstood as the closest to the view of neoclassical economics. However, it is the same as logic in the category 1 too. There is no stable or same equilibrium point to which a price backs again. Probably, J.S. Mill's system using the numerical example for explanation of trade theory may be evil for the followers to understand his *process analysis* correctly.

A model of Jenkin is just product by misleading use of a term. In the result, economics is deprived of the room where adjustment time is considered. It is said that Jenkin perverts the view of classical economics, in which cybernetic method is adopted and which Robinson states its feature as "the very process of moving has an effect upon the destination of the movement," to the view of neoclassical economics.

To make matters worse, Jenkin was a stepping stone between J.S. Mill and Marshall in a bad sense. Firstly, Mill's scientific methodology influences Marshall's pure theory. In the result, not only international value theory but also domestic value theory is not analyzed in terms of money. Secondly, a view on economics transformed. Marshall emphasizes the concept of stable equilibrium and considers it as the theme that economics should pursue. Mill's purpose is to explain the equalization *process* between the quantity of demand and the quantity of supply through time. This method belongs to the *process analysis.* So, the historical process of economy that Robinson and Leijonhufvud point out can no longer be analyzed in Marshall.

J.S. Mill's system itself has a cause of inducing misunderstanding and transformation of a view on economics, but the misreading by Jenkin and Marshall is the turning point of the great transformation in economics. At the same time, it is the turning point of view on economics from plutology to catallactics.

References

Aldrich, J. (2004). The discovery of comparative advantage. *Journal of the History of Economic Thought, 26*, 379–399.

Blaug, M. (1962). Economic theory in retrospect, Richard D. Irwin, inc.

Brownlie, A. D., & Prichard, M. F. L. (1963). Professor Fleeming Jenkin, 1833-1885, pioneer in engineering and political economy. *Oxford Economic Papers, 15*(3), 204–216.

Cournot, A. (1838). *Recherches sur les principes mathématiques de la théorie des richesses*. Paris: Chez L. Hachette.

Dupuit, J. (1844). De la mesure de l'utilité des travaux publics, *Annales des Ponts et Chaussées*, s. II, 2nd semester, 332–375, English translation by R.H. Barback, On the Measurement of the Utility of Public Works, *International Economic Papers*, 1952, n. 2, 83–110.

Faccarello, G. (2015). A calm investigation into Mr Ricardo's principles of international trade. *The European Journal of the History of Economic Thought, 22*(5), 754–790.

Hollander, S. (1987). *Classical economics*. Oxford: Basil Blackwell.

Howitt, P. (2002). In M. De Vroey & J. Lallement (Eds.), *On Keynesian economics and the economics of Keynes, A dictionary article translated and publisehd in French in Dictionnaire des grandes œuvres économiques*. Xavier Greffe: Edition Dalloz.

Hutchison, T. W. (1972). The Marginal Revolution and the decline and fall of english classical political economy. *History of Political Economy, 4*(2), 442–468.

Jenkin, H. C. F. (1868). Trade union: How far legitimate. *North British Review, 9*, 1–62.

Jenkin, H. C. F. ([1870] 1931). The graphic representation of the laws of supply and demand, and their application to labour, In: Sir Alexander Grant (ed.), *Recess Studies*, Edinburgh: Edmonston and Douglas, 1870, 151–185. Reprinted in *Series of Reprints of Scarce Tracts in Economic and Political Science*, No. 9, London: The London School of Economics and Politcal Science.

Jevons, W. S. (1871). *The theory of political economy*. London: Macmillan and Co..

Keynes, J. M. ([1930] 1971). *A treatise on money: The pure theory of money, The collected writings of John Maynard Keynes*, vol. V. London: Macmillan.

Keynes, J. M. ([1933] 1972). *Essays in biography, The collected writings of John Maynard Keynes. Vol. X*, London: Macmillan.

Keynes, J. M. ([1936] 1973). *The general theory of employment, interest and money, The collected writings of John Maynard Keynes.* vol. VII, London: Macmillan.

Leijonhufvud, A. (1968). *On keynesian economics and the economics of keynes*. Oxford: Oxford University Press.

Leijonhufvud, A. (1993). Towards a Not-Too-Rational macroeconomics. *Southern Economic Journal, 60*(1), 1–13.

Maneschi, A. (2004). The true meaning of David Ricardo's four magic numbers. *Journal of International Economics, 62*, 433–443.

Marshall, A. (1876). On Mr. Mill's Theory of Value, *Fortnightly Review*, April, reprinted in A. C. Pigou (ed.) *Memorials of Alfred Marshall*, London, 1925.

Marshall, A. (1879). *Pure Theory* (*Foreign Trade-Domestic Values*). No.1 in series of reprints of scarce tracts in economic and political science: London, 1930.

Marshall, A. ([1890] 1961). *Principles of Economics*, 9th (variorum) edition, with annotations by C. W. Guillebaud, Vol. I, London: Macmillan.

Mawatari, S. (1997). *J. S. Mill's Economics*, Ochanomizu-shobo (in Japanese).
Meoqui, J. M. (2014). Reconisilling Ricardo's comparative advantage with Smith's productivity theory. *Economic thought, 3*(2), 21–37.
Mill, J. S. (1843). System of logic: Ratiocinative and inductive, 1843; 1846; 1851; 1856; 1862; 1865; 1872. In: *Collected Works of John Stuart Mill, Vol. VII, VIII*, textual editor J. M. Robson, introduced by R. F. McRae, Toronto: University of Toronto Press, 1973.
Mill, J. S. (1844). *Essays in some unsettled questions of political economy*. London: John W. parker, West Strand.
Mill, J. S. (1848). *The principles of political economy: With some of their applications to social philosophy* (Vol. I and II). Boston: C.C. Little & J. Brown.
Mill, J. S. (1869). Thornton on Labour and its claims, *Fortnighty Review*, n.s. Vol. 5, May and June, in *Collected Works of John Stuart Mill, Vol. V*, Toronto: University of Toronto Press.
Ricardo, D. ([1817] 1951). On the principles of political economy and taxation. In: Piero Sraffa with the collaboration of M. H. Dobb (Eds.), *The works and correspondence of David Ricardo, Vol. I*, Cambridge: Cambridge University Press.
Robinson, J. (1953). Imperfect competition revisited. *Economic Journal, 63*(251), 579–593.
Ruffin, R. J. (2002). David Ricardo's discovery of comparative advantage. *History of Political Economy, 34*(4), 727–748.
Schumpeter, J. A. (1954). *History of economic analysis*. London: George Allen & Unwin Limited.
Schwarz, J. P. (1972). *The new political economy of J.S. Mill*. Durham: Duke University Press.
Thornton, W. T. ([1869] 1967). *On labour: Its wrongful claims and rightful dues, its actual present possible future*. London: Macmillan.
Uemiya, S. (1981). Jevons and fleeming jenkin. *Kobe University Economic Review, 27*, 45–57.
Yoshii, S. (2014a). Great transformation to simultaneous determination system: The turning point of a view on economics. In Y. Shiozawa, & Y. Aruka (Eds.), *Rebuilding of economics* (Chapter 8, pp. 269–349). Chuo University Press (in Japanese)
Yoshii, S. (2014b). Great transformation to simultaneous determination system: The turning point of a view on economics, Annual conference paper of the European Society for the History of Economic Thought, University of Lausanne.
Yoshii, S. (2016). "What is the criterion which discriminates between Classical and Neo-Classical?: the true meaning of J. S. Mill's law of supply and demand." *mimeo*.
Yukizawa, K. (1974). The original meaning and the deformed interpretation of Ricardo's 'theory of comparative costs.' *Shogaku Ronsan* (Chuo University), *15*(6), 25–21. (in Japanese)

Comparative Advantage in the Light of the *Old* Value Theories

Taichi Tabuchi

Abstract The chapter examines the historical process of how the comparative advantage theory developed from James and John Stuart Mill to the modern theory, by way of Viner's real cost approach, Haberler's opportunity cost approach and Ohlin's factor endowment approach, in the light of old value theories. Since J. S. Mill, the theory of values had met with a succession of modifications, while the doctrine of comparative costs remained unchanged. Thus, the divergence between the general theory of values and the value theory used in the theory of international trade widened. The debate between Viner's real cost approach and Haberler's opportunity cost approach in the 1930s was an important turning point and resulted in the emergence of the new mainstream theory – the HOS model. Its unrealistic assumptions were derived from Haberler's opportunity cost approach.

Keywords Comparative advantage • Opportunity cost • Haberler • Viner • Samuelson

1 Introduction

This chapter presents a historical analysis of how the doctrine of comparative costs has developed from Ricardo to modern economics from the viewpoint of the new international value theory (Shiozawa 2007, 2014), focusing on the controversy in the 1930s. At that time, there were three competing approaches to the theory of comparative advantage: Jacob Viner's real cost approach, Gottfried Haberler's opportunity cost approach and Bertil Ohlin's factor endowment approach. It has been said that the last two quickly overwhelmed the first, and the Heckscher-Ohlin-Samuelson model emerged as the new mainstream theory of international trade after World War II through the efforts of Paul Samuelson.

T. Tabuchi (✉)
Faculty of Commerce, Doshisha University, Kyoto, Japan
e-mail: ttabuchi@mail.doshisha.ac.jp

Y. Shiozawa et al. (eds.), *A New Construction of Ricardian Theory of International Values*, Evolutionary Economics and Social Complexity Science 7,
DOI 10.1007/978-981-10-0191-8_9

However, a closer look at this process reveals problems and puzzles:

1. Viner and Haberler were rivals but shared the misconception that Ricardo's theory of comparative costs was based on the labour theory of value. They insisted on removing the labour theory of value from the doctrine of comparative costs and replacing it with their own value theories – Viner's real cost theory and Haberler's opportunity cost theory.
2. Viner failed to find a common unit with which to measure the subjective costs of labour and capital (i.e. the 'irksomeness' of labour and 'abstinence' or 'waiting'). On the other hand, Haberler's approach created a theory of international values with multiple factors, introducing the substitution curve based on Austrian value theory (now known as the 'transformation curve' or 'production possibility frontier'), which, as Viner pointed out, had the fatal flaw of assuming fixed factor supplies and factor indifference as to alternative uses.
3. Contrary to popular perception, Ohlin showed disdain for the concept of 'comparative advantage' and gave weight to money costs against real costs.
4. It is curious that Samuelson, who later synthesized the three approaches into the HOS model, insisted that the real costs and opportunity cost approaches could go together in 1938, but he criticized the product of their compromise as a 'rather awkward mumbo jumbo' in 1948.

As a result of the synthesis of the three approaches, the framework of pure exchange economy was formulated and established under the two-country two-commodity two-factor model. This chapter will try to provide a critical perspective on these problems and puzzles.

2 The Starting Point for Arguing: The Original Meaning of Ricardo's Theory of International Trade

Recent studies have revealed that the international trade theory first presented in David Ricardo's *Principles* (1817–1821) was entirely different from the standard interpretation, the so-called 'Ricardian model', still found in textbooks on international economics and even on the history of economic thought. The inevitable question that follows is 'Who was responsible for the reconstruction of Ricardo's theory?' Quite naturally, the answer depends on what is regarded as the original meaning of Ricardo's theory of international trade.

2.1 *The 'Sraffa-Ruffin Interpretation' and Yukizawa*[1]

Several recent publications (e.g. Ruffin 2002; Maneschi 2004) have provided the true meaning of the famous 'four magic numbers' (Samuelson 1969) in Chap. 7 of the *Principles*. According to Ruffin, Sraffa (1930) also understood their true meaning; thus, the new interpretation was named the 'Sraffa-Ruffin interpretation' by Andrea Maneschi. The essence of the new interpretation is as follows: the four numbers are not unit labour coefficients in the production of wine and cloth in England and Portugal but a representation of the labour needed to produce the amounts of wine and cloth actually traded; each country's gains from trade are simply given by the difference between the two numbers, without the need of any knowledge of the other country's labour inputs.

As early as the 1970s, Japanese Marxian economist Kenzo Yukizawa presented an interpretation essentially identical to Ruffin's (Yukizawa 1974, 1978; Tabuchi 2006, 2014). Yukizawa criticized the standard interpretation as a 'deformed interpretation' derived from John Stuart Mill (1844, 1848) and insisted that Ricardo's theory of comparative costs should be understood *as it was,* in the same way as in Ricardo's original logic.

Interestingly, both Yukizawa and Ruffin identified John Stuart Mill as the one who shaped the standard interpretation. Ruffin claimed the following:

> John Stuart Mill was responsible for the rational reconstruction of Ricardo in which the labor cost coefficients were interpreted as the amounts used in a unit of each good produced rather than Ricardo's labor cost of producing the amounts contained in a typical trading bundle. (Ruffin 2002: 742–43)

However, a close look at James and John Stuart Mill's discussions of the comparative costs examples reveals that it is not clear that they did not understand the original meaning of Ricardo's four numbers. Indeed, James Mill's statement in the 'Colony' article (James Mill 1818) certainly shows that he understood Ricardo's original presentation of the four numbers almost perfectly. It is not known whether John Stuart Mill read the article (Tabuchi 2006, 2014).

Rather, it was Viner and Haberler who misunderstood Ricardo's original theory and mixed it up with John Stuart Mill's exposition.

Haberler (1936) and Viner (1937) established the standard interpretation of the four numbers as follows: 'In chapter VII of his *Principles*, he gives the following celebrated example: In England a unit of cloth costs 100 and a unit of wine 120 units of labour; in Portugal a unit of cloth costs 90 and a unit of wine 80 units of labour' (Haberler 1936: 128) (Table 1).

[1]I discuss roughly the same issues in more detail in my book in Japanese (see Tabuchi 2006, Chaps. 3–5).

Table 1 Viner

	Amount of labour required for producing *a unit of*	
	Wine	Cloth
Portugal	*80*	*90*
England	*120*	*100*

Viner (1937: 445), emphasis added

2.2 Faccarello's Reading

Another epoch-making and far-reaching insight into Ricardo's theory of international trade appeared more recently (Faccarello 2015a, b). Faccarello criticizes Ruffin, Maneschi and the subsequent literature for focusing on only a few pages (about 15%) of Chap. 7 of the *Principles* and sticking to the traditional approach in *real* terms. According to Faccarello, Ricardo's work is like a jigsaw puzzle: no part can be analysed independently of the rest of the work. Through a careful reading of all of Chap. 7 (especially the discussion on money), the *Principles* as a whole and other writings by Ricardo, Faccarello draws three conclusions (among other important insights)[2]:

1. For Ricardo, there are no significant differences between domestic and international exchanges. Though Ricardo's analysis may seem to occur at the macro level, individuals, not countries, are the agents of trade, and every exchange is a monetary exchange. There are no specific international prices: in both domestic and international trade, micro-agents act in their self-interest, and the prices they pay will tend to be natural prices.
2. Thus, a country's 'gains from trade' are just an *unintended consequence* of the dynamics of individual agents in a competitive market, and the 'principle of comparative advantage' does not explain the flows of trade. Ricardo did not use the phrase 'comparative advantage' at all in explaining international trade.
3. The characteristics of an international equilibrium and the nature and impact of destabilizing shocks are analysed, and through changes in the distribution of precious metals, these shocks make the value of money differ among countries.

Faccarello's reading would urge us to radically review the history of the theory of international trade since Ricardo. Here again, we will see that the debate in the 1930s was an important turning point, alongside James and John Stuart Mill's discussions.

[2]Faccarello's seminars on Ricardo's theory of international trade, held at Tokyo University on March 23, 2016, and at Doshisha University on April 9, 2016, were also helpful on these points.

3 Real Cost Versus Opportunity Cost Debate Between Viner and Haberler[3]

3.1 Theory of Values and the Doctrine of Comparative Costs Prior to the 1930s

James and John Stuart Mill began to treat the doctrine of comparative costs independently of the monetary discussions. James Mill, who showed a thorough understanding of Ricardo's original presentation of the four numbers in the 'Colony' article, mentioned money only slightly there and left all the explanations about the manner in which 'this class of transactions are affected by the intervention of the precious metals' (James Mill 1818: 269) to Ricardo himself. James and John Stuart Mill's subsequent discussions on the comparative cost examples and international exchange were made almost solely in real terms, though John Stuart Mill, in his chapter on 'International Values', faintly suggested that considering the problem in money terms would be possible (James Mill 1826; John Stuart Mill 1844, 1848).

John Stuart Mill also began to separate the theory of specific international values from the domestic theory of values. International values were explained with reference not to their costs of production but to a 'principle anterior to that of cost of production... namely, the principle of demand and supply' (John Stuart Mill, 1844, I.19). Thus, the problems of international trade were reconstructed as two questions: (1) What commodities are exported, and (2) what are the limits within which the terms of trade are determined?

From the era of John Stuart Mill to the 1920s, the theory of values had met with a succession of modifications, while the doctrine of comparative costs remained unchanged. The divergence between the general theory of values and the value theory used to explain the doctrine of comparative costs also widened (Mason 1926).

In Cairnes, Bastable and Marshall's theory of international trade, the doctrine of comparative costs occupied an important position, and they persisted in using the trade examples indicated in labour costs, or *real costs*. They defined and used a unit of real cost in quantitative terms in their explanation of the doctrine of comparative costs: the difference in value between two commodities was measured by the difference in the number of units of the real costs required for their production. On the other hand, in the general theory of values, they almost gave up the attempt to relate value to units of real cost in quantitative terms due to the analytical problems posed by the multiplicity of productive factors and different qualities of labour. Mason explained the situation as follows:

> Therefore, altho considerable and careful description may be given of the different forms which real costs may take, such as physical fatigue, monotony, social odium, or disesteem connected with various kinds of labor, and altho attention may be paid to the way in which

[3]See Bloomfield (1994, Chap. 7), Gomes (1990, Chap. 7) and Maneschi (1998, Chap. 8) for useful surveys of this debate.

these costs affect the distribution of labor among various occupations, and the remuneration received in these occupations, very little attempt is made to reduce these costs to measurable form. (Mason 1926: 65–6)

Cairnes, for example, attempted to reduce labour costs to three factors: '1st, the duration of the exertion, or quantity of labor; 2nd, its severity or irksomeness; and 3rd, the risk attending' (Cairnes 1874: 80). Cairnes still regarded these costs as something objective and measurable, but his explanation of the doctrine of comparative costs became ambiguous: 'a country will export those commodities in the production of which its relative average costs, in terms of labor time, irksomeness, and risk, are least' (Mason 1926: 79). Furthermore, Cairnes, as did Taussig (1927), introduced the ambiguous idea of 'non-competing groups' to explain the immobility of different groups of labour, which further complicated his theory.

Faced with the analytical difficulties of the multiplicity of productive factors, Bastable (1887) introduced the concept of 'units of productive power' to compare among 'sacrifices', and Marshall (1923: 157) introduced representative 'bales' as a common unit in terms of labour of various qualities and capital, both of which had little explanatory power.

The divergence between the general theory of values and the value theory used to explain the doctrine of comparative costs was most remarkable in Marshall (1890, 1923). While, in his *Principles* and elsewhere, Marshall deliberately conceded that values do not correspond to real costs, he showed only the crude labour cost examples with no reservations in explaining the doctrine of comparative costs when considering international trade in Appendix H of *Money Credit and Commerce*.[4]

3.2 The Failure of Viner's Real Cost Approach

It was therefore an unpromising project for Viner to rehabilitate the real cost approach in the 1930s. It was fairly clear that units of real costs provided no explanation of values or prices.

According to Bloomfield (1994: 155–7), Viner's real cost approach is as follows: it assumed the proportionality of market prices to quantities of the services of the various factors (i.e. 'real costs') instead of labour-time costs. The 'real costs' were defined as the subjective disutilities attaching to the irksomeness of labour and the abstinence, or 'waiting', in providing the services of capital. In this approach, prices should be determined by the preferences among various occupations and between employment and unemployment. Viner explained his approach as follows:

[4]Mason (1926: 86) made an ironic remark about Marshall: 'It seems strange, upon the appearance of his last book, to find him clinging to a branch of the tree which he had himself already cut off some years before'.

> I understand by a 'real-cost theory of value' a theory which holds that there is at least a strong presumption of rough proportionality between market prices and real costs, and that therefore propositions which depend for their validity on the existence of such rough proportionality are not for that reason to be regarded as invalid unless and until evidence is produced tending to show that in the particular situation under examination no such approach to proportionality between prices and real costs exists . . . But even if no presumptions as to proportionality of prices to 'real Costs' can be established, general value theory must, of course, take account of 'real costs' in so far as they exist and influence relative prices in any manner. Demolition of the 'real-cost theory of value,' therefore, does not have as an appropriate sequel abandonment of 'real-cost' analysis. (Viner 1937: 491–2)

Viner admitted that the real costs theory of value suffered from failures of reasoning:

> It must be conceded, therefore, that the existence of variable proportions between labor costs and capital costs and the absence of any procedure by which a bridge can be built between real labor costs and the subjective costs connected with capital or 'waiting' makes it impossible to postulate a close relationship between prices and real costs. (Viner 1937: 514–15)

To sum up, Viner failed to find a unit with which to measure the subjective costs of labour and capital.

3.3 Haberler's Opportunity Cost Approach and Its Fatal Flaws

Viner and Haberler were rivals but shared the misconception that Ricardo's theory of comparative costs was based on the labour theory of value. They insisted on subtracting the labour theory of value from the doctrine of comparative costs and replacing it with their own value theories: Viner's real cost theory and Haberler's opportunity cost theory (see Tabuchi 2006, Chap. 5; Bloomfield 1994, 154–62; Maneschi 1998, Chap. 8):

> The association of the comparative-cost doctrine with the labor-cost theory of value is a historical accident, a result merely of the fact that Ricardo, in his pioneer exposition of it, expressed real costs in terms of quantities of labor. (Viner 1937: 490)

Haberler eliminated the labour theory of value from the doctrine of comparative costs in a more elegant way, by introducing the substitution curve based on the Austrian opportunity cost theory of value (Haberler 1930[1985], 1936).[5] He insisted that 'this latter doctrine [the labour theory of value] holds goods, as a special case of

[5]Haberler (1929: 381) mentioned Mason (1926) as the source of the 'objection that the comparative-cost theory builds on an old-fashioned and abandoned labor theory of value' and claimed that 'cautiously formulated, our doctrine may escape even this dangerous criticism without losing importance, as I shall try to show on another occasion'. The reference to 'another occasion' may have been to Haberler (1930).

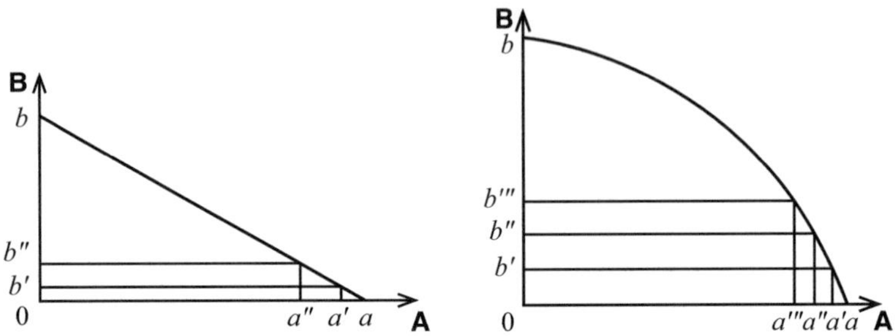

Fig. 1 Haberler's elimination of labour theory of value from the doctrine of comparative costs (Haberler 1936: 176)

general theory, if there is only factor of production: homogeneous labor' (Haberler 1936: 175). He continued (Fig. 1):

> It is now obvious that we have no further need of the Labor Theory of Value. We can derive the conditions of substitution between the two commodities, and express them in the form of a substitution-curve, when many different factors of production are available. (Haberler 1936: 177)

The substitution curve was a diagrammatic representation of a country's maximum attainable set of output for two commodities from a given supply of productive factors. The rate at which commodities could be substituted for each other was described under constant, increasing and decreasing opportunity cost conditions. Under increasing costs, the production substitution curve is drawn strictly concave to the origin. This provided an effective solution to the analytical problems posed by the multiplicity of productive factors that faced not only Bastable and Marshall but also Taussig and Viner (Haberler 1930; Gomes 1990: 102).

The substitution curve is now called the 'transformation curve' or the 'production possibility frontier'. It is so popular that essentially identical diagrams appear in virtually all textbooks on the theory of international trade. Haberler's presentation was the first exposition of the 'Ricardian Model', positioning it as a 'special case' of the more general case of the concave production possibility frontier with multiple factors.

Chipman was generous with compliments about Haberler's contribution:

> Undoubtedly Haberler's most significant contribution was his reformation of the theory of comparative costs (Haberler 1930), which revolutionized the theory of international trade. Prior to this paper, the Ricardian theory still held sway, but had been so amended with ill-defined concept such as 'real-cost' and 'unit of productive power' taking the place of labour allocation that it had lost all its simplicity and elegance. Haberler introduced the production 'substitution curve' (now usually known as the production-possibility frontier), allowing for several factors of production, and taken to be concave to the origin as a result of diminishing returns. This laid the foundations for Ohlin's theory, as well as Lerner's and Samuelson's. (Chipman 1987: 581)

Haberler's approach seemed able to help build a theory of international trade with multiple factors. However, it had fatal flaws, which Viner (1937, chap. 8) accurately pointed out[6]:

1. It assumed fixed factor supplies: factors that physically exist in the economy are utilized for production at the maximum level at any moment.[7]
2. It assumed factor indifference as to alternative uses: factors are homogeneous and can be transferred in no time to different uses.

If the two assumptions were dropped, it would prevent equality between the price ratio of two products and their transformation ratio (i.e. opportunity cost).

A more fundamental problem with Haberler's approach, which Viner did not mention, might be that the transformation ratio is determined not by the absolute amount of labour and other factors required to produce some output but by the alternative output forgone (Haberler 1936: 177). The 'cost' of a commodity is the amount of another commodity that might have been produced by the same resources. Comparative advantage is explained as the divergence between the transformation ratios of two countries,[8] but the theory 'does not explain why these divergent ratios of exchange should exist in the countries' (Mason 1926: 73).

Viner also pointed out that assuming community indifference curves in the opportunity cost approach contradicts the analysis of changes in income distribution.[9] Surprisingly, the first such diagram was presented in a lecture given by Viner at the

---

[6]It is suggestive that in the epigraph of Chap. 8 of his book *Gains from Trade: The Doctrine of Comparative Costs*, Viner cited a passage from Jevons' *Principles of Science*: 'It is always to be remembered that the failure of an argument in favour of a proposition does not, generally speaking, add much, if any probability, to the contradictory proposition' (Viner 1937: 437).

[7]It is a misrepresented statement that 'a point on the production possibility frontier shows that the economy is at the full employment level'. It expresses the maximum level of full employment, not full employment in general. In the labour markets, for instance, a point on the production possibility frontier, based on the assumption of inelastic factor supply, shows that the economy attains the 'saturation level' of employment, that is, each and every member of the society who has labour ability, say, at the age from 15 to 65 years old, *is* employed. On the other hand, Keynes' full employment means not a certain level of employment but variable amount of employment that depends on wage level.

[8]'We should have to set out, instead of a series of absolute labour-costs, . . . a series of relative prices or exchange-ratios, using any one commodity as a *numeraire* with which to measure prices. This series would equally represent the substitution-ratios, since these are the same as the exchange ratios. Each country would specialise in those branches of production in which it had a comparative advantage or, in other words, would produce those goods whose costs were relatively lowest. For example, if in country I one unit of A exchanged against one-and-a-half units of B, and in country II one unit of A exchanged against one of B, country I would plainly have a comparative advantage in the production of B' (Haberler 1936: 182).

[9]Assuming community indifference curves implicates that the economy's consumption decisions may be based on the tastes of a single representative individual. Viner criticized this as follows: 'The opportunity-cost approach encounters, therefore, on the income side, the same type of difficulty of weighting in the absence of knowledge of the proper weights as does the real-cost approach on the cost side' (Viner 1937: 523).

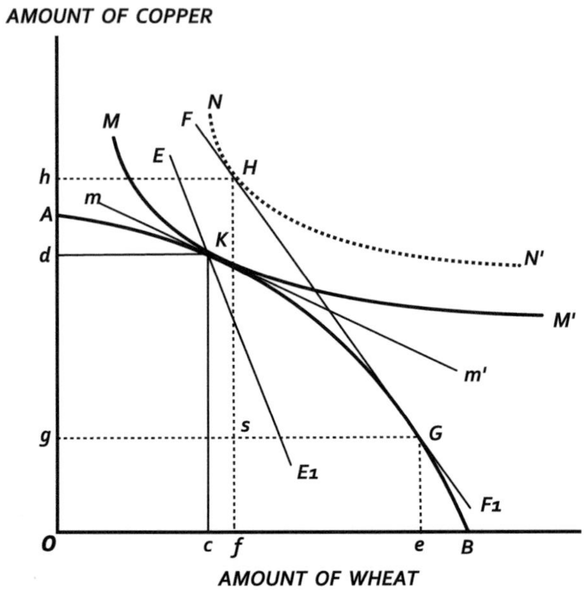

Fig. 2 Viner's diagram in January 1931 (Viner 1937: 521)

London School of Economics in January 1931. Here, a strictly concave production possibility curve with community indifference curves was applied to the problem of international trade for the first time, to show the gain from trade in a comparison between pretrade and posttrade equilibrium for a country. Viner claimed that his purpose in presenting the diagram was 'to stress the limitations rather than the possibilities of this [opportunity-costs] approach' (Viner 1937: 521) (Fig. 2).

Though Haberler deliberately avoided the use of community indifference curves,[10] Lerner (1932, 1934) and Leontief (1933) soon presented essentially identical but more elaborate ones, which remain among the most popular in textbooks on international trade theory.

Haberler later accepted some of Viner's critique, and a compromise was reached between the two approaches:

> I grant you, of course, that disutility of work, irksomeness or even abstinence . . . have to be considered. The only differences is that I prefer to put these things on the income side, while you put them on the cost side. That seems to me a difference in using words and not a material difference. (Haberler, letter to Viner, March 10, 1955, cited in Bloomfield 1994: 158).

Based on this compromise, Haberler (1950) tried to analyse the case of free trade with Keynesian unemployment using the production possibility frontier with elements of the real cost approach. In the real cost approach, prices should be

[10]'It must not be forgotten, however, that a shift in production will usually be accompanied by a redistribution of income. This precludes the uncritical application of community indifference curves, either as an explanatory device . . . or as a criterion of welfare' (Haberler 1950: 226).

determined by preferences among occupations and between employment and unemployment. Haberler adopted this idea of unemployment, dropping the assumptions of the original opportunity cost approach that factor supplies are fixed and that factor is indifference as to alternative uses. However, the supply curve of labour assumed in the real cost approach is essentially identical to Keynes' 'second postulate of classical economics';[11] therefore, the case Haberler tried to analyse was not the case of Keynesian unemployment but that of a kind of voluntary unemployment. It seems inevitable that the theory of international trade based on either approach would be insensitive to the problem of Keynesian unemployment (see Tabuchi 2006, Chap. 5).

4 Samuelson, Ohlin and the HOS Model

4.1 Samuelson's Discontents About the Compromise Between Viner and Haberler

A closer look at the discussions reveals that it was not as easy to build the new mainstream theory as was assumed, contrary to the popular perception that Haberler's opportunity cost approach and Ohlin's factor endowment approach quickly overwhelmed the real cost approach, and consequently the HOS model emerged after World War II through the efforts of Paul Samuelson.

In the middle of the debate, in 1938, Samuelson was quick to insist that both of the approaches could go together:

> It will be seen from the above that the doctrine of opportunity cost, properly stated, in no way contradicts the so-called pain-cost theory of value. In fact, when stated with full qualifications, the doctrine of opportunity cost inevitably degenerates into the conditions of general equilibrium. (Samuelson 1938: 777)[12]

However, after a compromise was reached between the two approaches, Samuelson reversed his stance and criticized the product of the compromise in 1948:

> Professor Viner has steadfastly maintained the more general equilibrium approach of Walras, Pereto and Marshall against his opponents Knight, Haberler and Robbins. And one by one they have either had to maintain an empirically gratuitous position (that all factors must be perfectly *inelastic* in total supply and indifferent between different uses) or else have had to reformulate the opportunity cost doctrine so that it becomes not only a rather awkward mumbo-jumbo, but loses all novelty and distinctiveness as well... But when Viner seems to argue that normative proposition in international trade cannot be deduced from a full general equilibrium analysis in much the same way that they can be

[11] 'The utility of the wage when a given volume of labour is employed is equal to the marginal disutility of that amount of employment' (Keynes 1936: 5).

[12] Samuelson was initially sympathetic to Viner's approach. Samuelson was one of Viner's students at Chicago University, and Viner was in turn a student at Harvard under Taussig (see Samuelson 1972).

> from the inadmissibly simple classical real costs comparative advantage, I part company with him. (Samuelson 1948a: 866)

It is difficult to comprehend why, despite the harsh criticism on both sides of the debate, Samuelson assumed essentially the same unrealistic assumptions when he built the HOS model (Samuelson 1948a, 1949; Stolper and Samuelson 1941). Later, Baldwin ironically advocated Haberler's approach against Samuelson's criticism:

> It seems no more 'empirically gratuitous' (to use Samuelson's expression) to assume that factors are fixed in supply and indifferent among various uses than to assume that production functions for any good are identical among all countries or that tastes are not only identical among nations but homothetic. Indeed, the usual simplified form of the Heckscher–Ohlin proposition requires (among others) the same assumptions that Haberler made. (Baldwin 1982: 143–4)

4.2 Samuelson's Oversimplification of Ohlin's Theory

Similarly, Ohlin's theory had not been well evaluated by Samuelson. In 1948, Samuelson inveighed against Ohlin's assumptions of immobile factors, and he dared to compare Ohlin to a 'murderer who returns again and again to the scene of his crime' (Samuelson 1948a: 167). It was not until the 1970s that Samuelson acknowledged that Ohlin was right and credited Ohlin's theory as a 'seminal proposition' (Samuelson 1971b: 365), though he referred to Ohlin (1933) and praised his achievement in the first edition of *Economics*, also published in 1948:

> The person who has most clearly emphasized how commodity trade partially relieves the scarcity in all countries of the less abundant factors of production is the Swedish economist and public financier, Ohlin (pronounced O'Lean). He has made the following important addition to the classical doctrine of comparative cost:
>
> Free movements of labor and capital between countries will tend to equalize wages and factor prices. *However, even without any movements of productive factors across national boundaries, there will result a partial (but not necessarily complete) equalization of factor prices from the free movement of goods in international trade.* (Samuelson 1948b: 557)

Samuelson integrated Ohlin's theory into the HOS model of comparative advantage almost solely from the point of view of factor endowment and factor price discussions. However, Samuelson neglected important aspects of Ohlin's theory. Contrary to popular perception, Ohlin (1924, 1933) showed disdain for the 'doctrine of comparative costs' and the concept of 'comparative advantage' (see, e.g. Ohlin 1933: 571–90). He was regarded as a prominent opponent of the doctrine of comparative costs.

Ohlin also gave weight to money costs against real costs:

> If, as some writers suggest, real costs are as a rule proportional to money costs, there is every reason for building up the theory in terms of the latter, thus avoiding many difficulties, and then 'translating' the conclusions into real cost terms, when questions of economic policy are discussed. If, on the other hand, real costs are not proportional to money costs, it is difficult to believe that the former concept is a practical tool for the study of trade and price problems. (Ohlin 1933: 590)

One of his contemporaries observed as follows:

> Ricardo himself in his analysis of the distribution of the precious metals demonstrated that it was quite possible to approach the whole question in terms of money values . . . But with all this Ohlin is the only economist who has avowedly adopted this method, apparently believing that he has thereby made a revolution in international trade theory. (Bowley 1937: 202)[13]

Bowley's claim may be somewhat exaggerated; it seems strange that such an important aspect of Ohlin's theory has hitherto been almost neglected.

5 Concluding Remarks

In this chapter, we re-examined the process by which the theory of international trade developed from James and John Stuart Mill to the modern theory by way of Viner's real cost approach, Haberler's opportunity cost approach and Heckscher-Ohlin's factor endowment approach in light of the value theories underlying the doctrine of comparative costs. Although Ricardo's theory of international trade was intrinsically linked to the theory of money, the doctrine of comparative costs was extracted from the unified body of Ricardo's original theory.[14] Afterwards, the divergence between the general theory of values and the value theory used in the theory of international trade widened. The debate in the 1930s was an important turning point of international trade theory since Ricardo, alongside J. S. Mill's discussions, and resulted in the emergence of the new mainstream theory – the HOS model.

Samuelson eventually formulated, accompanied by confusions and inconsistency, the HOS model with the unrealistic assumptions. As we saw before, these unrealistic assumptions were derived from *Haberler*'s opportunity cost approach.

We are now so accustomed to using 'opportunity costs', 'production possibility frontier' and 'community indifference curves' in the analysis of international trade. However, we need to eliminate these concepts and radically review the history of the international trade theory since Ricardo.

We are thus back where we went wrong in understanding Ricardo.

---

[13] I thank Prof. Y. Shiozawa for pointing me to this book.

[14] Samuelson (1971a) is exceptional in that it showed how it is possible to integrate Ricardian comparative advantage, Marshall' s partial equilibrium schedules of supply and demand and Hume's gold-flow mechanism for exchange rate equilibrium with the neoclassical general equilibrium model of international trade.

References

Baldwin, R. E. (1982). Gottfried Haberler's contributions to international trade theory and policy. *The Quarterly Journal of Economics, 97*(1), 141–148.
Bastable, C. F. (1887). *The theory of international trade with some of its applications to economic policy*. Dublin: Hodges, Figgis.
Bloomfield, A. I. (1994). *Essays in the history of international trade theory*. Aldershot: Edward Elgar.
Bowley, M. (1937). *Nassau senior and classical economics*. London: George Allen & Unwin Ltd.
Cairnes, J. E. (1874[2004]). *Some leading principles of political economy newly expounded*. Reprinted as Boylan, Tom & Tadhg Foley (Eds.), *Collected works of John Elliot Cairnes,* vol. 5. London: Routledge.
Chipman, J. S. (1987). Haberler, Gottfried (born 1900). In J. Eatwell, M. Milgate, & P. Newman (Eds.), *The new palgrave: A dictionary of economics* (Vol. 2, pp. 581–582). London: Macmillan.
Faccarello, G. (2015a). A calm investigation into Mr Ricardo's principles of international trade. *The European Journal of the History of Economic Thought*, *22*(5), 754–790. It will also be published as chapter 6 in Senga, S., Fujimoto, M. and T. Tabuchi (Eds.), 2017. *Ricardo and international trade*. London: Routledge.
Faccarello, G. (2015b). Comparative advantage. In H. D. Kurz & N. Salvadori (Eds.), *Elgar companion to David Ricardo*. Cheltenham: Edward Elgar.
Gomes, L. (1990). *Neoclassical international economics: An historical survey*. London: Macmillan Press.
Haberler, G. (1929). The theory of comparative cost once more. *The Quarterly Journal of Economics, 43*(2), 376–381.
Haberler, G. (1930[1985]). Die Theorie Der Komparativen Kosten Und Ihre Auswertung Für Die Begründung Des Freihandels. *Weltwirtschaftliches Archiv, 32*, 349–370. Translated into English in Koo, A. Y. C. (Ed.) *Selected essays of Gottfried Haberler* (3–19). Cambridge, MA: The MIT Press.
Haberler, G. (1936). *Theory of international trade*, trans. from the German 1933 edition by A. Stonier and F. Benham. London: William Hodge.
Haberler, G. (1950). Some problems in the pure theory of international trade. *The Economic Journal, 60*(238), 223–240.
Keynes, J. M. (1936). *The general theory of employment, interest and money.* Reprinted as *Vol. VII of the collected writings of John Maynard Keynes*. London/Basingstoke: Macmillan, 1973.
Leontief, W. W. (1933). The use of indifference curves in the analysis of foreign trade. *The Quarterly Journal of Economics, 47*(3), 493–503.
Lerner, A. P. (1932). The diagrammatical representation of cost conditions in international trade. *Economica, 12*(37), 346–356.
Lerner, A. P. (1934). The diagrammatical representation of demand conditions in international trade. *Economica, 1*(3), 319–334.
Maneschi, A. (1998). *Comparative advantage in international trade: A historical perspective*. Cheltenham: Edward Elgar.
Maneschi, A. (2004). The true meaning of David Ricardo's four magic numbers. *Journal of International Economics, 62*(2), 433–443.
Marshall, A. (1890[1997]). *Principles of economics*. Reprinted as C. W. Guillebaud (Ed.), *Collected works of Alfred Marshall*. Bristol: Overstone.
Marshall, A. (1923[1997]). *Money credit and commerce.* Reprinted as Groenewegen, Peter (Ed.), *Collected works of Alfred Marshall.* New York: A.M. Kelley.
Mason, E. S. (1926). The doctrine of comparative cost. *The Quarterly Journal of Economics, 41*(1), 63–93.

Mill, J. (1818). Colony. In *Encyclopaedia Britannica, Supplement to the 4th, 5th and 6th eds.*, reprinted in 1989, vol. 3 (pp. 257–273). Tokyo: Meicho Fukyu Kai.
Mill, J. (1826). *Elements of political economy, Third Edition Revised and Corrected*. London: Baldwin, Cradock and Joy, reprinted in 1844, London: H. G. Bohn, and in 1965, New York: Augustus Kelley.
Mill, J. S. (1844). *Essays on some unsettled questions of political economy*. London: John W. Parker.
Mill, J. S. (1848). *Principles of political economy*. London: John W. Parker, reprinted in 1973. New York: Augustus M. Kelley.
Ohlin, B. (1924[1991]). *The theory of trade*, in Heckscher, Eli F. and Bertli Ohlin, *Heckscher–Ohlin Trade Theory*. Translated, edited, and introduced by Harry Flam and M. June Flanders (pp. 71–214). Cambridge, MA: The MIT Press.
Ohlin, B. (1933). *Interregional and international trade*. Cambridge, MA: Harvard University Press.
Ruffin, R. J. (2002). David Ricardo's discovery of comparative advantage. *History of Political Economy, 34*(4), 727–748.
Samuelson, P. A. (1938). Welfare economics and international trade. *American Economic Review, 28*. Reprinted in Joseph E. Stiglitz (Ed.), *The collected scientific papers of Paul A. Samuelson*, vol. 2. Cambridge, MA: MIT Press, 1966.
Samuelson, P. A. (1948a). International trade and equalisation of factor prices. *Economic Journal, 58*(230), June. Reprinted in Joseph E. Stiglitz (Ed.), *The collected scientific papers of Paul A. Samuelson,* vol. 2. Cambridge, MA: MIT Press, 1966.
Samuelson, P. A. (1948b). *Economics: An introductory analysis*. New York: McGraw-Hill.
Samuelson, P. A. (1949). International factor price equalisation Once Again. *Economic Journal, 59*(234), June. Reprinted in Joseph E. Stiglitz (Ed.), *The collected scientific papers of Paul A. Samuelson,* vol. 2. Cambridge, MA: MIT Press, 1966.
Samuelson, P. A. (1969). The way of an economist. In Paul A. Samuelson (Ed.), *International economic relations.* London: Macmillan. Reprinted in Robert C. Merton (Ed.), *The collected scientific papers of Paul A. Samuelson,* vol. 3. Cambridge, MA: The MIT Press, 1972.
Samuelson, P. A. (1971a). An exact Hume–Ricardo–Marshall model of international trade. *Journal of International Economics, 19*(1), 1–18. Reprinted in Joseph E. Stiglitz (Ed.), *The collected scientific papers of Paul A. Samuelson,* vol. 3. Cambridge, MA: MIT Press, 1972.
Samuelson, P. A. (1971b). Ohlin was right. *The Swedish Journal of Economics, 73*(4), 365–384.
Samuelson, P. A. (1972). Jacob Viner, 1892–1970. *Journal of Political Economy, 80*(1), 5–11.
Shiozawa, Y. (2007). A new construction of Ricardian trade theory: A many-country, many-commodity case with intermediate goods and choice of techniques. *Evolutionary and Institutional Economics Review, 3*(2), 141–187.
Shiozawa, Y. (2014). *A final solution of Ricardo Problesm on interantional values*. Tokyo: Iwanami Shoten (in Japanese).
Sraffa, P. (1930). An alleged correction of Ricardo. *Quarterly Journal of Economics, 44*(3), 539–545.
Stolper, W. F., and Samuelson, P. A. (1941). Protection and real wages. *The Review of Economic Studies, 9*(1), November, 58–73 Reprinted in Joseph E. Stiglitz (Ed.), *The collected scientific papers of Paul A. Samuelson,* vol. 2. Cambridge, MA: MIT Press, 1966.
Tabuchi, T. (2006). Trade, money & power: Critique of international economics. Tokyo: Housei University Press (in Japanese).
Tabuchi, T. (2014). *Yukizawa's interpretation of Ricardo's 'theory of comparative costs.'* Paper presented at paper the Ricardo Conference organized by the Ricardo Society, Tokyo. September, 2014. Revised version will be published as chapter 4 in Senga, S., Fujimoto, M. and T. Tabuchi (Eds.), 2017. *Ricardo and international trade*. London: Routledge.
Taussig, F. W. (1927). *International trade*. London: Macmillan, reprinted in 1966, New York: Augustus M. Kelly.

Viner, J. (1937). *Studies in the theory of international trade.* Reprinted in 1955, London: George Allen & Unwin.

Yukizawa, K. (1974). The original meaning and the deformed interpretation of Ricardo's 'theory of comparative costs.' *Shogaku Ronsan* (Chuo University), *15*(6), 25–51. Reprinted in Morita, Kiriro (Ed.), 1988. *Classical theory of international trade* (pp.111–136). Tokyo: Dobunkan (in Japanese).

Yukizawa, K. (1978). Formation of classical theory of international trade: Ricardo and James & John Stuart Mill. In Yukizawa, Kenzo and Masaharu Tanaka et al. (Eds.), *On method and history of social science* (pp. 203–224). Kyoto: Minerva Shobo (in Japanese).

An Overview of Research into International Values in Japan

Hideo Sato

Abstract In Japan, research into international values has been conducted vigorously since the latter half of the 1940s. From a global perspective, studies that emphasize demand factors in the determination of international values have been dominant; however, this has not been the case in Japan. Neoclassical studies of this subject have not been as vitalized. Rather, many of the studies have succeeded the works of Ricardo, Marx, Graham, and Sraffa who placed a high priority on supply factors in the determination of commodity prices. Research on this topic is divided roughly into two periods owing to its contents and characteristics. One is the period until the 1980s and the other is that since the 1990s. Research in the first period was chiefly carried out by Marxian economists, and that in the second period, based on Graham and Sraffa, has led to the birth of the new theory of international values developed in this book. In this chapter, we provide an overview of research into international values in Japan. In addition, we explain Graham's relatively unknown theory of international values and show the fundamental structure of the Graham-type model (a modified version of Graham's original model and a multi-country multi-commodity Ricardian trade model). Furthermore, we present a way in which to derive an equilibrium solution of this model practically.

Keywords Multi-country multi-commodity • Link commodity • Frank D. Graham

1 Introduction

In Japan, research into international values has been conducted vigorously since the latter half of the 1940s. From a global perspective, studies that emphasize demand factors in the determination of international values have been dominant; however, this has not been the case in Japan. Neoclassical studies of this subject have not been as vitalized. Rather, many of the studies have succeeded the works of Ricardo, Marx,

H. Sato (✉)
Tohoku University, Sendai, Japan
e-mail: vermeer@qit.ne.jp

Y. Shiozawa et al. (eds.), *A New Construction of Ricardian Theory of International Values*, Evolutionary Economics and Social Complexity Science 7,
DOI 10.1007/978-981-10-0191-8_10

Graham, and Sraffa who placed a high priority on supply factors in the determination of commodity prices. Research on this topic is divided roughly into two periods owing to its contents and characteristics. One is the period until the 1980s and the other is that since the 1990s.

Research in the first period was chiefly carried out by Marxian economists, although a few modern economists joined the discussion in the very early stages. Two of the many issues were central. Firstly, how is Marx's *labor theory of value*, which is constructed under the assumption of a single market with the free movement of labor and capital, modified in the world market without the free movement of labor? Secondly, which principles determine the patterns of the international division of labor (or international trade) and world commodity prices? The first problem relates to "the modification of the determination of value by the quantity of labor," while the second is the theory of the international division of labor or international trade.

These problems are virtually one for the following reasons. From the standpoint that affirms the labor theory of value, the second problem is solved automatically if the first is solved, while from the standpoint that denies the theory, the first does not exist at all.

The argument in the first period was polemical and appeared to be highly complex because many issues were connected. Researchers involved in the argument were strongly influenced by Marx's writings and debated the interpretations of the Marx's words. As Marx himself offered no definite or coherent writing dealing with international values or international trade, many researchers have relied on fragments found here and there in *Capital*, *Outlines of the Critique of Political Economy*, *Theories of Surplus-Value*, and so on in order to show their own views to be right.

Although the argument produced a large number of works, research diminished rapidly in the 1990s. Following the collapse of the Soviet Union and the communist states of Eastern Europe around 1990, the influence of Marxian economics waned. The natural result was that research relying on Marx's writings lost its force with a decline in Marxian economics. However, simultaneously, a new type of research into international values has begun to grow, although the number of researchers involved remains low. This new research stream, based on Graham and Sraffa, has led to the birth of the new theory of international values developed in this book.

In this chapter, we provide an overview of research into international values in Japan. The remainder of the chapter is organized as follows. Section 2 outlines the research until the 1980s. Although the issues are many, we focus our attention on the most central and important. Section 3 describes the new research conducted since the 1990s. In addition, we explain Graham's relatively unknown theory of international values and show the fundamental structure of the Graham-type model (a modified version of Graham's original model and a multi-country multi-commodity Ricardian trade model). Furthermore, we present a way in which to derive an equilibrium solution of this model practically. Finally, we refer to what we consider to be the remaining challenges.

2 Research into International Values Until the 1980s

2.1 Nawa's Research Motivation and the Theory of Key Commodity

Research into international values in Japan was started by Toichi Nawa. His research results related to this topic are contained in Nawa (1949). Nawa, a Marxian economist, was convinced that the labor theory of value was right. This theory is constructed under the assumption of a single market where labor and capital move freely. Is, then, the theory invalid in the world market where labor does not move? For him, this was an important matter of concern.

About this problem, David Ricardo also pondered, ultimately developing *the theory of comparative costs*. In Chapter 7 of *On the Principles of Political Economy and Taxation*, he stated that the "same rule which regulates the relative value of commodities in one country, does not regulate the relative value of the commodities exchanged between two or more countries" (p. 133). In addition, he argued that equal quantities of labor are exchanged in the same country, while unequal quantities of labor are exchanged between different countries. He did not explain, however, how international exchange ratios of labor are determined. The example that the labor of 100 Englishmen was given for the labor of 80 Portuguese was only deduced from the given commodity terms of trade between cloth and wine. For him, the labor theory of value is valid only in domestic trade and invalid in international trade.

Karl Marx also examined this problem. He had stronger conviction about the labor theory of value than Ricardo, suggesting that it was not invalid but rather modified in the world market in two ways. The first is related to the intensity of labor: "In every country there is a certain average intensity of labor below which the labor"[1] is not considered to be labor of normal quality in each country. In domestic market, "only a degree of intensity above the national average affects the measure of value by the mere duration of the working time." This is not the case in the world market. The average intensity of labor differs from country to country. "These national averages form a scale, whose unit of measure is the average" of the national averages. "The more intense national labor, therefore, as compared with the less intense, produces in the same working time more value." The second is related to the productivity of labor. In the world market, more productive national labor is considered to be more intense, producing more value in the same working time as compared with less productive national labor.

Marx described the above in Chapter 20 (Chapter 22 in the English edition published in 1887) of *Capital* Volume I. The description was qualitative rather than quantitative and lacking in concreteness. Indeed, it could not be proven that the labor theory of value is also valid in the world market unless international exchange

[1]Marx (1887, p. 396). The following quotations from Marx are on the same page.

ratios of labor are defined quantitatively. He, however, wrote nothing about this to the last. For Marxian economists, theory of international values was the missing link of Marx's theory of values.

Nawa attempted to discover the link, writing that in the world market, the specific national labor that produces a globally important commodity or a *key commodity* is the measurement standard to determine the weight of national labor; further, national labor that produces other commodities is also evaluated according to the same standard. He offered the following numerical example. Suppose that the key commodity (e.g., cotton yarn) is P and the other commodity (e.g., agricultural products) is Q. Further, developed country A requires each one working day to produce each unit of P and Q, while developing country B requires 12 working days to produce a unit of P and two working days a unit of Q. Then, according to each country's labor productivity of P, the evaluation that one working day of country A is equal to 12 working days of country B is given for the labor of both countries. Hence, one unit of P produced in country A is exchanged for six units of Q produced in country B.

Nawa's *theory of key commodity* was not supported by researchers, however. His numerical example is a two-country two-commodity model in the same way as Ricardo's. Both examples are also the same in that the more productive country in both commodities exports the commodity with the higher degree of productivity advantage and imports the commodity with the lower degree. However, Nawa did not explain why country A exported P, even though country A's P produced in one working day had an equal international value to country B's P produced in 12 working days. Moreover, in his numerical example, country B received no gain from trade and, therefore, should have no incentive to trade in the first place.

2.2 Theory of National Productivity Differentials

International values were a topic of the plenary sessions in the second (1950) and third (1951) academic conferences of the Japan Society of International Economics, which was founded in 1950. The argument was developed with a central focus on the review of Nawa's theory, and heated discussion took place among Marxian economists and modern economists. Is an international exchange of unequal quantities of national labor an unequal exchange of value? What is the relation between Ricardo's trade theory and Marx's? How should John Stuart Mill's *theory of reciprocal demand* be evaluated? What does the introduction of currency into trade theory bring about? While the active argument[2] continued after the conferences, the views of researchers still did not converge, but an exception was the so-called *theory of national productivity differentials*.

[2]See Kinoshita (1960) and Naruse (1985) about the outline of the argument.

This theory, which replaced the theory of key commodity, insisted that the measurement standard to determine the weight of national labor was the national productivity differentials obtained by averaging the productivity differentials of individual sectors. Based on a two-country two-commodity model, the national productivity differential is determined somewhere between the respective productivity differentials of two commodities. Using Nawa's example above, the national productivity differential is somewhere between the productivity differential of commodity P (12 versus 1) and that of commodity Q (2 versus 1), for example, 6 versus 1. One working day of country A produces the same international value produced by six working days of country B. Country A exports P and imports Q. The commodity terms of trade are thus "1 unit of $P = 3$ units of Q"; consequently, both countries gain from trading.

This was the same composition as Ricardo's theory of comparative costs. However, although many researchers accepted this theory, problems remained. For example, there was no convincing explanation about a way to average the productivity differentials of individual sectors. If some kind of weighted average were to be used, we would need to specify the weights based on three options. The first would be to use trade values as weights. However, we would immediately understand this to be wrong if we recollect that the theory of international values is also the theory of clarifying trade values, trade volumes, and other factors relating to trade. Selecting the trade values as weights is to put the cart before the horse.

The second would be to use output values as weights. This also has a drawback. Output values are prices multiplied by output volumes, and prices and output volumes (therefore output values) vary according to the patterns of the international division of labor. The weights could not be determined without first determining the patterns. If we need to specify the weights in order to determine the patterns, this is circular logic. Then, can we use demand values as weights? Demand values are prices multiplied by demand volumes. Prices vary by pattern, and the demand volumes vary by price. It is not until the patterns are determined that the weights can be. Again, the same problem as in the case of the output values arises.[3]

The shortcomings of the second and third weights were barely noticed. The reason of this was concerned with the fact that research into international values in Japan was carried out exclusively with two-country two-commodity models. In these models, the patterns of the international division of labor are already decided. Hence, researchers did not consider the abovementioned relation between the patterns and the weights in depth. If multi-commodity (at least three or more) models had been used, the situation would have changed a little.

[3]Although most researchers did not give a quantitative definition of national productivity differentials, a number of them, we think, adopted the second standpoint, as Japanese Marxian economists believed that demand was never involved in the determination of values. Exceptionally, Kihara (1986) adopted the first standpoint and Yukizawa (1957) the third.

More serious problems emerge in multi-country multi-commodity models. Regardless of which weights we use, it is impossible to calculate a weighted average if we do not know the productivity differentials of individual sectors in all countries. Can we know the productivity level of the car industries in developing countries or crude oil industries in non-oil-producing countries? Nevertheless, the theory of national productivity differentials has continued to reign as a popular theory[4] with such problems neglected.

One main reason for this is that no alternative was presented until the 1980s. Another is that real-world statistics seemed to show the labor productivity level of a whole country. The data are obtained by dividing GDP by the total working population or the total working hours in a country, being the labor productivity per person or per hour. It might be possible to assert that the very differentials of these levels were the national productivity differentials; indeed, many researchers asserted so. The existence of these practical data, however, never removed the theoretical problems of the theory; it only concealed them.

2.3 Aspects as Trade Theory

Here, we compare Ricardo's theory of comparative costs (Ricardo 1817, Chapter 7), Mill's theory of reciprocal demand (Mill 1848, Chapter 18), and the theory of national productivity differentials and summarize features as a trade theory. The basic models of all three theories are two-country two-commodity models, but each has a different logical structure to determine the commodity terms of trade (CTT), or world relative prices, and the double factorial terms of trade (DFTT), or international exchange ratios of national labor. Ricardo determined DFTT by taking CTT as given, Mill determined CTT and DFTT by adopting reciprocal demand, and the theory of national productivity differentials determined CTT and DFTT by adopting national productivity differentials.

It is an important feature of the labor theory of value that relative prices are determined regardless of demand, making the theory incompatible with the theory of reciprocal demand. While, in Mill's trade theory, a change in demand causes immediately a simultaneous change in price and quantity supplied, in the theory of national productivity differentials, a change in demand brings only a change in quantity supplied since prices are already determined by fixed production costs. This is why many Marxian economists support the theory.

Thus, in Japan, two trade theories with the same origin as Ricardo's trade theory, namely, neoclassical trade theory, which was formed through Mill, and

[4]There was criticism of the theory by Marxian economists. Sasaki (1989) pointed out that it was fundamentally impossible to average the productivity differentials of different sectors, while Motoyama (1982) wrote that we could never know the productivity level of the car industry in developing countries.

Marxian trade theory, which was formed through Marx, confronted each other. Marxian economists continued to criticize *the Heckscher-Ohlin-Samuelson model* (HOS model), which was recognized as the core of neoclassical trade theory. They criticized the assumption of the model that both countries have identical production technologies as unrealistic. They also suggested that factor price equalization, the logical consequence of the model, does not occur in the real world; rather, there are very large wage differentials among countries. It appeared as if the theory of national productivity differentials had a huge advantage in explaining wage differentials, because, to explain these, only the existence of national productivity differentials (the productivity differentials of the key commodity in case of Nawa's theory) is needed. They further stated that the neoclassical production function that assumes smooth substitutability between capital and labor is impractical.

On the contrary, the mainstream economists claimed that Ricardo's and Marx's models are primitive one-factor models in which there is no capital and thus that these models are lacking in reality. This argument is wrong. Labor input coefficients in the labor theory of value consist not only of "direct labor" expended by workers to produce commodities but also of "indirect labor" embodied in intermediate goods and the consumption of capital goods. In these models, capital is not nonexisting; it is merely converted into labor. If there is capital, there are ordinarily profits. In an economy with profits, the proportional relationship between the labor input coefficients and prices is lost. Japanese Marxian economists were fully aware of this matter, and therefore, the "transformation problem," or transformation from labor values into production prices, became an important subject of debate.

However, researchers interested in the transformation problem were not working in the field of international values, while researchers interested in international values did not tackle transformation problems. They both went on their own paths independently without crossing each other. Researchers of international values, although they knew the existence of the transformation problem, constructed a trade theory not in terms of prices but in terms of labor values, which they considered to be permissible as an approximate approach.

Roughly speaking, Marxian trade theory followed Ricardo's theory of comparative costs. What the former added to the latter was the determination of world relative prices and wage-rate differentials by using the productivity differentials of the key commodity or national productivity differentials. The addition, as stated above, met with little success.

Despite many similarities, there were two great differences between Ricardo's trade theory and Marxian trade theory. One is the relation between foreign trade and foreign direct investment (FDI). Ricardo indicated "the difficulty with which capital moves from one country to another, to seek a more profitable employment" (Ricardo 1817, pp. 135–136) and did not deal with FDI. The HOS model, because of the factor price equalization theorem, was also indifferent to FDI. In Marxian trade theory, some studies arguing for a logical relation between foreign trade and FDI existed,[5] though a few in numbers. From the viewpoint of the present-day world economy, it is clear that they both have a close connection. The pioneering of such studies cannot be denied.

The other is the dynamic aspect of trade. As is well known, Ricardo thought that all trading countries gain from trade, meaning that a peaceful and harmonious world society should be realized through the international division of labor based on the comparative costs in each period. Marxian economists opposed this thought. They argued that all trading countries gaining from trade is only right from a static short-run viewpoint. From a dynamic long-run viewpoint, another aspect emerges: according to industries specializing in the international division of labor, the future economic growth of each country is greatly affected. A typical example is the international division of labor in agriculture and manufacturing: while a future high growth rate is expected in countries specializing in manufacturing, economic development may be restrained in countries specializing in agriculture. Consequently, the per capita income differentials between manufacturing countries and agricultural countries will widen over time. From such a viewpoint, they strongly supported Alexander Hamilton, who advocated developing manufacturing in the United States after independence, and Friedrich List, who argued in favor of adopting protective trade policy toward England to progress the industrialization of Germany. This dynamic viewpoint concurs with the history of the world economy and shows an advantage of Marxian trade theory over neoclassical trade theory.

2.4 Kojima's and Negishi's Interpretations of Ricardo

At the last of this section, we introduce two exceptional theories. Although most mainstream economists accepted the theory of reciprocal demand, there were two exceptions, Kojima (1951) and Negishi (1982, 1996). Both these representative modern economists in Japan, from a completely different viewpoint to Marxian economists, insisted that Ricardo's trade theory could determine the terms of trade without considering reciprocal demand.

Kojima (1951) constructed a two-country three-commodity model in which the third commodity (gold) was added to two ordinary commodities (cloth and wine) and determined the terms of trade by combining the specie-flow mechanism with the comparative costs structure. We explain his theory with a little arrangement. His model is divided into two cases. In the first case, gold is produced in both countries and does not move internationally. The real production costs of one unit of cloth, wine, and gold are the labor of 100, 110, and 100 men in England and 90, 80, and 80 men in Portugal. If it is assumed that one unit of English money (£e) and Portuguese money (£p) both contain 1/45 units of gold (the official prices of gold are £e 45 and £p 45), gold parity is £e 1 = £p 1. Then the English wage rate per person is £e 0.45

[5]See Muraoka (1968) and Sasaki (1998).

(45/100) and that of Portugal is £e 0.5625 (45/80). Accordingly, England exports cloth at a price of £e 45, and Portugal exports wine at a price of £e 45. Thus, the terms of trade are determined without demand conditions.

In the second case, the starting point is the situation in which the labor costs of wine fall from 110 men to 100 men after an improvement in winemaking in England, but all other factors remain unchanged. Through this improvement, the price of wine made in England falls from £e 49.5 to £e 45, and the export of wine made in Portugal stops. An English trade surplus thus occurs, which, in turn, causes appreciation of English money and an outflow of gold from Portugal into England. As a result, English prices rise uniformly (e.g., by 3.3%), and Portuguese prices fall uniformly (e.g., by 3.3%). Then, the English prices of cloth, wine, and gold all rise from £e 45 to £e 46.5, and the Portuguese prices of cloth, wine, and gold fall from £p 50.6 to £p 48.9, from £p 45 to £p 43.5, and from £p 45 to £p 43.5, respectively.

Through these adjustments, it becomes possible for Portugal to export wine and gold, and the trade equilibrium is restored. Under this new equilibrium, the official prices of gold change from £e 45 to £e 43.5 and from £p 45 to £p 43.5 in England and Portugal, respectively, and the exchange rate settles into gold parity. The English wage rate per person changes from £e 0.45 to £e 0.465 (46.5/100) and that of Portugal from £e 0.5625 to £e 0.54 (43.5/80). Kojima, after having provided the explanation above, insisted that the terms of trade were determined without demand conditions in this second case as well as in the first case.

However, Kojima's use of numerical examples to provide this explanation was lacking in clarity. In particular, he offered no explanation about why the trade equilibrium occurred in the second case when the fluctuation band of prices was not 2% or 4%, but 3.3%. Being aware of such gaps, Negishi (1996) aimed to compensate for the weaknesses of Kojima (1951). While he disagreed with Kojima's idea in the early days,[6] he later altered his view and became a supporter with some reservations.

Negishi constructed a two-country three-commodity model in which England specializes in the production of cloth and Portugal in the production of wine. In the model, P_c and P_w represent the prices of cloth and wine in terms of gold after starting trading, L_e and L_p are the supply of labor in England and Portugal, a_{ce} and a_{ge} are the unit labor cost of cloth and gold in England, a_{wp} and a_{gp} are the unit labor cost of wine and gold in Portugal, V_e and V_p are the constant velocity of circulation of money in England and Portugal, and G and M are the world stock of gold and its distribution to England. Given that gold is used exclusively for money, the international distribution of gold can be explained by the quantity theory of money:

$$P_c * L_e / a_{ce} = V_e * M \tag{1}$$

$$P_w * L_p / a_{wp} = V_p * (G - M) \tag{2}$$

[6]See Negishi (1982, p. 202).

If conditions (1) and (2) are satisfied, trade between England and Portugal is balanced, and there is no movement of gold between the countries. Suppose that gold is produced only in Portugal. Then,

$$P_w = a_{wp}/a_{gp} \tag{3}$$

From the three equations, the following is obtained:

$$P_c/P_w = \left(a_{gp} * V_p * G - L_p\right) V_e * a_{ce} / \left(V_p * L_e * a_{wp}\right)$$

The left-hand side represents the equilibrium terms of trade, which are determined without introducing reciprocal demand for cloth and wine provided the values of the exogenous parameters of the model meet several conditions (not stated here). Further, if gold is produced only in England, the terms of trade are different from those above, becoming $P_c/P_w = V_e{}^*L_p{}^*a_{ce}/(a_{ge}{}^*V_e{}^*G - L_e)V_p{}^*a_{wp}$. Furthermore, if gold is produced in both countries, $P_c/P_w = a_{gp}{}^*a_{ce}/(a_{ge}{}^*a_{wp})$.

There are some problems in the argument above, however. While he stated that trade between England and Portugal is balanced if conditions (1) and (2) are satisfied, the reason is unclear. It seems that the left-hand side of the equations represents national income and the right-hand side national expenditure, since the trade equilibrium is realized when national income equals national expenditure. It is, however, not convincing to regard the product of the gold stock and the velocity of circulation of money as national expenditure. Moreover, it is hard to understand that the gold stock is given, even though there are three patterns of gold production and the terms of trade are different according to these patterns. His attempt to compensate for the shortcomings of Kojima's work does not succeed.

Negishi (1982) presented another way in which to determine the terms of trade without recourse to demand factors. Here, his model is a compact one that uses Ricardo's numerical example of comparative cost theory. According to him, Ricardo had the notion that the wage rate of laborers equals the amount necessary to purchase the commodities required to support of themselves and their families. Therefore, in the Ricardian two-commodity model, the wage rate is expressed by

$$w = c_1 p_1 + c_2 p_2$$

where c_1 and c_2 denote the given quantities of cloth and wine, w is the wage rate, and p_1 and p_2 are the prices of cloth and wine. To simplify, it is assumed that c_1 and c_2 are identical among different countries.

Next, let us introduce a profit rate into Ricardo's numerical example in which cloth in England and wine in Portugal require 100 and 80 units of labor, respectively; further, we assume that England specializes perfectly in the cloth industry and Portugal in the wine industry.[7] Then, prices are expressed by

[7]Although Negishi (1982) did not exclude the case in which one country produces two commodities, we omit it here to explain the case.

$$p_1 = (1+r)\,100\,(c_1 p_1 + c_2 p_2)$$
$$p_2 = (1+r')\,80\,(c_1 p_1 + c_2 p_2)$$

where r and r' denote the profit rates of England and Portugal.

The difference in labor productivity brings about a difference in the profit rate, given that the wage rate is identical in both countries. If the gap in the profit rate is very large, international movements of capital occur and the gap diminishes. However, the profit rates of both countries do not equalize fully because, as emphasized by Ricardo, most men of property are satisfied with a lower profit rate in their own country rather than seeking a higher profit rate in foreign nations.

So, let us assume $R' = aR(a < 1)$, where $R = 1/(1+r)$ and $R' = 1/(1+r')$.Then, the relative prices of p_1 and p_2 or the terms of trade between cloth and wine are determined regardless of the demand factors.

Despite the strengths of Ricardo's interpretation by Negishi, the wage differential is nowadays very large, and capital moves around the world swiftly and easily compared with period of Ricardo. Negishi (1982) should be understood as a paper in the field of the history of economics.

3 Research into International Values Since the 1990s

3.1 Rediscovery of Graham's Theory of International Values

Research into international values which is within the scope of Marxian economics has tapered off since the 1980s, and the number of researchers interested in the subject has also decreased. Instead, new research outside the framework of Marx has started. In this section, this new body of research is addressed.

First, we describe the "rediscovery" of Graham's theory of international values. Frank D. Graham (1890–1949), a mainstream US economist, researched international values from the 1920s and published his major book *The Theory of International Values* in 1948.[8] His research, however, has not been praised within mainstream economics and has been almost forgotten. The reason is that while the origin of mainstream trade theory is Mill's theory of reciprocal demand, Graham criticized Mill's theory thoroughly. On the contrary, Marxian economists, who are critical of Mill's theory, have also refused to accept Graham's theory as excellent and have ignored it entirely. The reason is that, while the labor theory of value is the most important foundation for Marxian economics, Graham regarded the theory as a stumbling block and refused it. However, Graham's theory was decisively important for the new theory of international values. We explain the features of his theory in some detail.

[8]His related works are Graham (1923, 1932, 1948). The following explanation mainly relies on Graham (1948).

(1) *Graham was the first to present the existence of an equilibrium solution in a multi-country multi-commodity trade model.*

We can sum up the fundamental structure of Graham's model as follows:

1. There are many countries and many commodities.
2. There are no intermediate goods and no profits. All commodities are for consumption.
3. For each country, constant opportunity costs, economic scales, and demand structures are given.
4. Full employment and trade equilibrium (or national expenditure equals national income in each country) are fulfilled.
5. There are no transport costs and no trade barriers.

Under these assumptions, the patterns of the international division of labor, international values, and each country's volumes of production, export, import, and consumption are determined uniquely.

Graham explains the above, while providing no mathematical treatment,[9] by using many numerical examples. In earlier trade theories, although there was the example that an equilibrium solution is derived in a two-country multi-commodity case,[10] some possible patterns of the international division of labor were only shown at best in a multi-country multi-commodity case.[11] Indeed, Graham was the first to present the existence of an equilibrium solution in a multi-country multi-commodity (four-country three-commodity or ten-country ten-commodity) case.

(2) *To explain domestic values and international values by the same logic, Graham expresses production techniques of commodities not by labor costs (inputted labor), but by opportunity costs.*

According to Graham, each country's production techniques differ in every sector. While the labor theory of value expresses the difference in these techniques by using the difference in labor input coefficients, he expresses it by using the difference in the opportunity cost of each commodity. Concretely, he designates a specific commodity as a benchmark commodity (the opportunity cost of this commodity is one) and expresses the production techniques of other commodities by the number of units producible by giving up production of one unit of the benchmark commodity. The opportunity costs are essentially constant,[12] as distinct from those

[9]McKenzie (1954a) presented a mathematical treatment for Graham's model, and McKenzie (1954b) tried to prove the existence and uniqueness of the equilibrium solutions in the model. Shiozawa (2014), however, indicated that the proof was wrong because the demand functions assumed by McKenzie were different from Graham's (p. 290).

[10]See von Mangoldt (1975).

[11]See Section 4, Chapter 8 of Viner (1937).

[12]Graham refers to the case of variable opportunity costs too and indicates that the number of commodities produced in common in more than one country would grow under the increasing opportunity costs (Graham 1948, pp. 146–151).

of neoclassical trade theory which are increasing. Graham describes the reason for using opportunity costs as follows:

> When we think in terms of opportunity cost it can be conclusively demonstrated that Ricardo, Mill, and the neo-classicists, were wholly wrong in supposing that the same rule which regulates the relative value of commodities in one country does not regulate the relative value of the commodities exchanged between two or more countries. (Graham 1948, p. 333)

We also explain the other two given conditions. The economic scale of each country is expressed by the production volumes of the benchmark commodity which is realized when each country specializes in the commodity. Although full employment is supposed, the volumes of production factors and absolute productivity levels are not shown. Therefore, differentials in per capita income or wage rates among countries are not argued in the theory of international values directly and are treated as another problem.[13] The demand structures of each country are given by the expenditure coefficients of each commodity (amount expended on each commodity divided by national income). The sum of the coefficients is one (i.e., all income is expended) in every country.

(3) *International values are determined by the opportunity costs in each country and link commodities.*

International values or the world relative prices of commodities are determined not by reciprocal national demand, but by the opportunity costs in each country just like domestic values. What is important in this determination is the existence of commodities produced in common in more than one country, termed *link commodities*. This link commodities link the opportunity costs of countries that produce the same link commodities, meaning that the relative prices of all the commodities produced in these countries are determined uniquely. In principle, every country has at least one link commodity, suggesting numerous link commodities in the world at large. As a result, a body of link commodities links the opportunity costs of all countries and thus determines the international values of all the commodities in the world. The link commodities are, in turn, determined by the interaction among the opportunity costs, economic scales, and demand structures in each country. According to Graham, the link commodity was the missing link of the classical theory of value.

(4) *In the face of changes in demand, international values are highly stable.*

International values, formed once, are highly stable in the face of changes in demand. Such changes are adjusted through changes in production volumes

[13]Graham is not indifferent to the problem. For example, he writes that national prosperity (per capita income or wage rate) is a function of two variables, per capita physical productivity and the terms of trade, and the former is more important (ibid., p. 50, pp. 212–213, p. 233). He also refers to money wages (ibid., p. 261, p. 307).

and export-import volumes without price changes. If drastic changes in demand occur, prices might change slightly. In this case, the price changes are necessarily accompanied by changes in the pattern of the international division of labor. Newly formed international values are also based on the linkage of the opportunity costs in each country.

However, depending on the three given conditions of 3 in the above (1), the linkage of opportunity costs may be disconnected. Graham calls such a state of disconnection *limbo* (see the next subsection for more details) and regards this state as highly improbable. In the limbo case, a small change in demand brings about an immediate change in international values. Exemplified by using a two-country two-commodity model, the limbo case is a situation in which each country specializes in a commodity with a comparative advantage, an ordinary case used in textbooks. According to him, however, a situation in which one country produces two commodities and the other country produces either commodity with a comparative advantage has a far higher probability. Then, international values are determined by the opportunity cost of the former country, and reciprocal demand plays no part.

Graham's theory of international values was also introduced to Japan by several researchers from the 1950s to the 1980s. However, these introductions were either critical of his theory or only partial offerings.[14] Sato (1990) gave high acclaim to the theory and introduced the whole picture of it to Japan. Furthermore, Sato (1994) presented a two-country multi-commodity model, which was a modified version of Graham's model in three ways: production techniques are expressed not by opportunity costs, but by labor input coefficients, volumes of usable labor are given, and not only the linkage case but also the limbo case is treated. Hereafter, we refer to this modified model as a *Graham-type model* to distinguish from Graham's original model. In the next subsection, we explain a way in which to obtain an equilibrium solution of the Graham-type model.

Another stream of research into multi-country multi-commodity model that started from Jones (1961) aimed to solve the patterns of each country's specialization in the setting in which countries and commodities are equal in number. Several modern economists in Japan also wrote related papers.[15] However, although they typically referred to Graham, they did not understand the importance of link commodities.

[14]Kojima (1949) and Minabe (1956) were critical. Noguchi (1987) introduced the two-country two-commodity case affirmatively.

[15]About this research stream, see Sect. 2, Chapter "Analysis of Production Efficient Patterns of Specialization Allowing Intermediate Inputs: The Meaning of Shiozawa's Model with a Viewpoint of Modern Economics" of this book.

3.2 *Derivation of the Equilibrium Solution in the Graham-Type Model*

Graham's attempt to present a general equilibrium in a multi-country multi-commodity Ricardian trade model had several weaknesses. First, his model did not include intermediate goods and profits. Second, he did not show how to derive the equilibrium solution from the given conditions: he showed only the calculation results of his numerical examples. Third, he virtually ignored the limbo case, which he regarded as highly improbable. However, as McKenzie (1954a) indicated, this was wrong. The probability of the limbo case, while certainly small, is not negligible. We also have to derive an equilibrium solution about the limbo case as long as we cannot ignore the limbo. Fourth, he did not address underemployment case, although his model was essentially compatible with an underemployment.[16]

As stated later, the first problem was solved by Shiozawa (2007, 2014), and the content of the solution is shown in Chapter "The New Theory of International Values: An Overview" of this book. In this subsection, the second and third problems are addressed on the basis of Sato (2016). Although Sato (2016) also presented an underemployment version of the Graham-type model, this is not covered here.

Model Setting and Definition of Terms There are M countries and N commodities (M, N: an integer of ≥ 3 and $M < N$). The labor input coefficients, volumes of usable labor, and expenditure coefficients in each country are given. Conditions 2, 4, and 5 of (1) in the previous subsection are adopted, with another condition that domestic wage rates are equal in all sectors added.

Given the international division of labor, some sectors in each country continue the production activity, and other sectors cease it. The former is called *active point* and the latter *non-active point*. The patterns of the international division of labor have to be *reasonable*. Here, "reasonable" means a situation in which both the "production costs of active points = prices of commodities" and "production costs of non-active points > prices of commodities" are fulfilled.

The patterns of the international division of labor (hereafter, the patterns) can be classified into two types. One is when all countries are linked through link commodities. We refer to this as *the linkage type*. In this type, there are $M + N - 1$ active points, and all the active points are linked (see McKenzie, 1954a). Hence, by taking a commodity as the numéraire, the prices of all commodities and wage rates of all countries can be expressed by the labor input coefficients (see three-country four-commodity case mentioned later). In other words, once the patterns are determined, all the relative prices and wage rates (hereafter, the prices/wage rates) are determined by the patterns themselves, or there is a one-to-one correspondence between the patterns and the prices/wage rates.

[16]See Sect. 4 for the reason.

```
        linkage type                    limbo type with two disconnections
x-----x-----x-----x-----x-----x         x-----x     x-----x-----x     x
     active points: N+5                        active points: N+3
```

Fig. 1 An example of the two types of the international division of labor

The second type is called *the limbo type*. In this type, the patterns have fewer active points than M + N − 1. Here the linkage of countries and active points is not perfect, and one or more disconnections of the linkage occur. Therefore, determining all the prices/wage rates by the patterns only is not possible. As mentioned above, Graham called such a situation limbo and virtually ignored the limbo type; however, we cover this type. The disconnection is not always one. Theoretically, the disconnection can occur in the range from 1 to M − 1, and the number of active points decreases according to the number of disconnection. If a pattern of the international division of labor has two disconnections, the active points of this pattern are M + N − 3. Figure 1 illustrates these two types in a six-country N-commodity case (commodities are not shown in the figure).

Six countries (expressed by x) are all linked in the linkage type, whereas in the limbo type, the linkage is disconnected in two places, and countries are divided into three groups within which they are linked.

Process of the Derivation of the Equilibrium Solution An equilibrium solution is derived through the following process. First, we have to search for and identify reasonable linkage-type patterns, which are determined only by the labor input coefficients. The number of the reasonable linkage-type patterns is (M + N − 2)!/{(M − 1)!(N − 1)!} in an M-country N-commodity case.[17]

Second, for all reasonable linkage type patterns, we calculate the production volumes of the active points and prices/wage rates. As the prices/wage rates are determined according to each pattern, only the production volumes are unknown. Since the number of active points in the linkage type is M + N − 1, the number of unknowns is also M + N − 1. On the contrary, the number of equations is also M + N − 1, where the M equations express the conditions of full employment in each country and the N − 1 equations the conditions of the supply-demand balance for each commodity.[18] As the unknowns and equations are equal in number, we can solve all the equations mathematically. However, whether the solutions obtained mathematically are valid economically is another problem, leading to the next process.

[17]See Shiozawa (2014, p. 372). If M and N are large, it is very difficult even to identify the reasonable patterns. Including the rest of the process, the support of computer program would be needed in order to calculate actually.

[18]Although the number of conditions (therefore, equations) is N, one is invalid owing to Walras' law.

Third, from the $(M + N - 2)!/\{(M - 1)!(N - 1)!\}$ set of solutions, we select a set that has all positive solutions, as the production volumes must be positive economically. If there is such a set, the solutions of this set are the equilibrium solutions required. The pattern, production volumes, and the prices/wage rates are determined. The consumption volumes and export-import volumes in each country are also able to be calculated easily.

Fourth, when no set of solutions is all positive, we must expand the search range to find the equilibrium solution to the patterns of the limbo type. For all reasonable patterns of the limbo type,[19] we have to calculate the production volumes of the active points and prices/wage rates. We explain the process by assuming that the number of disconnections is l. Then, "$l + 1$" country groups are formed (see Fig. 1). The relative prices of the commodities produced in each country group and the relative wage rates of countries that belong to the same group are determined by the pattern itself, while the prices/wage rates among country groups are not determined only by the pattern. To determine the prices/wage rates among groups, we have to add the wage rates of a country in each country group which does not produce a numéraire commodity as unknowns. The number of additional unknowns is l. On the other side, the number of active points is $M + N - 1 - l$. Eventually, regardless of the number of disconnection, the total unknowns are still $M + N - 1$, and we can solve all the equations mathematically.

Lastly, we have to select a set of solutions that fulfills the following two conditions: all the solutions are positive and the solution set passes a competitive test. The test is to check whether non-active points are competitive by comparing the production costs of non-active points with the prices of the commodities. As all the prices/wage rates are already obtained by the fourth process above, the test itself is, though laborious, simple. If at least one non-active point is competitive, the set is disqualified. Only one set satisfies these two conditions and this set is the equilibrium solution.

Case of a Three-Country Four-Commodity We now give an example of the abovementioned in the case of a three-country four-commodity. There are the three countries of *A*, *B*, and *C* and the four commodities of 1, 2, 3, and 4. We define a_{ij}, b_{ij}, L_i, p_j, and w_i as commodity j's labor input coefficient in country i, commodity j's expenditure coefficient in country i, volumes of usable labor in country i, commodity j's price, and wage rate of country i, respectively. Consumption volumes are expressed as $w_i L_i b_{ij}/p_j$. The numéraire is commodity 1. The six unknowns are expressed as x_h (h= 1, 2, ..., 6).

Let us begin with the linkage type. The unknowns are all the production volumes of the active points. The unknowns' subscript number is assigned in order from the commodity of the smaller number in country *A* to the larger number in country *C*.

[19]The number of the reasonable patterns is $\sum(M + N - l - 2)!/\{(M - l - 1)!(N - l - 1)!l!\}$ $(l = 1, 2, \ldots, M - 1)$, where l is the number of disconnections (suggested by the description of Shiozawa, 2012, p. 50).

For example, in the pattern that country *A* produces commodities 1 and 2, country *B* commodities 2 and 3, and country *C* commodities 3 and 4, the prices/wage rates and system of equations are expressed as follows:

Prices and wage rates:

$$p_1 = 1$$
$$p_2 = a_{A2}/a_{A1}$$
$$p_3 = a_{B3}/a_{B2} * p_2 = a_{B3}/a_{B2} * (a_{A2}/a_{A1})$$
$$p_4 = a_{C4}/a_{C3} * p_3 = a_{C4}/a_{C3} * (a_{B3}/a_{B2}) * (a_{A2}/a_{A1})$$
$$w_A = 1/a_{A1}$$
$$w_B = a_{A2}/a_{B2} * w_A = a_{A2}/\,(a_{B2} * a_{A1})$$
$$w_C = a_{B3}/a_{C3} * w_B = (a_{B3} * a_{A2})\,/\,(a_{C3} * a_{B2} * a_{A1})$$

Conditions of full employment:

$$a_{A1} * x_1 + a_{A2} * x_2 = L_A$$
$$a_{B2} * x_3 + a_{B3} * x_4 = L_B$$
$$a_{C3} * x_5 + a_{C4} * x_6 = L_C$$

Conditions of supply-demand balance (only three of the four are valid):

$$x_1 * p_1 = w_A L_A b_{A1} + w_B L_B b_{B1} + w_C L_C b_{C1}$$
$$x_2 * p_2 + x_3 * p_2 = w_A L_A b_{A2} + w_B L_B b_{B2} + w_C L_C b_{C2}$$
$$x_4 * p_3 + x_5 * p_3 = w_A L_A b_{A3} + w_B L_B b_{B3} + w_C L_C b_{C3}$$
$$x_6 * p_4 = w_A L_A b_{A4} + w_B L_B b_{B4} + w_C L_C b_{C4}$$

Although we have to rewrite these in the case of other patterns, this is easy and would be sufficient for exemplification.

Next is the limbo type. In the pattern that country *A* produces commodities 1 and 2, country *B* commodities 3 and 4, and country *C* commodity 4 only, the prices/wage rates and system of equations are expressed as below. Here, the production volumes of five active points (x_1–x_5) and country *B*'s wage rate (x_6) are unknowns.

Prices and wage rates:

$$p_1 = 1$$
$$p_2 = a_{A2}/a_{A1}$$
$$p_3 = a_{B3} * x_6$$
$$p_4 = a_{B4} * x_6$$
$$w_A = 1/a_{A1}$$
$$w_B = x_6$$
$$w_C = a_{B4}/a_{C4} * x_6$$

Conditions of full employment:

$$a_{A1} * x_1 + a_{A2} * x_2 = L_A$$
$$a_{B3} * x_3 + a_{B4} * x_4 = L_B$$
$$a_{C4} * x_5 = L_C$$

Conditions of supply-demand balance (only three of the four are valid):

$$x_1 * p_1 = w_A L_A b_{A1} + x_6 * L_B b_{B1} + a_{B4}/a_{C4} * x_6 * L_C b_{C1}$$
$$x_2 * p_2 = w_A L_A b_{A2} + x_6 * L_B b_{B2} + a_{B4}/a_{C4} * x_6 * L_C b_{C2}$$
$$x_3 * p_3 = w_A L_A b_{A3} + x_6 * L_B b_{B3} + a_{B4}/a_{C4} * x_6 * L_C b_{C3}$$
$$x_4 * p_4 + x_5 * p_4 = w_A L_A b_{A4} + x_6 * L_B b_{B4} + a_{B4}/a_{C4} * x_6 * L_C b_{C4}$$

The above is one way of practically deriving an equilibrium solution in the Graham-type model.

3.3 *Sraffian Trade Theory*

Sraffa (1960), of which the Japanese translation was published in 1962, was well known among some nonmainstream economists in Japan. Japanese Sraffians were greatly interested in the re-switching of techniques and capital reversing being concerned with the Cambridge capital controversies. Studies of trade theory using Sraffa's price system, however, were delayed considerably, as it was difficult to incorporate intermediate goods or profits into a trade model, especially a multi-country multi-commodity trade model.

The first wide-ranging study was Takamasu (1991), which, based on the studies of Ian Steedman and John Stanley Metcalfe (Steedman 1979), examined Ricardian trade theory in detail. The author claimed that the introduction of profits or intermediate goods into the Ricardian trade model might cause losses from trade[20] and that the Heckscher-Ohlin theorem and factor price equalization theorem are not always valid even without the factor intensity reversal, as long as capital is not the given but commodities produced by means of commodities and human labor.[21] In addition, he proved that in a multi-country multi-commodity Ricardian trade model with intermediate goods and without profits, competitive equilibria exist.

[20] The probability that the existence of trade in intermediate goods brings about losses from trade is not zero but very small. On the contrary, the possibility for the existence to yield extended gains from trade is very large. (See McKenzie (1954a), Evans (1989), and Samuelson (2001).)

[21] However, his numerical example to show the invalidity of the theorems was in fact wrong, since the example did not satisfy the condition that there is no factor intensity reversal. Kurose and Yoshihara (2016) indicate this and give a correct example to show the invalidity of the theorems.

Thereafter, there was no noticeable progress in research into the theory of international values or trade theory by nonmainstream economists in Japan. The publication of the research achievements of Shiozawa changed this situation markedly (Shiozawa 2007, 2014). He succeeded in constructing the world production frontier of the multi-country multi-commodity model with intermediate goods and profits, proving that international values including the wage rates of each country were determined uniquely by a combination of three factors, namely, the production techniques of countries, distribution of labor powers to countries, and world demand (not national reciprocal demand). His model incorporates intermediate goods and profits into the Graham-type model and is the world market version of the Sraffian price system. He calls the model *the Ricardo-Sraffa trade economy*. Further explanation is omitted here since the model is described in detail in this book.

4 Remaining Challenges

Stimulated by Shiozawa's works, studies of trade theory by nonmainstream economists were revitalized since several years ago. In 2014, the workshop on the theory of international values, led by Shiozawa, began, and now research presentations and discussions are conducted four times a year. The topic are various: Trade and unemployment, Graham's theory of international values re-examined, Marxian trade theory revisited, Trade in value added and networks of production and trade, Thinking about the prospects of the Ricardo-Sraffa-Shiozawa trade model, Roles of demand in the determination of international values, Dynamic industry in the light of new trade theory, and so on. Several of the pertinent research achievements are included in this book.

Although research into international values has advanced recently, the remaining challenges are many. First, researchers must clarify how international values are determined under the condition of underemployment. Most trade models, for reason of necessity to close the models, are structured under the assumption of full employment. In the real world, however, underemployment is a normal state. Needless to say, it is desirable that the assumptions of models reflect reality. Further, in the case of the new theory of international values, which places high priority on quantity adjustments over price adjustments, the model settings of underemployment are especially desirable because the movement of productive resources among domestic industrial sectors, which requires a protracted period of time, is indispensable for quantity adjustments under full employment, while changes in the operating rate and employment rate are sufficient for those under the condition of underemployment.

Second, researchers must combine the knowledge obtained from the new theory of international values with economic growth theory and development economics. The new theory has a considerably different logical structure from mainstream trade theory. Therefore, a prescription for economic growth or economic development may be different between the new theory and mainstream theory. For the new theory, it is necessary to build not only a logical but also a policymaking counterweight to the mainstream, by deepening our understanding of relations between markets or corporations and states.

Third, researchers must join the new theory of international values to the theory of the international movement of capital. Ricardo composed his trade theory on the premise that capital did not move internationally, although he knew that capital did really move between countries. As mentioned already, HOS theory does not deal with international capital movement. In Marxian trade theory, some attempts in this direction have been made, but it has been insufficient. The present world economy in which globalization is progressing is characterized by the fact that capital moves among countries vigorously. The new theory has to incorporate this fact.

Fourth, researchers must verify the relevance of the new theory. Until now, empirical testes have been performed on Ricardian trade theory and HOS theory. The test results of the former take the view that Ricardian theory can be generally supported (see Golub and Hsieh 2000 and Chapter 3 of Krugman et al. 2015). For the latter, the results are less good (see Trefler 1995 and Chapter 5 of Krugman et al. 2015). A theory must always be verified by reality. The new theory is no exception.

There may be other challenges, but we think the above four are the most important.

Acknowledgment We would like to thank Kazuhiro Kurose for his useful suggestions and helpful comments.

References

Evans, H. D. (1989). *Comparative advantage and growth: Trade and development in theory and practice*. New York: Harvester Wheatsheaf.

Golub, S. S., & Hsieh, C.-T. (2000). Classical Ricardian theory of comparative advantage revisited. *Review of International Economics, 8*(2), 221–234.

Graham, F. D. (1923). The theory of international values re-examined. *Quarterly Journal of Economics, 38*(1), 54–86.

Graham, F. D. (1932). The theory of international values. *Quarterly Journal of Economics, 46*(4), 581–616.

Graham, F. D. (1948). *The theory of international values*. Princeton: Princeton University Press.

Jones, R. W. (1961). Comparative advantage and the theory of tariffs: A multi-country, multi-commodity model. *Review of Economic Studies, 28*(3), 161–175.

Kihara Y. (1986). Essence of the surplus profits obtained by Export (19), *Tokyokeidai GakkaiShi* (Tokyo Keizai University), *147*, 217–258 (in Japanese).

Kinoshita, E. (Ed.). (1960). *Readings in the theory of international values.* Tokyo: Kobundo (in Japanese).

Kojima, K. (1949). Graham's theory of international values. *Hitotsubashi Ronso, 22*(4), 579–589 (in Japanese).

Kojima, K. (1951). Ricardo's theory of international balance of payments equilibrium. *Annals of the Hitotsubashi Academy*, *2*(1), 76–92. (Reprinted in *Trade, investment and Pacific economic integration: Selected essays of Kiyoshi Kojima*. Tokyo: Bunshindo, 1996).

Krugman, P. R., Obstfeld, M., & Melitz, M. J. (2015). *International economics: Theory and policy* (10th ed.). Boston: Pearson.

Kurose, K., & Yoshihara, N. (2016). *The Heckscher-Ohlin-Samuelson model and the Cambridge capital controversies* (TERG Discussion Papers, 346). Online: http://www.econ.tohoku.ac.jp/e-dbase/dp/terg.html

Marx, K. (1887) *Capital: A critique of political economy* (Vol. I), English edition of "Das Kapital". Online: http://www.marxists.org/archive/marx/works/1867-c1/index.htm

McKenzie, L. (1954a). Specialisation and efficiency in world production. *Review of Economic Studies, 21*(3), 165–180.

McKenzie, L. (1954b). On equilibrium in Graham's model of world trade and other competitive systems. *Econometrica, 22*(2), 147–161.

Mill, J. S. (1848). *Principles of political economy: With some of their applications to social philosophy*. New York: John W. Parker.

Minabe, N. (1956). International values and gains from foreign trade. *Keizaigaku Zasshi* (Osaka City University), *34*(1/2), 41–75 (in Japanese).

Motoyama, Y. (1982). *Toward a trade theory*. Tokyo: Yuhikaku (in Japanese).

Muraoka, S. (1968). Toward a theory of capital export. *Kenkyu-nenpo Keizaigaku* (Tohoku University), *29*(3/4), 51–61 (in Japanese).

Naruse, S. (1985). On the argument over International values. In Kinoshita, & Muraoka (Eds.), *State, international trade and world market.* Tokyo: Yuhikaku (in Japanese)

Nawa, T. (1949) *Studies in the theory of international values.* Tokyo: Nippon Hyoron Sha (in Japanese).

Negishi, T. (1982), The labor theory of value in the Ricardian theory of international trade. *History of Political Economy*, *14*(2), 199–210.

Negishi, T. (1996). Japanese studies on Ricardo's theory of foreign trade. *Japanese Economic Review*, *47*(4), 335–345.

Noguchi, A. (1987) Graham's critique of the classical theory of international trade. *Keizaigaku Ronsyu* (University of Tokyo), *53*(2), 25–44 (in Japanese).

Ricardo, D. (1817). *On the principles of political economy and taxation.* Edited by Piero Sraffa with the collaboration of M. H. Dobb, *The works and correspondence of David Ricardo*, vol. 1. Cambridge: Cambridge University Press, 1951.

Samuelson, P. A. (2001). A Ricardo-Sraffa paradigm comparing gains from trade in inputs and finished goods. *Journal of Economic Literature, 39*(4), 1204–1214.

Sasaki, T. (1989). Toward a theory of international values. *Keizaigaku Kenkyu* (Hokkaido University), *39*(2), 40–59 (in Japanese).

Sasaki, T. (1998). International division of labour and capital export in a Ricardian model. *Economic Journal of Hokkaido University*, *27*, 1–9. The Japanese article with the approximately same content was published in *Keizaigaku Kenkyu* (Hokkaido University), *41*(3), 1991.

Sato, H. (1990). F. D. Graham's theory of international values. *Artes Liberales* (Iwate University), *46*, 163–191 (in Japanese).

Sato, H. (1994). *The principles of international division of labor and foreign trade.* Tokyo: Sohusya (in Japanese).

Sato, H. (2016). *New development in Graham-type theory of international values* (TERG Discussion Papers, 343). Online: http://www.econ.tohoku.ac.jp/e-dbase/dp/terg.html (in Japanese).

Shiozawa, Y. (2007). A new construction of Ricardian trade theory: A many-country, many-commodity case with intermediate goods and choice of production techniques. *Evolutionary and Institutional Economics Review, 3*(2), 141–187.

Shiozawa, Y. (2012). *Subtropical convex geometry as the Ricardian theory of international trade.* Discussion Paper uploaded in Shiozawa's Contribution page in ResearchGate.

Shiozawa, Y. (2014). *A final solution of Ricardo problem on international values.* Tokyo: Iwanami Shoten (in Japanese).

Sraffa, P. (1960). *Production of commodities by means of commodities; Prelude to a critique of economic theory*. Cambridge: Cambridge University Press.

Steedman, I. (Ed.). (1979) *Fundamental issues in trade theory.* London: Macmillan.

Takamasu, A. (1991). *Neo-Ricardian trade theory.* Tokyo: Sobunsha (in Japanese).
Trefler, D. (1995). The case of the missing trade and other mysteries. *American Economic Review, 85*(5), 1029–1046.
Viner, J. (1937). *Studies in the theory of international trade*. New York: Harper & Brothers.
von Mangoldt, H. (1975). On the equation of international demand. *Journal of International Economics, 5*(1), 55–97. The German original was published in 1863 and the translation from German was carried out by Siegfried Schach and was edited by J. S. Chipman.
Yukizawa, K. (1957). *Toward an international economics.* Kyoto: Minerva Shobo (in Japanese).

MIX
Papier aus verantwortungsvollen Quellen
Paper from responsible sources
FSC® C105338

If you have any concerns about our products,
you can contact us on
ProductSafety@springernature.com

In case Publisher is established outside the EU,
the EU authorized representative is:
Springer Nature Customer Service Center GmbH
Europaplatz 3, 69115 Heidelberg, Germany

Printed by Libri Plureos GmbH
in Hamburg, Germany